ACTINS

3rd EDITON

PUBLISHED TITLES

Transcription Factors 2: Helix-loop-helix, 2nd Edn. *T. Littlewood and G. Evan.*
Calcium-binding Proteins 1: E F Hands, 2nd Edn. *H. Kawasaki and R. H. Kretsinger.*
Actin-binding Proteins 1: Spectrin, 2nd Edn. *J. Hartwig.*
Extracellular Matrix 1: Fibril-forming Collagens, 2nd Edn. *K. Kadler.*
Motor Proteins 1: Kinesins, 2nd Edn. *G. Bloom and S. Endow.*
Cell Adhesion Molecules 1: Ig-like, 2nd Edn. *T. Brümmendorf and F. G. Rathjen.*
Motor Proteins 2: Myosins, 1st Edn. *J. R. Sellers and H. Goodson.*
Phosphoprotein Phosphatases, 1st Edn. *B. J. Goldstein.*
GTP-binding Proteins: Heterotrimeric, 2nd Edn. *S. Pennington.*
Intermediate Filament Proteins, 2nd Edn. *R. Quinlan, B. Lane and C. Hutcheson.*
Proteinases 1: Lysosomal Cysteine, 1st Edn. *H. Kirschke, N. Rawlings and A. Barrett.*
Nuclear Receptors, 1st Edn. *H. Gronemeyer and V. Laudet.*
Leucine Zippers. Transcriptional Factors, 3rd Edn. *H. Hurst.*

FORTHCOMING TITLES

Helix-loop-helix Transcription Factors, 3rd Edn. *T. Littlewood and G. Evan.*
Tyrosine Phosphoprotein Phosphatases, 2nd Edn. *B. J. Goldstein.*
E F Hand Calcium-binding Proteins, 3rd Edn. *H. Kawasaki and R. H. Kretsinger.*
Heterotrimeric G Proteins, 3rd Edn. *S. Pennington.*
Network and Filament Forming Collagens, 1st Edn. *M. Van der Rest and B. Dublet.*
Serine Phosphoprotein Phosphatases, 1st Edn. *P. Cohen and D. Basford.*
Lysosomal Cysteine Proteinases, 2nd Edn. *H. Kirschke, N. Rawlings and A. Barrett.*
Fibrillar Collagens, 3rd Edn. *K. Kadler.*
Proline Isomerases, 1st Edn. *A. Galat.*
Myosins, 2nd Edn. *J. R. Sellers and H. Goodson.*
Nuclear Receptors, 2nd Edn. *H. Gronemeyer and V. Laudet.*
Kinesins, 3rd Edn. *V. Allan and R. Cross.*
Gelsolin Family, 1st Edn. *H. Hinssen.*
IgCAMS, 3rd Edn. *T. Brümmendorf and F. G. Rathjen.*

ACTINS

3rd EDITION

PETER SHETERLINE

Department of Human Anatomy
and Cell Biology,
University of Liverpool,
Liverpool L69 3BX, UK

JON CLAYTON

EMBL, Meyerhofstrasse 1,
D-69117 Heidelberg, Germany

JOHN C. SPARROW

Department of Biology,
University of York,
York YO1 5DD, UK

ACADEMIC PRESS
Harcourt Brace & Company, Publishers

London, San Diego, New York, Boston, Sydney, Tokyo, Toronto

ACADEMIC PRESS LIMITED
24–28 Oval Road
LONDON NW1 7DX

U.S. Edition Published by
ACADEMIC PRESS INC.
San Diego, CA 92101

This book is printed on acid free paper

1st Edition published as *Protein Profile* Volume 1, Issue 1: Actin, 1994
2nd Edition published as *Protein Profile* Volume 2, Issue 1: Actin, 1995

A catalogue record for this book is available from the British Library

ISBN 0-12-639985-9

Typeset in Great Britain by Alden, Oxford, Didcot and Northampton
Printed in Great Britain by Halstan & Co. Ltd, Amersham, Bucks.

ACTINS
3rd EDITION

PROTEIN PROFILE

CONTENTS

PETER SHETERLINE
Department of Human Anatomy and Cell Biology, University of Liverpool, Liverpool L69 3BX, UK
Tel: +44 51-794-5450
Fax: +44 51-794-5517
Email: shep@liverpool.ac.uk;

JON CLAYTON
EMBL, Meyerhofstrasse 1 Heidelberg, Germany.
Tel: +49 6221 387 241
Fax: +49 6221 387 306
Email: clayton@eros.embl-heidelberg.de

JOHN C. SPARROW
Department of Biology, University of York, York YO1 5DD, UK
Tel: +44 904-432826
Fax: +44 904-432860
Email: jcs1@ebor.york.ac.uk

LIST OF FIGURES

LIST OF TABLES

SERIES PREFACE

The *Protein Profile* series has been developed from a recognition that individuals find it increasingly difficult to readily access the enormous amount of information accumulated by the international research community; information which is crucial for the efficiency and quality of their activities.

The *Protein Profile* series aims to provide a practical, comprehensive and accessible information source on all major families of proteins. Each volume of *Protein Profile* is focused on a single family or sub-family of the proteins, and contains tables and figures presenting as comprehensive an accumulation of structural, kinetic and biochemical information available on that particular protein group, coupled to an extensive bibliography. Every volume will be refined and updated to provide users with a practical up-to-date single source of information by the publication of new editions approximately every two years.

CONTENT

The text provides a brief overview of the biological context of the function of the protein group followed by an overview of available information on:

- function
- kinetic and biochemical properties
- sequences, sequence relationships and sequence features
- domain structure
- mutations
- 3-D structure
- binding sites of protein
- ligand binding sites and interactions with drugs
- derivatization sites
- proteolytic cleavage sites

Each volume follows the same format but with different emphases depending on the protein family. The text is extensively supported by tables and figures listing key information gathered from the literature with comprehensive reference to primary sources.

Available protein sequences, or where there are very large numbers, representatives from each sub-group, are aligned on fold-out pages at the back of each issue so that alignments can be viewed in their entirety.

Available references pertinent to the properties, structure and function of the proteins are listed in a full bibliography. References are numbered for access in the text, but are also arranged alphabetically under headings to allow browsing. Reviews are listed at the beginning and key papers are identified.

ACTINS
3rd EDITION

GLOSSARY

^{125}I-HBE, Methyl-*p*;-hydroxybenzimidate iodoester

A15, *Dictyostelium* actin gene

AA, Acetic anhydride

ABP, Actin-binding protein

9AC, 9-Anthroyl choline

Act88F, *Drosophila* indirect flight muscle actin gene

ACT1, Yeast actin 1 (conventional actin)

ACT2, Yeast actin 2 (actin-like)

ADP-F-actin, F- (or G-) actin with bound ADP or ATP

ADP.Pi-F-actin. F-actin with bound ADP with Pi in the γ site

AMPCPP, $\alpha\beta$-Methylene-adenosine-5′-triphosphate

AMPPCP, $\beta\gamma$-Methylene-adenosine-5′-triphosphate

AMPPLP, Adenosine-5′-phospho-pyridoxal-5′-phosphate

AMPNPP, $\alpha\beta$-Amido-adenosine-5′-triphosphate

AMPPN, β-Amido-adenosine-5′-diphosphate

AMPPNP, $\beta\gamma$-Amido-adenosine-5′-triphosphate

ANB-NOS, *N*-5-Azido-2-nitrabenzoyloxy succinimide

APG, ρ-Azidophenylglyoxal

ATPγS, $\beta\gamma$-Sulpho-adenosine-5′-triphosphate

AZP, *N*-(4-Azidobenzoyl)-putrescine

BFP, 3-Bromo-1,1,1-trifluoropropanone

BNHS, Biotin-*N*-hydroxysuccinimide

BPM, 4-Maleimidobenzophenone

CHD, 1,2-Cyclohexanedione

CRIA, (2-Nitrophenyl)-1-methyl-*N*-(resorufin-4-carbonyl)-*N*′-iodoacetyl piperazine

DABM, 4-Dimethylaminophenyl-azophenyl-4′-maleimide

DABSYL-Cl, 4-Dimethylaminoazobenzene-4-sulphonyl chloride

DACM, *N*-[7-Dimethylamino-4-methyl-3-coumarinyl] maleimide

DBF, Mercuri-dibromofluorescein

DCQ, 1-(2-Hydroxyethyl)-6-[(2,2-dicyano)vinyl]-2,3,4-trihydroquinoline

DDPM, *N*-(4-Dimethylamino-3,5-dinitrophenyl) maleimide

DEPC, Diethylpyrocarbonate

DMS, Dimethylsuberimidate

DNase I, Deoxyribonuclease I

DNP-ATP, *S*-Dinitrophenyl-6-mercaptopurine riboside-5′-triphosphate

DNPG, 2,4-Dinitrophenyl glutathionyl disulphide

DNS-A, 5-(Dimethylamino)naphthalene aziridine

DNS-Cl, 5-(Dimethylamino) naphthalene-1-sulphonyl chloride

DNS-cysteine, 5-(Dimethylamino)naphthalene cysteine

DTAF, Dichlorotriazinyl aminofluorescein

DTNB, 5,5′-Dithio-*bis*-(2-nitrobenzoic acid)

DTNBC, 5,5′-Dithio-*bis*-(2-nitrobenzoic acid) cystinyl disulphide

DTNBG, 5,5′-Dithio-*bis*-(2-nitrobenzoic acid) glutathione

DZT, 5-Diazonium-(1H)-tetrazole

ε-ADP, 1,N^6-Etheno-adenosine-5′-diphosphate

ε-ATP, 1,N^6-Etheno-adenosine-5′-triphosphate

EDANS, *N*-(5-Sulpho-1-naphthyl) ethylenediamine

EDC, 1-(3-Dimethylaminopropyl)-3-ethyl carbodiimide

EDTA, Ethylene diamine tetra-acetic acid

EF-1α, Translation initiation factor

EF-2, Translation elongation factor

EGTA, Ethylene glycol tetra-acetic acid

EIA, Erythrosin iodoacetamide

EP-GP, Extra-parotid glycoprotein

FDAP, 3-(5-Fluoro-2,4-dinitroanilino) proxyl

FDNB, Fluoro-dinitrobenzene

FITC, Fluorescein-5-isothiocyanate;

FM, Fluorescein maleimide

FP, Fluorescence polarization

FRAP, Fluorescence recovery after photobleaching

FRET, Fluorescence resonance energy transfer

FTP, Formycin-5′-triphosphate

GA1, Stoichiometric gelsolin : actin complex

GCDFP-15, Gross cystic disease fluid protein-15

IA, Iodoacetic acid

IAE, Iodoacetamido eosin

IAEDANS, 5-{2-[(Iodoacetyl)-amino]ethyl}amino naphthalene-1-sulphonic acid

IAF, 5-iodoacetamido fluorescein

IAM, Iodoacetamide

IANBD, 4-[*N*-(Iodoacetoxy)ethyl-*N*-methyl]-amino-7-nitrobenz-2-oxa-1,3-diazole

IATR, Iodoacetamido tetramethylrhodamine

IAZB, 2,2′-Dicarboxy-4′-iodoacetamido-azobenzene

K_d, Equilibrium dissociation constant (M)

k_-, First order dissociation rate constant (s^{-1})

k_+, Second order association rate constant ($M^{-1} s^{-1}$)

MANS, 2(*N*-Methylanilino)naphthalene-6-sulphonic acid

MBB, Monobromobimane

MBS, *m*-Maleimidobenzoyl-*N*-hydroxysuccinimide ester

MNS-Cl, 2-(*N*-Methylanilino)naphthalene-6-sulphonyl chloride

NBD-Cl, 7-Chloro-4-nitrobenzo-2-oxa-1,3-diazole

NBDN, *n*-Nonylene-1,9-bis-[5-dithio-(2-nitrobenzoic acid)]

NBS, *N*-Bromosuccinimide

NEM, *N*-Ethyl-maleimide

nmr, Nuclear magnetic resonance

NSL-TP, Spin-labelled ATP, 6-mercapto-*N*-(1-oxyl-2,2,6-tetramethyl-4-piperidinyl)acetamido-$\alpha\beta$-D-ribofuranosylpurine-5′ di or triphosphate

2NTC, 2-Nitro-5-thiocyanobenzoic acid;

NVOC-Cl, [(Nitroveratryl)oxy]chlorocarbamate

PBM, *N*,*N*′-*p*-Phenylene-*bis*-maleimide

PCMB, *p*-Mercuribenzene sulphonate or *p*-hydroxy-mercuribenzoate

PFP-ITC, Pentafluorophenyl isothiocyanate

PIA, *N*-(1-Pyrenyl) iodoacetamide

PIP, *N*-(1-Pyrene)-3-iodopropionamide

PLP, Pyridoxal-5′-phosphate

PRODAN, 6-Acryloyl-2-dimethylamino naphthalene

PYM, *N*-(1-Pyrenyl) maleimide and *N*-(3-pyrenyl) maleimide

RITC, Rhodamine isothiocyanate

RNHS, 5-Carboxytetramethyl rhodamine-*N*-hydroxysuccinimide

RSMA, Rabbit skeletal muscle actin

S-1, Myosin subfragment 1

SA, Succinic anhydride

SABP, Secretory actin-binding protein

Salyrgan, *O*-Carboxymethylsalicyl[3-oxymercuri-2-methoxypropyl]amide. Hg

SH3 domain, src-homology region type 3

Spin-labels maleimide/IA; *N*-(1-Oxyl-2,2,6,6-tetramethyl-1,4-piperidinyl)maleimide or iodoacetamide, 3-(2-iodoacetamido)-2,2,5,5-tetramethyl-1-pyrrolidinyloxyl

TEA, Tartryl-*bis*-ε-aminocaprylazide

TG, Transglutaminase-catalysed amidation with primary amines (derivatized with a variety of groups)

TMB, Trimethylammoniobromobimane bromide

TNBS, Trinitrobenzenesulphonic acid

TNM, Tetranitromethane

TNP-ADP, 2′(or 3′)-*O*-(2,4,6-Trinitrophenyl)-adenosine-5′-diphosphate

To Helen, Tom, Polly and Nancy

ACTINS
3rd EDITION

INTRODUCTION

BIOLOGICAL CHARACTERISTICS OF ACTIN

An introduction to actin

Conventional actins appear to constitute a highly conserved family of cytoplasmic proteins found in all eukaryotes; but appear not to be present in prokaryotes. The functional characteristics originally used to identify actins include their ability to self-assemble into filaments which have a characteristic morphology (in the presence of magnesium ions and at physiological salt concentrations) and the ability of these filaments to stimulate the ATPase activity of myosin. Most, but not all actins, can also bind cytochalasins, phallotoxins and deoxyribonuclease I (DNase I). Whilst the majority of conventional actins fulfil these criteria, there is emerging a subfamily of highly divergent actins which share significant sequence homology but limited functional similarity. In addition, it is now clear that there is a large group of proteins (which includes the sugar kinases and ATPase heat-shock proteins) that share very limited sequence homology but similar three-dimensional molecular structures to actin and that share with actins the ability to bind and hydrolyse ATP at a structurally equivalent site. Since this latter group are also expressed in prokaryotes, actins may represent only one of several groups of descendants from some common ATP-binding ancestor.

Conventional actins have relative molecular masses close to 43 000 due to the high degree of conservation of sequence and amino acid number. Many different isoforms of actin exist in higher eukaryotes coded for by even larger numbers of genes. The significance of several different isoforms appears to be functional diversity. Isoforms are differentially expressed in different specialized tissue cells and within a single cell there may exist several isoforms (three in vertebrate skeletal muscle for example) which segregate to functional regions of the cell and can be demonstrated to have subtly different properties *in vitro*. Many different isoforms of actin have now been purified from animal, plant, protozoan and fungal sources and sequence data are available for many more (see Table 3).

Actin is predominantly found in the cytoplasmic compartment of cells. Actin monomers (G- or globular actin) assemble in cells to form actin polymers (actin filaments, F-actin or microfilaments). The location, extent, polarity and timing of assembly is regulated by a large number of actin-binding proteins which are ultimately regulated by signal transduction pathways initiated at the cell surface. Filaments in turn, are organized into different, but characteristic three-dimensional organizations by a further group of actin-binding proteins that can cross-link, cap and bundle actin filaments and also attach filaments to membranes, often via transmembrane solute transporters, adhesion- or signal-receptors (see Figure 3). The actin cytoskeleton is highly dynamic in most cells, with measured half-lives for filament populations of the order of minutes. This turnover is a consequence of the intrinsic dynamics of actin filaments themselves which in turn is the consequence of the actin ATPase. Each actin monomer catalyses the hydrolysis of one molecule of ATP during a single cycle of assembly and disassembly. Thus, whilst the characteristic pattern of actin filaments in particular cell types exists continuously (to be observed in fixed cells by immunocytochemistry), the individual filaments that comprise these structures are continually changing. Evolution of new structural organizations of actin filaments occurs by cycles of assembly of new structures from the products of disassembly of the old.

The turnover of actin filaments and the properties of particular actin-binding proteins that determine the organization of the actin cytoskeleton are modulated by environmental signals via hormone, growth factor, and extracellular matrix receptors, and also from internal differentiation signals. Known modulators of cytoskeletal architecture include calcium ions, pH, inositol lipids and phosphorylation. It is likely that the targets of these signal pathways are the actin-binding proteins. However, there is evidence for the *in situ* phosphorylation and ADP-ribosylation of actins and the actin ATPase cycle can be modulated by monomer-binding proteins which catalyse or inhibit nucleotide exchange and by changes in pH or inorganic phosphate concentrations.

There are many reports of the presence of actin in the nucleus [1583, 1654, 1657, 1802, 1830, 2365].

Actin filaments have been localized by both HMM-labelled colloidal gold and by phalloidin close to nucleoli of neurons [1543, 1642]. Actin may only be accessible after treatment of nuclei with endonucleases (6592), accounting for the apparent conflict in reports. Ribonucleoprotein particles from parainfluenza virus both bind to and are activated to transcribe in the presence of F-actin but not G-actin *in vitro* [629], and the export of mRNA from fungal nuclei is cytochalasin-sensitive [2339], suggesting the possibility of both structural and allosteric roles for actin in the nucleus.

The actin cytoskeleton

The actin cytoskeleton appears to contribute to two aspects of motility that together (or independently) lead to the motile responses characteristic of eukaryotic cells. The first class of function involves the assembly of actin filament architectures which constrain the polarity and spatial organization of filaments such that they form tracks along which myosin can generate force in an appropriate direction. The most obvious example of this is contraction of striated muscle, where parallel interdigitating sets of actin filaments are made to slide past each other by myosin filaments, both shortening the contractile units (the sarcomeres) and generating significant forces along the common axes of the actin filaments. This subclass of myosin-based motility results from the activities of myosins of class II. More recently, myosins of class I (up to class VIII now – see Myosin issue of *Protein Profile*, Vol. 2, issue 12) have been identified that appear to interact with membranes and other structures and to move these relative to actin filaments. The functions of this class of myosin presumably contribute to the movement of organelles (e.g. chloroplasts) along actin filaments and the movement of the leading lamellipodium over the substrate in locomoting animal cells. In myosin-based motility, the polarity of actin filaments determines the direction of movement, the myosins moving in a pointed to barbed-end direction (see below). Myosin-based motility accounts for many cellular phenomena including muscle contraction, contraction of the collar that forms between daughter cells during cytokinesis, the exertion of force by cells on the extracellular matrix, morphogenic movements during development and clot retraction by platelets. The myosins are covered in another issue of *Protein Profile*.

The actin cytoskeleton may also be able to generate shape change and motility in the absence of myosin, by induced changes in the structure of cross-linked actin filament gels and by polymerization against a surface. In the former, swelling and shrinkage of gels (or any complex three-dimensional shape change) can be induced by selective breakage or formation of cross-links, releasing or storing entropic or osmotic energy. For the latter, polymerizing filaments may progressively consolidate space under a membrane formed as a result of random thermal movements or, more likely, by pumping of ions to induce local influx of water. Ruffling of leading lamellipodia, phagocytosis and shape changes of cells may include such mechanisms of motility. The complex process of cell locomotion clearly involves aspects of both polymerization-driven protrusion and myosin-based motility.

A relatively small proportion of cytoplasmic proteins (<50%) is solubilized if the plasma membrane is breached by hydrophilic non-ionic detergents (e.g. Brij 58). These data imply that many cytoplasmic proteins are not soluble but are bound in some way to cytoplasmic structures. Many glycolytic enzymes are bound in this way to the actin cytoskeleton of non-muscle animal cells and to the thin filaments of skeletal muscle cells. Furthermore, binding of many such proteins to actin filaments can be demonstrated *in vitro*. It is as yet unclear whether the catalytic efficiency of these bound enzymes, or of allosteric effects between them differs between their bound and soluble forms.

Protein synthesis and targeting of actin

A significant proportion of both ribosomes [812] and mRNAs [2088] co-localize with the actin cytoskeleton. More detailed morphological studies [2340] show that poly-A mRNAs localize close to filamin and α-actinin within a few nanometres of intersections between actin filaments. These observations suggest that the two major locations for translating ribosomes are endoplasmic reticulum membrane for mRNAs carrying an export signal sequence and the actin cytoskeleton for cytoplasmic proteins. Whether localization of mRNA on the actin cytoskeleton is a necessary prerequisite for translation of certain messengers is as yet uncertain, but the observation that both the translation initiation factor EF-1α and the elongation factor EF-2 bind directly to actin [2199] implies a close relationship between the actin cytoskeleton and translating ribosomes.

Furthermore, different actin mRNAs coding for different isoforms of actin are localized differentially about the actin cytoskeleton; often close to the cellular location of that particular isoform. β and

γ cytoplasmic actins have a different location in cells which is broadly matched by the localization of their respective mRNAs [1601, 2350]; similarly, β cytoplasmic actin and α cardiac actin mRNAs are segregated in myocytes [2355]. Even more surprising is the observation that the differential distribution of mRNAs can be modulated by signal transduction processes [2350, 2360] similar to those which cause redistribution of the translated protein isoforms. The relationship between translation and assembly into structures remains unclear. Unassembled actin appears to be distributed both throughout the cytoplasm and in discrete foci [2102]. It is conceivable that newly assembled actin filaments containing particular isoforms are serviced from specific synthetic foci. Disruption of the localization process for β-actin mRNA also disrupts the actin-rich lamellipodium [2356], suggesting a close functional link between translation and assembly for actins.

The untranslated 3′ end of mRNAs is responsible for the differential locations of actin isoforms [2355, 2356]. Swapping 3′ ends or adding an actin 3′ end to some carrier protein leads to localization determined entirely by this untranslated region. Further analysis of the 3′ end has identified two 'zip code' regions, one more dominant than the other, which appear to be conserved and whose function can be abolished by mutation [2356].

Newly synthesized actins (including the actin-related proteins) appear to be folded, in common with tubulins, on a toroidal chaperonin complex, TCP-1. Pulse labelling and morphological studies point to a transient ATP-dependent interaction of the unfolded nascent actin with the pore of TCP-1, [2147, 2362, 2367]. It may be interesting to note that the actins share a similar peptide motif with TCP-1, which may bind to some complementary site in actin to prevent folding [2342].

In animal cells there are at least three characteristic levels of organization of actin filaments: firstly, antiparallel arrays (stress fibres and less obvious bundles) that are homologous to the myofibrillar organization in muscle; secondly, parallel arrays that form dynamic protrusive structures (microspikes or filopodia) at the cell surface; and thirdly, isotropic arrays of filaments underlying the plasma membrane and attached via ABPs to transmembrane proteins in the bilayer. Each of these particular levels of organization is associated with particular (overlapping) sets of ABPs. The characteristic stress fibre organization of animal cells in culture is shown in Figure 1.

In plants, the actin filaments are also arranged in bundles that form helical arrays under the plasma membrane (Figure 1). Filaments in plant cells also appear to share the dynamic properties of filaments in animal cells as they reorganize significantly during the cell cycle. At present the amount of information available for the identification and characterization of plant actins and their associated ABPs is scant, but what evidence there is suggests that similarly complex cytoskeletal functions occur in both plant and animal cells [127].

The actin cytoskeleton also appears to interact with other polymer systems in the cytoskeleton. For example, the collapse of intermediate filaments around the nucleus in cells deprived of their microtubule network is inhibited by disruption of the actin cytoskeleton and some microtubule-associated proteins (in particular MAP 2s) appear to function to link actin filaments and microtubules. More extended discussions of these general functional properties of the actin cytoskeleton can be found in some of the books and reviews collected in the bibliography.

Actin in muscle

By far the most well studied actin is the α-skeletal muscle isoform from vertebrates. This isoform of actin is the major myofibrillar actin and represents some 20% or more of the total protein in these cells. Routine methods for the preparation of gram quantities of highly purified actin from muscle are available. Striated muscle myofibrils represent the most extensively studied contractile network in eukaryotic cells. The exquisite symmetry of the arrangement of filaments and the repetition and registration of identical contractile structures within and between each myofibril have allowed the use of powerful averaging techniques to study structure at high resolution and have enhanced the interpretation of other morphological and localization studies. Although muscle myofibrils are the most highly evolved contractile structures, less highly ordered examples of contractile structures using similar principles of organization and function are common in all eukaryotes.

In non-muscle cells and, to a lesser extent, smooth muscle cells, the contractile structures are less obviously organized, and perhaps more significantly, they retain a high degree of plasticity, both in terms of the apparent direction of force generation and in their location within the cell in response to different physiological situations. Thus, whilst striated muscle cells have evolved to select and develop those potential properties of contractile proteins suitable for generating large forces in a single dimension, non-muscle cells appear to have retained or developed the means to regulate force production *per se* and to exert both spatial and temporal control over the

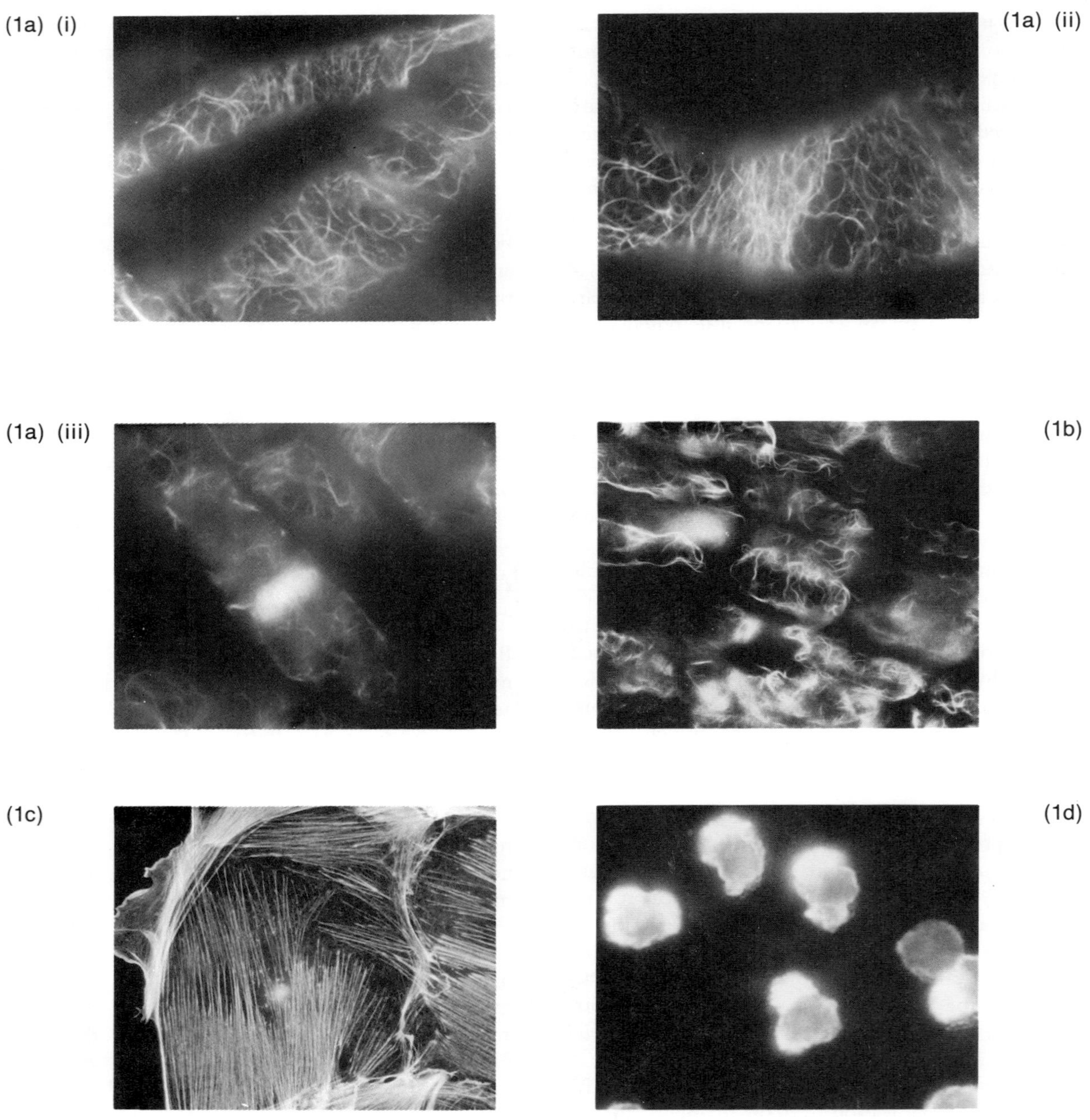

FIGURE 1 Localization of F-actin in plant and animal cells. *Fluorescent micrographs of bound rhodamine phalloidin show the location of polymerised actin in (a) carrot cells (Dauca carrota) in suspension, (b) lower plant rhizoids (liverwort Riccia flutans), (c) PtK2 cells in culture, (d) neutrophil leucocytes. Carrot cells in a(i) are in interphase; a(ii) and a(iii) showing increased filament density at presumptive cleavage plane and at the phragmoplast after mitosis respectively. Images in (a), (b) and (c) kindly provided by Dr. Clive Lloyd, Prof. Jeff Duckett and Dr. Sue Handel respectively.*

three-dimensional contractile networks appropriate to their differentiated functions. The mechanism of turnover of monomers in sarcomere filaments is at present poorly understood; however, micro-injection of labelled monomers into muscle cells suggests exchange into both the A-band and distal ends of filaments [263].

Purification of actin

Actin has been isolated from metazoan muscle [1878, 1879, 1735, 770, 967, 1890, 1752, 1759, 1764, 1767, 1769, 1947], metazoan non-muscle [1757, 75, 1823, 1923, 1924, 1902, 1733, 1745, 1759, 1781, 1786, 1795, 1796, 1812, 1814, 1829,

1842, 1847, 1870, 1894], plants [1875, 1931], fungi [1791, 1825, 1773, 1922, 875–878] and protozoa [1769, 1785, 1799, 1801, 1807, 1853]. Actin-like protein has been isolated from vertebrates [780]. The vast majority of structure and function experiments with purified actin have been performed using the rabbit skeletal muscle isoform of actin (identical to other mammalian and some other vertebrate skeletal muscle actins). Actin comprises about 20% by weight of skeletal muscle proteins and yields of 0.4–0.5 g of highly purified actin are possible from muscle obtained from a single rabbit. The most widely used approaches for purification, which differ only in detail, are those based on the method of Feuer developed in the late 1940s. Most of the soluble proteins are removed by homogenizing muscle and extensively extracting it in high, followed by low ionic strength buffers. The protein mass remaining (predominantly the myofibre proteins) are then dehydrated with acetone to form what is known as an acetone powder. The dehydration step serves both to denature many of the other proteins and to allow the drying of the powder into a form that can be stored indefinitely. The acetone powder is then extracted with low ionic strength buffers containing ATP to dissociate any myosin remaining and to sustain nucleotide binding to actin (which denatures in its absence) and calcium ions to displace any bound magnesium. The Ca.ATP monomer has a high critical concentration (about 6 mg ml^{-1}) and so can be

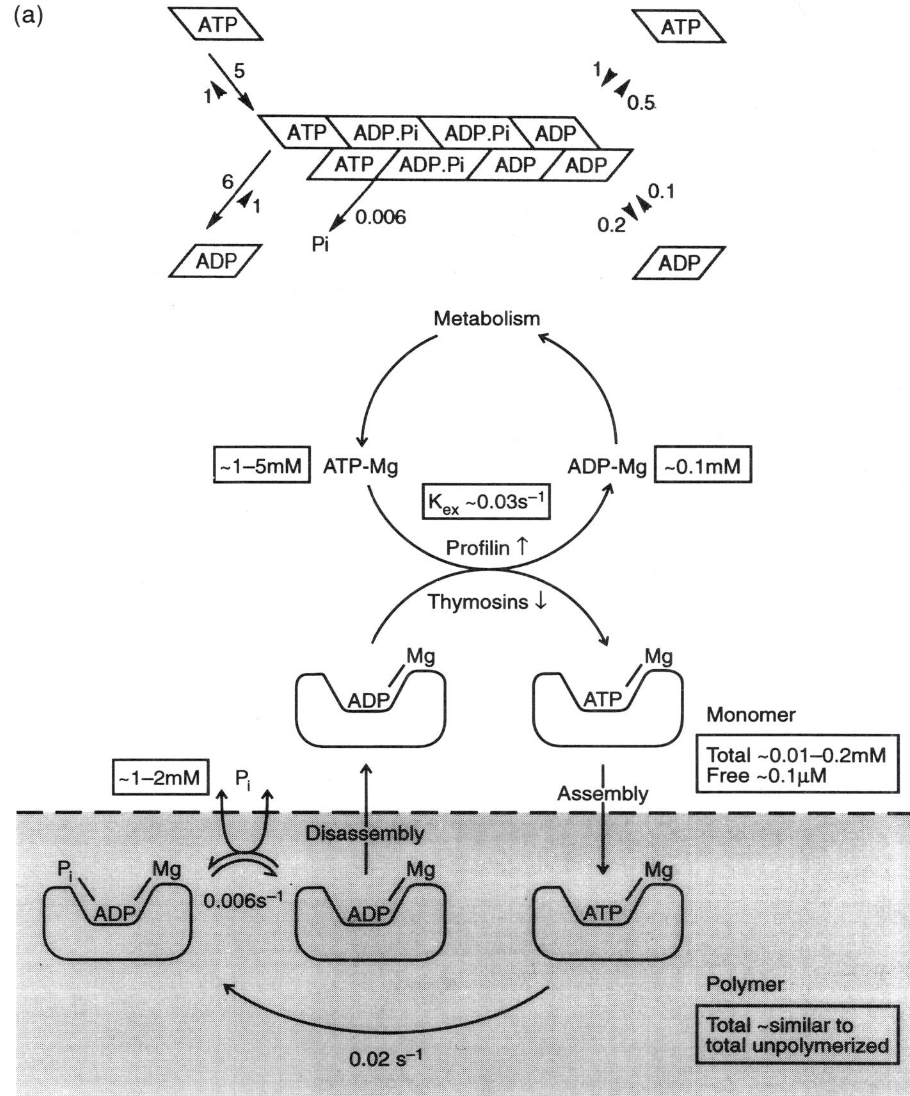

FIGURE 2 Assembly and disassembly reactions. *(a) This scheme shows the possible assembly and disassembly reactions during polymerization of actin. The length of the arrows is roughly proportional to the value of the rate constants (see Table 2), which are also indicated as numbers beside the arrows. (b) The ATPase and actin assembly/disassembly cycles combined with relevant concentrations and rates where not given elsewhere. Rate constants are in the following units:* k_+ *in* $\mu M^{-1}.s^{-1}$; *for* k_- *in* s^{-1}.

TABLE 1 Non-covalent actin-binding ligands

Ligand	References
Small actin-binding ligands	
Cations	34
Adenine nucleotides	401
Actin-binding proteins	
Gelsolin family	
Gelsolin; brevin	159, 661, 821
Adseverin	742
Earthworm 40 kDa protein	2215
Fragmin; cap42a; Severin; 45 kDa	1418, 873, 643, 775
Mbh1; macrophage-capping-factor; MCP; gCap39; scinderin	785, 815, 587, 859
Villin; 100 kDa *Dictyostelium*; dematin	626, 689
Capping proteins	
β-actinin	843
Acumentin	814a
Aginactin; Hsc70	798, 641, 678
Band 4.1; ezrin; moeisin; radixin; EM10	795, 794, 653, 2265
Caenorhabditis CP	578, 842
Cap100	688
CapG; gCap39; macrophage capping protein	2247
CapZ; β-actinin; cap32; cap34; cap37	159, 605, 678
DNase I	359
Schwannomin, merlin	793, 835
Tropomodulin	2270
Monomer-binding proteins	
Actobindin	148
Actophorin; ADF; actin-depolymerizing factor; destrin; cofilin depactin; COF1; ABP1p; dematin	741, 573, 532, 679, 638, 740, 697, 711, 485
ASP-56	659
Thymosin β4; thymosin β9; thymosin β10; Fx	768, 860, 451, 2231
Profilin	665, 805
Vitamin-D-binding protein; VDBP; group-specific component; Gc-globulin, Gc	614
α-Actinin family	
α-Actinin; actinogelin; gelactin	9
ABP-120	613
ABP-240	687
Adducin	594
Dystrophin; DMDR protein; DRP; dystrophin-related protein; utrophin; DP-71	782, 648, 647, 649
Filamin; ABP-280; ABP; actin-binding protein; gyronemin	668, 2219
Fimbrin; L-plastin; T-plastin; ABP-67; SAC6; macrophage 70 kDa protein	769, 779, 2290
Spectrin; fodrin; calspectin; TW 240–260	56, 660
Miscellaneous ABPs	
ABP-50; LAV3-4; ABP-46	630, 642, 818
Actobindin	148, 2205
Aldolase	773
Angiogenin	695, 694
Annexin I; calpactin II; lipocortin I; p35; GIF; chromobindin 9	661
Annexin II; calpactin I, lipocortin II; p36; PAP-IV; protein I; chromobindin 8	701, 739
Annexin VI; lipocortin VI; protein III; p70; p68; 67K calelectrin; chromobindin 20	692
c-Abl	751, 836, 2266
Caldesmon	671, 2249
Calponin; gizzard p34K	830
Calvasculin	2268
Chaperonins; TCP1	1285, 756, 819
Coactosin	627
Comitin; p24	849
Coronin	628
Cortactin; p80; p85	853
Crystallins, α-A-crystallin; α-B-crystallin	667
Dematin; Band 4.9	813
Deoxyribonuclease I, DNase I; DNaase I	396
Dictyostelium 34 kDa	650
Drebrin	2198
EF-1α	778

TABLE 1 Continued

Ligand	References
EF-2	2199
EGF receptor; EGFR	630
EP-GP	2260
Fascin; singed; *Xenopus* 53 kDa protein; lymphocyte 55 kDa protein	2203, 2226, 2250
gag; p58	2235
GAP-43; B-50	683
GCDFP-15	2260
Gelactins I–IV	747
Glued	780
Hisactophilin	676
HSP27; HSP28; HSP25	763
HSP90	772
HSP100	719
Inositol-1,4,5-trisphosphate receptor	2213
Insertin	847
LSP1; lymphocyte-specific protein	702
Lysozyme	720
MAP1A; MAP1B	656
MAP2	855
MARCKS	577
Mitochondrial protein	2361
Myelin basic protein	792
Myosin (motor site)	610
Myosin I tail (2nd site)	791, 709, 2263
Myosin alkali light chain 1	2223
Myosin light chain kinase	680, 706
Nebulin	700, 607
17β-Oestradiol dehydrogenase; hydratase-dehydrogenase-epimerase	2239
Physarum 210 kDa protein	698
PIK-A49	856
Ponticulin	611
Ribonucleoprotein complexes (RNPs)	629
SABP	2260
Scruin; Kelch; spe-26	632, 1314, 2267
Synapsin I	588, 606, 1651
Synapsin IIa	2208
Talin	709, 801, 2217
Tau protein	855, 540
TCP-1 chaperonin	2362, 2367
Tensin	847, 2240
Transgelin; C4h; SM22; WS3-10; p27	715, 809, 2254
Tropomyosin	793a
Troponin I	735
unc-60	748
Vimentin	2207
Vinculin; metavinculin	820, 2234, 2245
Actin-binding toxins, drugs and other organics	
9AC (hydrophobic probe)	603
Auramine-O (cation analogue)	429
Cannabinols	2013
Chaetoglobosins	2031
Cytochalasins	117
Cyclosporin A	2038
DCQ (molecular rotor)	1256
cis-Diaminodichloro-platinum(II) (binds actin to DNA)	759
Doxorubicin	1970, 1971, 1969
Goniodomin A	1995
Jasplakinolide	2380
Latrunculin	1975
Lipids	811, 816–17, 737, 832, 658, 2214
Mycalolide B	2390
Scytophysins, tolytoxin	2047, 2392
Swinholide A	2381

TABLE 1 Continued

Ligand	References
MANS (hydrophobic probe)	603
Microcystis aeruginosa toxin	2055
Phallotoxins	2036
Phosphate esters	476
Reserpine	2039, 1470
Salyrgan (binds sulphydryls)	1447
Virotoxins	2036
Yersinia YopE toxin	2054, 2053

extracted at high concentration without polymerizing. The extracted monomer is then polymerized by addition of 0.1 M KCl and magnesium to displace the calcium. The polymer is freed from tropomyosin by increasing the KCl concentration and collected by centrifugation. The polymer pellet is then depolymerized by dialysis against a low ionic strength buffer containing calcium. It has become standard practice to gel-filter the monomer to remove some unidentified actin-binding proteins. This approach has been used for several different muscle tissues and for some non-muscle tissues where large amounts of material are available. Actin can also be purified by directly exploiting its affinity for DNase or by indirectly exploiting its affinity for profilin. DNase I affinity chromatography relies on the ability of almost all monomeric actins to bind DNase I. Actins can be eluted by chelating Ca^{2+} ions necessary for DNase binding. This approach is suitable for small amounts of material. Actin bound to profilin (profilactin) is retained on poly-proline affinity media and can be eluted by ammonium sulphate. The binding of profilactin to poly-proline appears to be via the profilin, and recent data on the function of profilactin in yeast suggest that this site on profilin (which is homologous to the SH3 domain) may function in the recognition of poly-proline sequences on signal transduction proteins. This method is only suitable for some non-muscle tissues where the profilin concentration is sufficiently high to sequester a significant proportion of the cellular actin (in most cases thymosin $\beta 4$ is the major monomer-sequestering protein). The profilactin purified is a mixture of the two isoforms with usually the β-form predominating (reflecting the tissue composition). The actin can be liberated from the profilin by treatment with high magnesium concentrations that leads to actin para-crystal formation. Actin has a lower than average pI and can be bound to anion exchangers like DEAE cellulose. The actin can be eluted by conventional salt gradients. Purification usually requires further purification steps on gel-filtration and hydroxyl-apatite columns or by polymerization cycles.

FUNCTIONAL PROPERTIES OF ACTIN

Actin polymerization *in vitro*

The 'functional' form of actin in cells appears to be the filament. F-actin is a non-covalent polymer that assembles spontaneously *in vitro* under the ionic conditions believed to exist in cells, that is, about 50–100 mM KCl, pH 7.0–7.1 and 1 mM free Mg. Actin is usually prepared with bound Ca at low ionic strength. Under these conditions, actin has a high critical concentration (C_c, the concentration above which extra actin assembles into filaments). The C_c under purification conditions (in the presence of Ca and at low ionic strength) is about 0.1 mM, whereas under physiological conditions it falls to about 0.1 μM (see Table 2) and the majority of the actin spontaneously assembles. *In vitro* there is a lag phase before net assembly reaches a maximum rate, which is due to the slow rate of exchange of bound Ca by Mg allowing the formation of oligomers to nucleate filament assembly. Spontaneous nucleation is inhibited in cells by monomer sequestering proteins.

The turnover of actin filaments in cells is rapid; individual actin filaments assemble and disassemble over a period of minutes. If actin polymerization were an equilibrium system, then the filaments would assemble to equilibrium (which would leave the critical concentration of actin unassembled) and remain there, apart from some random monomer exchange at filament ends. To maintain the actin filament system in flux, a chemical switch is required that allows one state for actin that will assemble to form stable filaments, then a change in state of monomers within the filament such that it is now 'unstable' and will disassemble. The lifetime of a filament is then determined by how quickly the switch operates. This is necessarily a steady-state system and requires a continuous influx of energy to prevent its eventual collapse into the 'low energy' equilibrium state. The chemical switch for actin is the hydrolysis of

TABLE 2 Summary of actin biochemical data

Parameter	Data	References
Actin chemical and physical data		
Relative molecular mass	~43 kDa	
Number of amino acids	375 (−8 to +5)	See alignments
Isoelectric point	5.4–5.9	1947, 1763, 1931
Diffusion coefficient (G-actin, D_{w20}, $cm^2\,s^{-1}$)	5.7×10^{-7}, 5.3×10^{-7}, $4.9–6.1 \times 10^{-7}$	338
Structural data		
Monomer atomic structure	DNase I:, profilin: and gelsolin segment 1:actin crystals, myosin:F-actin structure	1260, 1261, 1262, 746, 1329, 804, 1350, 803, 788, 1320
Monomer dimensions (nm)	6.7 × 4.0 × 3.7	1260
Polymer molecular structure	Various ultrastructural techniques	487, 1265, 1296, 126, 1336, 1343, 1304, 1363, 1371, 1321, 1320, 69, 1296
Polymer diameter (nm)	7–10	1296, 1242
Helix periodicity (2-start, nm)	36–39 half-pitch	1343, 1280
1-start genetic left-hand helix	−166.2° translation per monomer 2.75 nm rise	74
2-start long-pitch right-hand helix	13–14 monomers per half-turn	1222
Angular disorder of monomers	5–6°	1222, 1201
Mean length at steady state (μm)	$\langle L \rangle n = 1.7–5.9$, $\langle L \rangle w = 2.6–7.2$	539, 246, 249, 250, 371, 296
Persistence length (μm)	10.0, 17.7	1783, 2170
Cooperativity	—	467, 1175, 1319
Stiffness		
F-actin ($pN\,nm^{-1}$)	65.3	2315
F-actin + tropomyosin, ($pN\,nm^{-1}$)	43.7	2315
Flexural rigidity	$7.3 \times 10^{-26}\,Nm^2$	1783, 499
Standard volume change for assembly ($ml\,mol^{-1}$)	−720 or +74 for Mg-F-actin, +79 for Ca-F-actin, +328 for K-F-actin	1239, 317
Binding data		
Nucleotide	Review	712
Relative affinities		
K_d (ADP/ATP) Mg-G-actin	4–12	712
K_d (ADP/ATP) Ca-G-actin	165–198	712
K_d (ADP/εATP) Mg-G-actin	(3), 2–5	530, 567, 712
K_d (ADP/εATP) Ca-G-actin	(30), 81–91	530, 567, 712
K_d (ATP/εATP) Ca- or Mg-G-actin	(0.19–0.87), 0.3–0.5	530, 555, 564, 567, 712
K_d (ATP/NSL-TP)	0.5	463
K_d (ADP/NSL-TP)	4	463
K_d (AMPPN/NSL-TP)	5	463
K_d (AMPPNP/NSL-TP)	12.5	463
K_d (AMPCPP/NSL-TP)	25	463
K_d (AMPPCP/NSL-TP)	25	463
K_d (AMP/NSL-TP)	10 000	463
ATP affinity		
(K_d, μM) cation-free G-actin	(0.5–300), 1	413, 555, 434, 1883, 1349, 570, 712
(K_d, nM) Ca- or Mg-G-actin	(0.07–100), 0.07	514, 462, 1883, 471, 315, 530, 565
(K_d, nM) DNase-G-actin	0.01	315
ADP affinity		
(K_d, nM) Mg-G-actin	0.3 (calculated from ratio)	712
(K_d, nM) Ca-G-actin	12 (calculated from ratio)	712
εATP affinity (K_d, nM) Ca- or Mg-G-actin	(5–22) 0.2 (calculated from ratio)	1883, 530
AMP.PNP (K_d, μM)	15	514, 462, 463
Inorganic phosphate and analogues		
At the γ-site		
Pi affinity (K_d, mM)	$H_2PO_4^-$: 0.47, Pi: 1.5 at pH 7	454, 568, 453
Pi stoichiometry	1 mol/mol	453
Pi dissociation ($k-$, s^{-1})	0.006	446
BeF_3 affinity (K_d, μM)	2	460, 461, 525
BeF_3 kinetics ($k+$, $M^{-1}\,s^{-1}$)	4	460
BeF_3 kinetics ($k-$, s^{-1})	8×10^{-6}	460
AlF_4 affinity (K_d, μM)	25	460, 461

TABLE 2 Continued

Parameter	Data	References
Exchange rates		555, 564, 1883, 551, 418, 475, 491, 530
ATP/ADP (Mg-G-actin, k_{-ADP}, s^{-1})	(3.3×10^{-3}–1×10^{-2}), 3.3×10^{-3} at 1 mM-Mg	712
ATP/ATP (Mg-G-actin, k_{-ATP}, s^{-1})	(5×10^{-4}–2×10^{-3}), 5×10^{-4} at 1 mM-Mg	712
ATP/ADP (Ca-G-actin, k_{-ADP}, s^{-1})	(3×10^{-2}–1.1), 1.1 at 0.1 μM-Ca	712
ATP/ATP (Ca-G-actin, k_{-ATP}, s^{-1})	(5×10^{-4}–1.5×10^{-2}), 1.5×10^{-2} at 0.1 μM-Ca	712
ATP/ADP (F-actin, $t_{1/2}$)	(2–20 h), 14 days	569, 499, 524, 506, 531
Association rates		534, 564, 555
ATP.Mg-G-actin (k_{+ATP}, M^{-1} s^{-1})	($1.1 - 6 \times 10^6$), 2×10^6 (calculated from $k_{+N}/k_{+\epsilon ATP}$)	712
ATP.Ca-G-actin (k_{+ATP}, M^{-1} s^{-1})	2×10^6	712
ADP.Mg-G-actin (k_{+ADP}, M^{-1} s^{-1})	($1.2 \times 10^6 - 3.6 \times 10^4$), 3×10^5	712
ADP.Ca-G-actin (k_{+ADP}, M^{-1} s^{-1})	1.2×10^6	712
Cations		
High affinity site	Review	34, 436, 416
Affinity		
Magnesium (K_d, nM)	(4–40), 5	376, 377, 385, 391, 34, 2182
Calcium (K_d, nM)	(1–50), 1	436, 376, 377, 391, 393, 413, 385, 418, 416, 34, 2182
Stoichiometry (cations : actin)	1	376, 391, 377, 423
Selectivity	Ca > Mn > Cd > Mg > Zn > Ni > Sr	424, 422, 551, 1277
Association rates		
Mg: ATP-G-actin (k_{+Mg}, M^{-1} s^{-1})	(2×10^5–2.5×10^5), 2.3×10^5	385, 393, 34
Mg: ADP-G-actin (k_{+Mg}, M^{-1} s^{-1})	2×10^4	(by analogy)
Ca: ATP-G-actin (k_{+Ca}, M^{-1} s^{-1})	(8×10^6–5×10^7), 2×10^7	385, 391, 393, 434, 34
Ca: ADP-G-actin (k_{+Ca}, M^{-1} s^{-1})	2×10^6	418, 393, 391, 385, 712
Dissociation rates		
Mg: ATP-G-actin (k_{-Mg} s^{-1})	(0.0012–0.0017), 0.0012	385, 416, 712, 34, 2182
Mg: ADP-G-actin (k_{-Mg} s^{-1})	(0.013–0.024), 0.018	418, 416, 712, 2182
Ca: ATP-G-actin (k_{-Ca} s^{-1})	(0.011–0.026), 0.014	391, 385, 413, 416, 393, 712, 1883, 418, 34, 2182
Ca: ADP-G-actin (k_{-Ca} s^{-1})	1.1	712
Intermediate-affinity sites	Review	34, 436, 416
Mg, Ca (K_d, mM)	(0.02–1.6), 0.15	377, 436, 49, 416
K (K_d, mM)	10	376, 377
Stoichiometry (cations : actin)	3–4	428, 436
Low-affinity sites	Review	34, 436
Mg, Ca (K_d, mM)	(5–41), 10	423, 416
K (K_d, mM)	100	416
Stoichiometry (cations : actin)	5	423
Selectivity	Ni > Mn > Ca > Mg = Sr > K	416, 423
Drugs		
Cyclosporin A		
Fluorescent cyclosporins (G-actin K_d, μM)	0.06, 0.57	2038
Cytochalasins		
Binding site	Barbed end	2065
Cytochalasin A (polymer K_d, μM)	0.15	2051
Cytochalasin B		
Binding affinities		
Monomer (K_d, μM)	51	1999
Polymer (K_d, μM)	(0.005–0.6), 0.05	2051, 2065, 2063
Stoichiometry (mol/mol actin)	1	1999
Cytochalasin C (polymer K_d, μM)	0.08	2051
Cytochalasin D		
Binding affinities		
Mg-G-actin (K_d, μM)	(2.6–4.6), 3	342, 2001
Ca-G-actin (K_d, μM)	18	
Polymer (K_d, μM)	(0.002–0.02), 0.01	1968, 2051
Stoichiometry (mol/mol G-actin)	1 (Ca), 2 (Mg)	342

TABLE 2 Continued

Parameter	Data	References
Cytochalasin E (polymer K_d, μM)	0.09	2051
Cytochalasin H (polymer K_d, μM)	0.15	2051
21,22-Dihydro-cytochalasin B (polymer K_d, μM)	0.15	2051
CB-γ-lactone (polymer K_d, μM)	0.2	2051
Chaetoglobosin B (polymer K_d, μM)	0.15	2051
Goniodomin A		
Stoichiometry (mol/mol actin)	1 : 1	1995
(F-actin, K_d, μM)	1	1995
Jasplakinolide		
K_d (F-actin nM)	15	2380
Phallotoxins		
Phalloidin (and various derivatives)		
Stoichiometry (P : A)	1 : 1	2036, 2385
K_d (polymer μM)	(0.01–0.4), 0.04	1967, 1987, 2079, 2385
k_+ (polymer M^{-1} s^{-1})	(420–520), 500	1967, 2385
k_- (polymer s^{-1})	8.3×10^{-5}–4.8×10^{-4}	1967, 2385
Tolytoxin		
(G-actin K_d, nM)	2–18	2047
Actin-binding proteins		
α-actinin (F-actin, K_d, μM)	(0.1–44), 2.4	488, 846, 663, 841, 672, 351, 2338
k_- (polymer s^{-1})	0.4, 0.7, 5.2, 9.6	663, 351, 2338
k_+ (polymer M s^{-1})	1.0–1.2×10^6	663, 351, 2338
Actobindin (G-actin, K_d, μM)	3.3 (each site 0.01 to actin dimer)	602, 2204
ADF (ATP-G-actin, K_d, μM)	0.2	862, 679, 485
ADP-G-actin (K_d, μM)	1.3	485
Calponin (F-actin, K_d, nM)	43 (smooth), 320 (skeletal)	851
CapZ (F-actin, K_d, nM)	0.2–0.6	605
Cofilin (G-actin, K_d, μM)	0.15	532
DNAseI (F-actin, K_d, mM)	0.1	515
DNAseI (G-actin, K_d, nM)	0.05–2	1403, 515
Drebrin (F-actin, K_d, μM)	0.12	2228
Stoichiometry (mol/mol actin)	1:5	2228
Dystrophin (F-actin, K_d, μM)	44, peptides: 1.3–5.7	844, 647
EF-2 (F-actin, K_d, μM)	0.85	2199
Stoichiometry (mol/mol actin)	1:8	2199
Filamin (F-actin, K_d, μM)	0.5	663
k_- (polymer s^{-1})	0.6	663
k_+ (polymer M^{-1} s^{-1})	1.3×10^6	663
Gelsolin ($G + GA_1$, K_d, nM)	1	802
Gelsolin (GA_2 + barbed end, K_d, nM)	0.07	807
MAP1 (F-actin, K_d, nM)	37	656
Myosin (S − 1) (G-actin, K_d, μM)	0.2, 0.8	624, 608, 708
Myosin light chain kinase (F-actin, K_d, μM)	13	706
Plastin (F-actin, K_d, μM)	1.8, 5.5 (+Ca)	779
Profilin (G-actin-β/γ, K_d, μM)	(0.4–1.1), 1.0	532, 727, 782
(fluorescent G-actin, K_d, μM)	3.6–30	212
(Mg.ATP-G-actin, K_d, μM)	0.5	2183
(Mg.ATP-G-actin, k_+, M^{-1} s^{-1})	4.5×10^7	2183
(Mg.ATP-G-actin, k_-, s^{-1})	10	2183
(Mg.ADP-G-actin, K_d, μM)	5	2183
SM22 (F-actin, K_d, μM)	70	715
Tensin (F-actin, K_d, μM)	0.1	2240
Barbed end (F-actin, K_d, μM)	20	2240
Stoichiometry (mol/mol actin)	1:10	2240
Thymosin β4 (Mg.ATP-G-actin, K_d, μM)	0.8, 1.7	681, 451, 2231
(Mg.ADP-G-actin, K_d, μM)	80	451, 2231
(Ca.ATP-G-actin, K_d, μM)	9	451
(Mg.ATP-G-actin, k_+, M^{-1} s^{-1})	1.5×10^6	2231
(Mg.ATP-G-actin, k_-, s^{-1})	2	2231
Thymosin β9 (Mg.ATP-G-actin, K_d, μM)	0.7	2231

TABLE 2 Continued

Parameter	Data	References
Thymosin β9 (Mg.ATP-G-actin, K_d, μM)	1.1	681
Tropomodulin (pointed end F-actin, K_d, μM)	0.2	2270
Tropomyosin (F-actin, K_d, μM)	0.18	2224
Hill coefficient	2.7	2224
Vitamin-D-binding protein (K_d, nM)	5.3	664
Assembly kinetics	Reviews	361, 508, 255, 457, 450, 448, 447
Critical concs		
Steady-state C_c		
ATP-F/G-actin (μM)	(0.07–0.36), 0.1	454, 508, 270, 538, 194
ADP-F/G-actin (μM)	(1.1–8.0), 1.0	313, 508, 454, 1968, 538, 464, 194, 365
ADP.Pi-F-actin (μM)	(0.05–0.18), 0.1	568, 538, 365
AMPPNP	0.8	194
In presence of phalloidin (μM)	0.037	1987
Barbed end		
ATP-F/G-actin (μM)	(0.07–0.23), 0.10	313, 315, 177, 195, 508, 269, 365
Pointed end		
ATP-F/G-actin (μM)	(0.20–4.0), 0.6	313, 315, 177, 192, 508, 269, 365, 538
ATPase kinetics		
F-ATP → F-ADP.Pi (s^{-1})	0.05, 0.02	184, 537
F-ADP.Pi → F-ADP (s^{-1})	0.006 ($t_{1/2} = 2$ min),	453, 452, 446
Polymerization rate constants		
Barbed end		
ATP-actin k_+ ($\mu M^{-1} s^{-1}$)	(1.4–11.6), 5	313, 315, 177, 192, 508, 269
ATP actin k_- (s^{-1})	(0.14–2.0), 1	313, 315, 177, 192, 508, 269
ADP-actin k_+ ($\mu M^{-1} s^{-1}$)	(0.75–3.8), 1	313, 508, 269
ADP-actin k_- (s^{-1})	(1.8–7.2), 6	313, 508, 269
Pointed end		
ATP-actin k_+ ($\mu M^{-1} s^{-1}$)	(0.07–2.2), 1	454, 313, 315, 177, 192, 508
ATP actin k_- (s^{-1})	(0.4–0.8), 0.5	313, 508
ADP-actin k_+ ($\mu M^{-1} s^{-1}$)	(0.05–0.16), 0.1	313, 508, 464
ADP-actin k_- (s^{-1})	(0.21–0.4), 0.2	313, 508, 464, 325
Treadmilling rates ($\mu m\ h^{-1}$)	(=2.7 nm.$k_{\text{-pointed}}$), 2.0	325
Annealing rate ($\mu M^{-1} s^{-1}$)	2.2	256
Fragmentation rate (s^{-1})	7×10^{-6}	256

Data and associated literature references are listed. Where there are many reported values for a particular parameter, the range is enclosed in brackets and the value following represents either a current consensus value or the mean of reported values. See Glossary for abbreviations

ATP. Actin assembly is accompanied by the stoichiometric hydrolysis of ATP. Each actin monomer under physiological conditions would be predicted to bind a molecule of Mg and one of ATP at their respective high-affinity sites. ATP bound to assembling monomers is hydrolysed not during the assembly step, but subsequent to incorporation within the filament to yield (eventually) an ADP-actin filament. Comparison between the rate of addition of subunits and the rate of hydrolysis of ATP in a population of rapidly assembling filaments shows that hydrolysis lags behind assembly [508]. The ATP hydrolysis step is thus uncoupled from the assembly step. Once the monomer is incorporated into the polymer (but not during the assembly step itself), there is a finite probability that the ATP will be hydrolysed. The rate constant for hydrolysis has been measured at about $0.02\,s^{-1}$ or a probability of 1 ATP hydrolysed per second per 50 monomers assembled [537]. The immediate product of hydrolysis is a monomer carrying ADP.Pi in the filament. The release of phosphate has a lower probability than hydrolysis, with a measured dissociation rate constant of about $0.006\,s^{-1}$ [446] or a probability of 1 Pi released per second per 170 monomers assembled. Thus, after a predictable period of time ($t_{1/2}$ = about 2 min), the filament which assembled as ATP-F-actin becomes ADP-F-actin via an ADP.Pi-F-actin intermediate. This is the conformational switch. Taken together, these observations predict that under conditions where the rate of assembly exceeds the rate of hydrolysis and phosphate release, an individual filament will consist of co-linear segments containing respectively ATP, ADP.Pi or ADP monomers in the same filament, in proportions according to the relative times monomers have been present since assembly and the rate constants of hydrolysis and Pi release.

There is also a difference between the C_c at opposite ends of the same filament (under physiological conditions) because of the different relative rates of monomer exchange at the pointed and barbed ends. Rates are some 20-fold faster at the barbed end than at the pointed end [313]. The slower rate of addition at the pointed end allows a high probability of ATP hydrolysis on the terminal monomer before the next subunit adds on, whilst the converse is generally true at the barbed end [464]. Thus, the pointed end will have a terminal ADP monomer, the barbed an ATP (or possibly ADP.Pi) monomer. The chemical difference at opposite ends resulting from the differences of exchange rates (and also possibly from differences in the rate of hydrolysis of the terminal subunits at opposite ends) confers different overall equilibrium constants at opposite ends. Since the equilibrium constant for the set of reactions is the reciprocal of C_c, the C_c is different at either end (see Table 2). Thus, since both ends of the same filament must necessarily share the same monomer pool and attempt to attain equilibrium at different concentrations of monomer (as determined by their respective C_c), there will be net loss of monomers from the pointed end and net gain at the barbed end leading to the effective shuttling of monomers assembled at the barbed end as ATP monomers towards the pointed end, to be released eventually at the pointed end as ADP monomers. This process is called head-to-tail assembly or treadmilling [2826] and its overall rate is determined by the slowest step in the process which is the release of ADP-actin at the pointed end [313]. In cells, conditions would predict a treadmilling rate of about $2\,\mu m\,h^{-1}$ (assuming no other constraints). Filament breakage and reannealing is also rapid *in vitro* and may contribute to filament reorganization [256].

The C_c at the barbed (ATP) end is about $0.1\,\mu$M and that at the pointed end about $0.7\,\mu$M under physiological conditions [538]. When ATP-actin assembles *in vitro* it will eventually reach the steady-state C_c, which is a concentration of monomer between the C_c at the barbed end and that at the pointed end since neither end can attain equilibrium in the same monomer pool. The greater rates of exchange at the barbed end determine that the steady-state C_c is close to $0.1\,\mu$M. It is at steady-state that treadmilling occurs. In cells, where ends may be capped, the actual C_c will depend on the relative proportions of free barbed and pointed ends. Figure 2 shows a summary of the possible assembly and disassembly reactions at both ends and their relative rates (see Table 2).

The release of phosphate from the γ-position in the nucleotide-binding site is reversible. The dissociation constant for binding of phosphate at this site is about 1mM, close to the phosphate concentration in the cytoplasmic compartment of cells. The C_c of ADP.Pi-actin ends is similar to that of an ATP-barbed end (about $0.1\,\mu$M) owing to a decrease in the off-rate constant for ADP.Pi-actin [538, 539]. Thus, changes in the concentration of Pi or pH [184] may also regulate the C_c at ends. In cells, monomers will be constrained to disassemble from high C_c ends and to assemble at low C_c ends. The spatial locations of capped ends presumably determine where in cells assembly and disassembly take place.

ADP-monomers dissociated from filaments do not significantly reassemble until the ADP has been exchanged for ATP since ATP monomers have a higher affinity for filament ends. The exchange rate is relatively slow ($0.003\,s^{-1}$ under physiological conditions) and so this step can become rate-limiting for assembly under certain conditions. However, the exchange rate can be modulated; positively (profilin) or negatively (thymosin) by several monomer-binding proteins (see nucleotide section) and this step might represent a major step for control of assembly in cells.

The monomer pool of actin in non-muscle animal cells is around, or in excess of, 50% of total cellular actin, giving a concentration for the unpolymerized actin pool greatly above the C_c (although free actin monomer will be necessarily at the C_c). This unpolymerized pool is maintained by binding of G-actin to a number of different monomer-binding proteins, the major contribution coming from thymosin β4. It is assumed that a large monomer pool diffused throughout the cytoplasmic compartment effectively uncouples assembly from disassembly by providing sufficient monomer for transient assembly anywhere in the cell and at any time. Data on the distribution and properties of the monomer pool are scarce, but attempts to visualize the differential distributions of monomer and polymer pools have indicated that monomer may also be organized within discrete structures that are associated with dynamic regions of the cell [37, 1564].

Interaction with actin-binding proteins

Although there are a large number of different actin-binding proteins in eukaryotic cells, the majority of those described fall into a smaller number of structural and functional families (see Table 1). The actin-binding proteins represent a good example of a diverse group of functionally

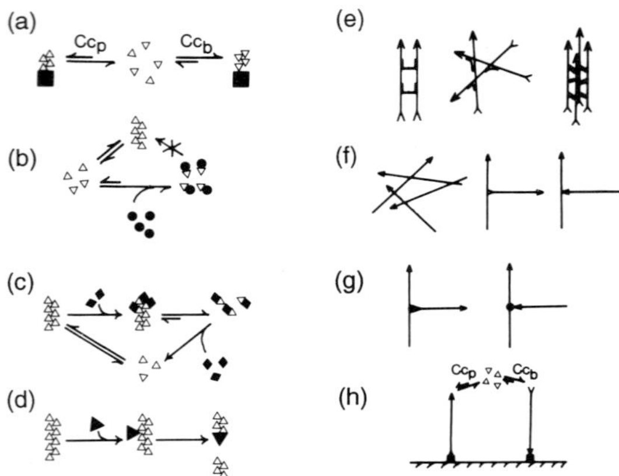

FIGURE 3. Scheme of functions of actin-binding proteins. *Actin monomers are indicated by triangles and filaments as lines with arrows indicating polarity. Actin-binding proteins are represented by solid symbols. (a) Capping proteins at opposite ends of the filament; (b) monomer sequestering proteins; (c) filament depolymerizing proteins; (d) severing/capping proteins; (e) cross-linking proteins: (f) self association of actin filaments; (g) nucleation by a cross-linking protein and (h) anchoring proteins.*

related proteins whose diversity of function is in part due to different combinations of functional domains assembled in individual proteins to provide various combinations of properties.

Functionally, the actin-binding proteins form the equivalent of the parts of some cellular construction kit (where actin is the girder) with sufficient numbers of different parts to allow a variety of unique structures to be assembled but with few enough to give a limited number of organizational patterns underlying the individual structures. The major categories of function include motor proteins (myosins), filament-stabilizing proteins (tropomyosins), filament-destabilizing proteins (cofilins), cross-linking proteins (parallel (villin, fimbrin), antiparallel (α-actinin) or disordered (filamin)) end-binding proteins (nucleation (gelsolin?), anchorage (talin) or capping (cap100)) severing proteins (gelsolin family) and monomer-binding proteins (sequestering (thymosins), regulatory (profilins)). These functions are schematized in Figure 3. In all eukaryotic cells examined to date, all categories are present. There are, in addition, a large number of actin-binding proteins whose function is poorly categorized and structures are unknown. Some of the actin-binding proteins will be covered in other issues of *Protein Profile*. The features of actin-binding proteins that seem most relevant to their function in cells are the spatial consequences of binding and the mechanical properties of the result, the relative affinities with which they bind actin, whether they share a binding site on actin with other ABPs (or whether they sterically or allosterically interfere with binding) and whether their function is regulated via signal pathways, Since ABPs and actin appear to share the same compartment and are free (transiently) to diffuse, the major experimental questions concern the preferential formation of one specific structure (rather than another) at a particular part of the cell and its function when it has assembled.

Actin Structure

THE ACTIN SEQUENCE

General sequence description

Actins form a family of eukaryotic proteins whose sequences are highly conserved, There have been several attempts to order actins into classes based on functional and sequence data. The variable sequence at the *N*-terminus has provided the simplest classification criteria, but there are other key amino acid positions that are characteristic of actins from different kingdoms and phyla. Using *N*-terminal sequence differences, actins can be divided into two classes, I and II. Class I actins have a Met-X terminus where X is an acidic residue, whilst class II actins usually have a Met-Cys-X sequence. In a few cases, the cysteine residue in class II actins is replaced by glycine or alanine [882, 981, 1092]. These differences can be discriminated by *N*-terminal processing enzymes, but apart from this do not appear to have any obvious functional correlate. Vertebrate actins have also been categorized into different classes based on the ordering of residues at the *N*-terminus [153]. Information on actin sequences, including a key for the OWL database abbreviations used in the alignment can be found in Table 3.

Actin sequences are more highly conserved than almost any other proteins. This high degree of conservation can be seen in an alignment (see Fold-out section) of the 123 unique actin protein sequences currently available Since the most extensively studied isoform of actin is that from vertebrate skeletal muscle, all other sequences have been compared with that. All numbering used in the text and figures refers to the amino acid position in skeletal muscle actin aligned with the residue in that particular sequence, as shown on the alignment. This choice of reference sequence gives some spurious apparent variation at positions where residues are only present in muscle actins; notably positions 6, 10, 16, 103, 162, 201, 225, 260 and 365. Despite this sequence property of muscle actins, there are very few amino acid positions that can be used to define precisely a particular group of actins, although there are somewhat more positions where a particular amino acid is indicative. Plants have an extra alanine at the *N*-terminus, proline at 7, an acidic at 228 and alanine at 252. Similarly, muscle actins tend to have four negative amino acids at the *N*-terminus rather than two or three (perhaps involved in more efficient activation of myosin) and the set of amino acids listed above. Echinoderms have a proline at 323 and arthropods a threonine at 234. Many other partial identifiers can no doubt be spotted.

Whilst amino acid substitutions are distributed throughout the actin sequences, there are clearly regions that are much more variable than others. The extreme *N*-terminus is highly variable, but is always acidic. This site is involved in myosin activation (see below) and is close to or part of the binding site for a number of other actin-binding proteins. For myosin, the degree of negative charge is important (see below) but the specific sequence apparently not. Other obvious regions of sequence variability are between residues 39–55 (the DNase-binding loop in subdomain IA), 187–199 (a helix in subdomain IIB which may be involved in tropomyosin binding), 223–235 (the small helix close to the filament axis), 260–280 (the proposed hydrophobic plug for interstrand interactions) and 316–325 (on subdomain IIA and proposed as a major self-assembly site). There is considerably more sequence variation in the lower eukaryotes and to a lesser extent plants. Some of this is genuine variability between organisms but these regions do include general differences between the chordate line organisms and the lower eukaryotes and plants. Very little is known about the properties of actin isolated from these variable groups save that all those looked at thus far both assemble and activate myosin. Where unusual properties exist, for example, inability to bind DNase or other actin-binding proteins, the actins tend to come from this group. Reviews on sequence comparisons can be found in Table 3. At only 13.1% of positions in the amino acid sequence are positions invariant in all actins; 25.1% of positions have substitutions of only one other amino acid (nearly always a conservative change), and in 14.1% of positions this substitution occurs

TABLE 3 Actin sequence data

Classification	Sequence ID	Common name/references	Comments
VERTEBRATES		1088, 123, 1110, 871, 24, 943, 40, 41, 116, 1255, 1062, 153, 1069, 154, 1010	Reviews
Mammalia		149, 151, 152, 450	Reviews
Homo sapiens		Human – 76, 96, 104	Reviews
Skeletal muscle	ACTS_HUMAN	1018, 980, 900, 1034, 1151, 1145, 1133, 931, 930, 874, 894, 946, 1169, 964, 1083, 1150, 1070, 929, 872, 932, 870, 954, 959, 982, 1054, 1080, 1111, *1114, *1117, *1118	Same protein sequence in mouse, rat, rabbit, chicken, cow
Cardiac muscle	ACTC_HUMAN	53, 893, 928, 894, 1019, 1150, 870, 1053, 934	Same as mouse, chicken, fetal
Vascular smooth muscle	ACTA_HUMAN	892, 719, 1141, 894, 1042, 1051, 1041, 990, 1150, 1038, 1123, 968	Same as mouse, rat, cow
Enteric smooth muscle	ACTH_HUMAN	96, 1001, 1055, 1042, 1151	Same as mouse, rat, chick gizzard
β non-muscle	ACTB_HUMAN [S38782]	1069, 965, 1087, 1145, 914, 1137, 871, 874, 894, 1007, 899, 927, 959, 968, 982, 76, 1034, 1080, 934, 1149, *1024, *1025, *1977, *875, *961, *989, *995, *1022, *1023, *1029, *1102, *1103, *1105, *1107, *1131, *1132, *1144, *960, *799, *1022	Same as mouse, rat, cow, chicken. *Mutation S38782 has Met at residue 139
γ non-muscle	ACTG_HUMAN	936, 937, 896, 888, 874, 894, 74, 959, 982, 1149, 1145, 1136, 1087, 1028, *896	Same in mouse, cow, rat, chicken. *Mutation
Oryctolagus cuniculis		Rabbit	
β non-muscle	ACTB_RABIT		
Bos bovis		Cow	
Vascular smooth muscle	ACTA_BOVIN		
Aves			
Gallus gallus		Chicken – 894	Reviews
Vascular smooth muscle	ACTA_CHICK	892, 1008	Pro not Gln at 359; so same as ACTA_HUMAN
Type 5	CHKACCY5 [A26559]	882, 894	Similar to β-isoform (same except for deletion of aas 356–363 and 368 to terminus)
Anser anser		Goose	
β non-muscle	ACTB_ANSAN	1005	
Amphibia			
Xenopus laevis		African clawed frog – 910	Reviews
α1 cardiac muscle	ACT1_XENLA	1124, 1060	
α2 cardiac muscle	ACT2_XENLA	1125	
α3 skeletal muscle	ACT3_XENLA	1124, 1058	Same as *X. tropicalis*
Type 5 non-muscle		1059	
Type 8 non-muscle		1059	
Xenopus borealis		Kenyan clawed frog	
β non-muscle	ACTB_XENBO	910	
Xenopus tropicalis		Western clawed frog	
α2 cardiac muscle	ACT2_XENTR	1125	
Pleurodeles waltii		Iberian ribbed newt	
Skeletal muscle (124 aas)	ACTS_PLEWA	1000	

TABLE 3 Continued

Classification	Sequence ID	Common name/references	Comments
Osteichthyes			
Cyprina carpio		Common carp	
β non-muscle	ACTB_CYPCA	1030, 1031	Same as grass carp *Ctenopharyngadon idella*
Skeletal muscle		887	
INVERTEBRATES			
Echinodermata		985	Reviews
Strongylocentrotus purpuratus		Purple sea urchin – 868	Reviews
Muscle (M)	ACTM_STRPU	906	
Non-muscle 1	ACTC_STRPU	905, 925, 974, 1109	In endoderm of larval stage only
Non-muscle IIa/IIb	ACT2_STRFN	890, 924, 945	Same as *S. franciscanus* actin 15B
Non-muscle IIIa/IIIb	ACT3_STRPU	868, 924, 944, 974, 1109	In embryonic ectoderm
S. franciscanus		Sea urchin	
Actin 15A	ACT1_STRFN	945, 985	
Lytechinus pictus		Sea urchin	
non-muscle I	T:LP09651_1	2101	
non-muscle II	T:LP09651_2	2101	
Pisaster ochraceus		Sea star	
Muscle	ACTM_PISOC	1009, 1011	
Non-muscle	ACTC_PISOC	1009, 1011	
Non-muscle, testicular (155 aas)	A60945	886a	Expressed in cells undergoing rapid cell division; actin FAT
Tunicata			
Styela clava		Sea squirt	
Muscle	ACTM_STYCL	1138	
Non-muscle, CA15 (213 aas)	A61043	1553	
Styela plicata		Sea squirt	
Muscle	ACTM_STYPL	Unpublished	
Non-muscle	ACTC_STYPL	Unpublished	
Halocynthia roretzi			
Muscle	ACTM_HALRO	1015	
Mollusca			
Aplysia californica		Californian sea hare	
Skeletal muscle	ACT_APLCA	915	Cells in muscle sheath around abdominal ganglia
Arthropoda – insecta		1064	Review
Drosophila melanogaster		Fruit fly – 45, 949, 979, 1036, 1044	Reviews
Larval muscle (3) 57A/B	ACT3_DROME	949, 951	
Adult muscle (4); 79B	ACT4_DROME	949, 1106	
Muscle (5); 87E	ACT5_DROME	949, 1036	
Indirect flight muscle (6); 88F	ACT6_DROME	639, 1106, 879, 949, 1135, *920, *922, *976, *993, *1104–5, *1119	*Mutations NB: 5aa errors in ACT6_DROME. JC1246; *D. simulans* is same as correct one
Non-muscle 5C	ACT1_DROME	886, 949	
Non-muscle 42A	ACT2_DROME	949	
Bactreocera dorsalis		Oriental fruitfly	
Muscle	BCCACTIN	2286	
Muscle	BCCACTINA	2286	
Muscle clone A3	BCCACTINB	2286	
Muscle clone A5	BCCACTINC	2286	
Bombyx mori		Silk moth	
Muscle A1	ACT1_BOMMO	1061	
Muscle A2	ACT2_BOMMO	1061	
Non-muscle A3	ACT3_BOMMO	1063	

TABLE 3 Continued

Classification	Sequence ID	Common name/references	Comments
Manduca sexta		Tobacco hornworm	
Actin	MOTACTINX	2300	
Anopheles gambiae		Mosquito	
Actin 1D	T:AGCTIDA_1	2298	
Arthropoda – crustacea			
Artemia sp.		Brine shrimp	
Actin 205	ACT1_ARTSX	1035, 1084	
Actin 211	ACT2_ARTSX	1035, 1084	
Actin 302	ACT3_ARTSX	992, 1035, 1084	Complete sequence, Sastre (*pers. comm.*)
Actin 403	ACT4_ARTSX	1035, 1084	
Procambarus clarkii		Crayfish	
Actin 1 (325 aas)	PRAACTIN	992	Lacks *N*-terminal 50 aas
Chelicerata			
Limulus polyphemus		Horseshoe crab	
Actin 3	T:LPACT3_1	Unpublished	
Actin 5	T:LPACT5_1	Unpublished	Acrosomal bundle
Actin 11	T:LPACT11_1	Unpublished	
Acoelomata – platyhelminthes			
Taenia solium		Tapeworm (pork)	
Actin	ACT_TAESO	891	
Echinococcus granulosus		Tapeworm	
Actin 1		912	Three genes; 1 and 2 sequenced
Actin 2		912	
Schistosoma mansoni			
Actin	T:SMACTIN	Unpublished	
Acoelomata – nematoda			
Caenorhabditis elegans		Roundworm – 939, *1012	Reviews
Actin 1	ACT_CAEEL	939, *1012, *1016	*Mutants; actins 1 and 3 identical
Actin 2	ACT2_CAEEL	939, *1012, *1016	
Actin 4	ACT4_CAEEL	939, *1012, *1016	
Onchocerca volvulus			
Actin 1	ACT1_ONCVO	1170	
[act1A]	[A48449]	1170	[1 change: V355 > Mt]
Actin 2	ACT2_ONCVO	1170	
Acoelomata – cnidaria			
Hydra attenuata		Hydra	
Non-muscle 6.2	ACT_HYDAT	943	
Podocaryne carnea			
Actin 1	JN0832	867	Only two actin genes; differ by 5 aas
Actin-4	JN0833	867	
PROTOZOA			
Ciliophora		957	Reviews
Oxytricha fallax			
Cytoplasmic (micronuclear)	ACT_OXYFA	957, 988	357 aas, complete
Macronuclear	OFAACTIN	956, 987, 988	357 aas, complete
Oxytricha nova			
Macronuclear	ACT_OXYNO	956	
Tetrahymena thermophila			
Macronuclear	ACT_TETTH	911	
Tetrahymena pyriformis			
Cytoplasmic	ACT_TETPY	977	Does not bind DNase, phalloidin, α-actinin or tropomyosin [237]
Euplotes crassus			
Cytoplasmic	ACT_EUPCR	966	

TABLE 3 Continued

Classification	Sequence ID	Common name/references	Comments
Apicomplexa			
Plasmodium falciparum		Malarial parasite	
Actin 1	ACT1_PLAFA	885, 1162	
	[S12628]		[1 change: I300 > Tr]
Actin 2	ACT2_PLAFA	1163, 1164	Only in sexual stage of life cycle – divergent sequence
Toxoplasma gondii	T:TG10429	Unpublished	
Sarcomastigophora			
Trypanosoma brucei		Sleeping sickness parasite	Identical to *T. congolese*, *T. cruzi*, *T. vivax*
Actin A	ACT1_TRYBB	880	
Actin B	ACT2_TRYBB	880	
Leishmania major			
Major actin gene	LEIACTIN	913	
Naegleria fowleri			No methyl-histidine in *N. gruberi*
Actin 1	ACT1_NAEFO	Unpublished	
Actin 2 (371 aas)	ACT2_NAEFO	Unpublished	
	NGRACTINI	Unpublished	
Entamoeba histolytica		Amoebic dysentery parasite	Does not bind DNaseI [657]
Actin A	ACT_ENTHI	926, 981	
Giardia lamblia			
Actin	T:GLACTI_1	Unpublished	
Physarum polycephalum		Slime mould – 864, 955, 963	Reviews
Plasmodial actin	ACT_PHYPO	873, 955, 963, 1065, 1146	ArdA, B and C genes all code for identical actin
Spherule actin	ACTD_PHYPO	864	ArdD gene expressed in spherules, less in *Plasmodium*
Acanthamoeba castellanii		Slime mould – 1092	Reviews
Actin 1	ACT1_ACACA	1072, 1143, 1158	
Dictyostelium discoideum		Slime mould – 1096	Reviews
Actin 1	ACT1_DIDCI	942, 1004, 1043–4, 1096, 1139, 1153, 1127, 984	
Actin 2 or A12	ACT2_DIDCI	942, 1004, 1043–4, 1096, 1139, 1153, 1127, 984	
Actin 3 or 3s1	ACT3_DIDCI	942, 1004, 1043–4, 1096, 1139, 1153, 1127, 984	
Actin 4 or 3s2	ACT4_DIDCI	942, 1004, 1043–4, 1096, 1139, 1153, 1127, 984	379 aas, complete
Actin 15 or 8	ACT8_DIDCI	942, 1004, 1043–4, 1096, 1139, 1153, 1127, 984	
PLANTS		91, 1046, 1086, 1113	Reviews
Coniferae			
Pinus contorta		Shore pine	
Actin (161 aas)	ACT_PINCO	998	
Fabeaceae			
Glycine max		Soybean – 975	
Actin 1	ACT1_SOYBN	129, 975, 1066	
Actin 3	ACT0_SOYBN	129	
Pisum sativum			
Actin 1	ACT1_PEA	Unpublished	
Actin 2	ACT2_PEA	Unpublished	
Apiaceae			
Daucus carrota		Carrot	
Actin 1	ACT1_DAUCA	1122	380 aas, complete
Actin 2	ACT2_DAUCA	1122	381 aas, complete

TABLE 3 Continued

Classification	Sequence ID	Common name/references	Comments
Cyperaceae			
Zea mays		Maize	
Actin 1	ACT1_MAIZE	1113	
[Actin 1A]	[MZEACT1G]	1113	[Gln insert at 354/355. Found in no other actin gene]
Oryza sativa		Rice – 1039, 1093	Reviews
Actin 1	ACT1_ORYSA	1039–40, 1093	
[actin 1]	[OSRAC1]	1040	[As Act1 but I51 > N]
Actin 2	ACT2_ORYSA	1039, 1093	
Actin 3	ACT3_ORYSA	1039, 1093	
Actin 7	ACT7_ORYSA	1039, 1093	
Caperaceae			
Arabidopsis thaliana		Mouse-ear cress	
Actin 1	ACT1_ARATH	1068	
Solanaceae			
Solanum tuberosum		Potato – 918	Reviews
Actin 58	ACT1_SOLTU	918	
Actin 71	ACT2_SOLTU	918	
Actin 75	ACT3_SOLTU	918	
Actin 85c (195 aas)	ACT4_SOLTU	918	
Actin 97	ACT5_SOLTU	918	
Actin 100	ACT6_SOLTU	918	
Actin 101	ACT7_SOLTU	918	
Nicotiana tabacum		Common tobacco	
Actin 25	ACT_TOBAC	1134	
Rhodophyta			
Chondrus crispus			
Actin	T:CC03676	Unpublished	
Chromophyta		Brown algae	
Costaria costata			
Actin (331 aas)	ACT_COSCS	883	
Chromophycota			
Fucus disticus			
Actin	T:FD11697_1	Unpublished	
Phycophyta		Green algae	
Volvox carterii		Volvox	
Actin	ACT_VOLCA	907	
Chlorophycota			
Acetabularia cliftonia			
Actin	T:ACACTIX_1	Unpublished	
Prymnesiophyta			
Emiliana huxleyi			
Actin 1 (365 aas)	A37431	884	Six genes – actins-1, 2, 4–6 encode same protein
Magnoliophyta			
Striga asiatica		Witchweed	
Actin	T:S68003	2305	
Sorghum vulgare			
Actin	T:SVSOAC1	Unpublished	
Actin 3		884	actin S189 > F
FUNGI			
Basidiomycotina			
Filobasidiella neoformans			
H99 actin	T:FNCTIN	Unpublished	
Histoplasma capsulatum			
Actin	T:HCCT_1	Unpublished	
Physinia graminis			
Actin	T:PGTRITACT_1	Unpublished	

TABLE 3 Continued

Classification	Sequence ID	Common name/references	Comments
Zygomycotina – oomycetes			
Achyla bisexualis			
Actin	ACT_ACHBI	883	
Phytophthora megasperma		Potato pink rot	
Actin	ACT_PHYME	923	
Phytophthora infestans		Potato late blight	
Actin 1	ACT1_PHYIN	1142	Made at all stages
Actin 2	ACT2_PHYIN	1142	Expressed at low levels
Zygomycotina – mucorales			
Absidia glauca			
Actin 1 (140 aas)	ACT1_ABSGL	Unpublished	
Actin 2	ACT2_ABSGL	Unpublished	
Deteromycotina – hyphomycetes			
Aspergillus nidulans		Mould	
Actin (γ)	ACTG_EMENI	938	
Thermomyces lanuginosus			Genus also known as *Humicola*
Actin	ACT_THELA	1165	
Candida albicans			
Actin	ACT_CANAL	1032	
Cryptosporidium parvum			
Actin	ACT_CRYPV	1002	
Ascomycotina – hemiascomycetes			
Saccharomyces cerevisiae		Budding (baker's) yeast	
Actin 1	ACT_YEAST	902, 904, 952–3, 983, 1071, 1073, *865, *1079, *1108, *1116	Same as actin 1 from *S. carlesbergensis.* *Mutation
Schizosaccharomyces pombe		Fission yeast	
Actin 1	ACT_SCHPO	1050	
Kluyveromyces lactis			
Actin	ACT_KLULA	917	
Unknown			
Pneumocystis carini			
Actin 1	PMACTINI	943a	
MISCELLANEOUS			
Amino acid methylation		930, 1143, 1158, 1475, 1398, 1117	
N-terminal processing		1098–9, 1143, 1114, 1480, 901, 1483, 1037, 1481, *904, *1092	*Review
Chromosome location		96, 951, 970, 1076, *1033	*Review
ACTIN-RELATED PROTEINS		2125	Review – suggests Arp1–3 classification
Mammalia			
Oryctolagus cuniculus		Rabbit	
Actin-RPV (376 aas)	O:ACTL_RABIT	730a	Arp1, Same as human and dog (T:HSACTINRP_1, O:S29075)
Homo sapiens		Human	
α-Centractin (376 aas)	T:HSACENT_1	898	Arp1
β-Centractin (376 aas)	T:HSBCENT_1	898	Arp1
Bos bovis		Cow	
Act2 (418 aas)	O:ACTL-BOVIN	1128	Arp3 (T:BTACT21_1)
Insecta			
Drosophila melanogaster		Fruit fly	
Arp87c (376 aas)	T:DMARP87C_1	2103	Arp1
Arp14D (395 aas)	T:DMARP14D_1	2103	Arp2, same as *A. castellani* 43 kDa, *C. elegans* ActC and *D. discoideum* 44 kDa

TABLE 3 Continued

Classification	Sequence ID	Common name/references	Comments
Arp66B	O:ACTL-DROME	2103	Arp3, same as *A. castellani* 49 kDa and *C. elegans* ActD (see also T:DMACTR66B_1)
Arp13E (398 aas)	T:DMARP_1	948	
Arp53D (376 aas)	T:DMARP53D_1	2103	
Acoelomata – nematoda			
Caenorhabditis elegans			
ActB (384 aas)		Unpublished	Arp1, same as *S. cerevisiae* Act5b
ActC (395 aas)		Unpublished	Arp2, same as *D. melanogaster* Arp14D, *A. castellani* 43 kDa and *D. discoideum* 44 kDa
ActD (418 aas)		Unpublished	Arp3, same as *D. melanogaster* Arp66B and *A. castellani* 49 kDa
PROTOZOA			
Sarcomastigophora			
Acanthamoeba castellani		Slime mould	
43 kDa (395 aas)		2145	Arp2, same as *D. melanogaster* Arp14D, *C. elegans* ActC and *D. discoideum* 44 kDa
49 kDa (418 aas)		2145	Arp3, same as *D. melanogaster* Arp66B and *C. elegans* ActD
Dictyostelium discoideum		Slime mould	
44kDa (395 aas)		Unpublished	Arp2, same as *D. melanogaster* Arp14D, *C. elegans* ActC and *A. castellani* 44 kDa
'Actin-like' (418 aas)		2296	
FUNGI			
Unknown			
Neurospora crassa			
Arp1 (380 aas)	O:ACT_NEUCR	2152	Arp1, same as *Pneumocystis* Arp1 (see also O:NEURO4CEN, T:NCROCEN_1)
? (332 aas)	T:NC14008_1	Unpublished	
Pneumocystis carini			
Arp1 (380 aas)		Unpublished	Arp1, same as *N. crassa Arp1*
ActII (385 aas)	O:PMCACTIN	Unpublished	Arp1
Ascomycotina – hemiascomycetes			
Saccharomyces cerevisiae			
Act2p (391 aas)	T:SCACT2G_1	1112	Arp2
Act3p (489 aas)	O:ACTR_YEAST	2285, 2278	(see O:S37563, T:SCACREL_1)
Act5p (384 aas)	T:SCH9315	2295	Arp1, same as *C. elegans* ActB
Schizosaccharomyces pombe			
Act2 (427 aas)	O:ACTL-SCHPO	1026	Arp3 (E:SPDNA)

This table lists actin and actin-like sequences presently in the current OWL sequence database with the exception of short sequence fragments (<100 amino acids) and indicates references for the various sequences. All the actin sequences (123) and actin-like sequences (7) were used in the actin and actin-like alignments, respectively; only complete, or nearly complete sequences were used to produce the phylogenetic trees (Figures 4a and 4b). The OWL database is version 23.2 (installed June 1994) (Bleasby, A. J. & Wootton, J. C. 1990, *Protein Engineering* **3:** 153–160).

TABLE 4 Summary of actin mutants and their effects

Position	Expression	Gene	Results	References
C0D,N,H	*in vitro*	Hs α	(−) Processing	1114
C0S,G,F,Y	*in vitro*	Hs α	Not processed	1114
del0	*in vitro*	Hs α	(+) Processed	1483
D1H, D1H/D4H	Dicty, pure	*ACT15*	<Motility; <V_{max}; <ATPase;	1126, 1127
D1/S2 to DED	Yeast, pure	*ACT1*	=Motility; >ATPase; =K_m	902
D1/S2 to DED	Yeast, pure	*ACT1*	=Polymerisation (caldesmon induced)	909
del1-12	*in vitro*	*Act88F*	>Profilin; =ATP; =DNase; zero ADP-ribosylation	639, 1238
D2N/E4Q	Yeast, pure	*ACT1*	<ATPase; =polymerization	901
D2N/E4Q	Yeast, *in vivo*	*ACT1*	(+) Growth; =processing; (−) F-actin	901
D2N/E4Q	Yeast, *in vivo*	*ACT1*	<Caldesmon; <caldesmon induced polymerization	909
del2-4	Yeast, pure	*ACT1*	<ATPase; =polymerization (<when caldesmon induced)	901, 909
del2-4	Yeast, *in vivo*	*ACT1*	(+) Growth; (−) processing; (−) F-actin	901
del2-4	Yeast, pure	*ACT1*	<Caldesmon inhibition of S1 ATPase; =caldesmon	909
D2V	Yeast, *in vivo*	*ACT1*	(+) Growth	983
D2A	Yeast, *in vivo*	*ACT1*	Recessive cold, heat sensitive; <spore viability	1160, 2289
D3A,H,N	*in vitro*	Hs α	=Polymerization; =S-1; =DNase; =NDG	1118, 1132
D3N	*in vitro*	Hs α	(−) Processed	1114
del3-4, D3K/D4K,D3A/D4A	Yeast, pure	Ch β	=DNase; =polymerization; (−) S-1	585
del3-4,D3A/D4A	Yeast, pure	Ch β	<ATPase; <velocity; =tropomyosin	877
D3K/D4K	Yeast, pure	Ch β	<ATPase; no velocity; =tropomyosin	877
E4V,A	Yeast, *in vivo*	*ACT1*	(+) Growth	983, 1160, 2289
D4E	Dros, *in vivo*	*Act88F*	(+) Structure	1095
D4H	Dicty, pure	*ACT15*	<Motility	1126
G6A	Dros, *in vivo*	*Act88F*	(−) Structure	1095
G6A/A7T	Dros, *in vivo*	*Act88F*	Flightless; (−) structure	1094
I10V	Dros, *in vivo*	*Act88F*	(+) Structure	1095
D11E,N, D11N/N12D	*in vitro*	Hs α	=Polymerization; =S-1; <DNase; <NDG; <processing	1118
D11H	*in vitro*	Hs α	<Polymerization; =S-1; <DNase; <NDG; <processing	1118
D11Q,K	Yeast, *in vivo*	*ACT1*	Dominant lethal	983
D11A	Yeast, *in vivo*	*ACT1*	Partial dominant lethal; <profilin	1160, 2233
D11E	Yeast, *in vivo*	*ACT1*	(+) Growth; viable spores	903
D11N	Yeast, *in vivo*	*ACT1*	(+) Growth; no viable spores	903
S14C, S14G	Yeast, *in vivo*	*ACT1*	Lethal	2188
S14T	Yeast, *in vivo*	*ACT1*	(+) Growth	2188
S14T	Yeast, pure	*ACT1*	(+) Polymerization	2188
S14A	Yeast, *in vivo*	*ACT1*	>Heat sensitivity in growth	2188
S14A	Yeast, pure	*ACT1*	>rate polymerization; (+) critical; <F-actin ATPase; <ATP affinity; <heat stability	2188, 2310
D24H/D25H	Dicty, pure	*ACT15*	Zero motility	984, 1126
D24A/D25A	Yeast, *in vivo*	*ACT1*	Recessive cold, heat sensitive	1160, 2289
R28C	Dros, *in vivo*	*Act88F*	Strong antimorph; weak hsp induction	1057, 994

TABLE 4 Continued

Position	Expression	Gene	Results	References
R28L	Cells	A^x	Causes melanoma cell-line	1103
P32L	Yeast, *in vivo*	*ACT1*	Temperature sensitive	1116
G36E/E83D	Dros, *in vivo*	*Act88F*	(−) Structure	1104
G36E/E83D/G245D	Dros, *in vivo*	*Act88F*	(+) Structure; flighted; hsp induction	1104
G36E/E83D/G245D	Cells	Hs β	(+) Morphology; <incorporation into structures; <DNase	1024, 1025
R37A/R39A	Yeast, *in vivo*	*ACT1*	Recessive cold, heat sensitive	1160, 2233
P38A	Yeast, pure	Ch β	<Polymerization; <motility; >profilin	878
K50A/D51A	Yeast, *in vivo*	*ACT1*	Recessive cold, heat sensitive	1160, 2289
D56A	Yeast, *in vivo*	*ACT1*	< > fimbrin binding; (+) heat sensitivity	865, 2290
D56A/E57A	Yeast, *in vivo*	*ACT1*	Recessive heat sensitive; =cold sensitive	1160, 2289
A58T	Yeast, *in vivo*	*ACT1*	Temperature sensitive	1116
K61N	Yeast, *in vivo*	*ACT1*	< > fimbrin binding; =heat sensitivity	865, 2290
K61A/R62A	Yeast, *in vivo*	*ACT1*	Partial dominant lethal; < profilin	1160, 2289
R68A/E72A	Yeast, *in vivo*	*ACT1*	(+) Growth; < profilin	1160, 2233, 2289
H73Y	*in vitro*	Hs α	<Polymerization; =DNase; =processing	1117
H73Y,R	Cells	Hs α	(+) Growth	1117
H73R	*in vitro*	Hs α	=Polymerization; =DNase; =processing	1117
I76F	Dros, *in vivo*	*Act88F*	Strongly antimorphic; weak hsp; induction; (−) Z discs	994, 1057, 1095
T79term	Dros, *in vivo*	*Act88F*	No actin accumulation	1082
D80A/D81A	Yeast, *in vivo*	*ACT1*	Recessive cold, heat sensitive	1160, 2289
E83A/K84A	Yeast, *in vivo*	*ACT1*	Recessive cold, heat sensitive	1160, 2289
H88Y	Yeast, *in vivo*	*ACT1*	< > fimbrin binding; >heat sensitivity	865, 2290
T89I	Yeast, *in vivo*	*ACT1*	< > fimbrin binding; >heat sensitivity	865, 2290
E93K	Dros, *in vivo*	*Act88F*	(−) Structure; no Z-discs	1119
E93A/R95A	Yeast, *in vivo*	*ACT1*	Recessive lethal	1160, 2289
E99H/E100H	Dicty, pure	*ACT15*	<Motility	984, 824
E99A/E100A	Yeast, *in vivo*	*ACT1*	Recessive heat sensitive; =cold	1160, 2289
R116A/E117A/KII8A	Yeast, *in vivo*	*ACT1*	Recessive heat sensitive; =cold; <profilin	1160, 2233, 2289
S129V	Dros, *in vivo*	*Act88F*	(+) Structure	1095
V139M	Cells, *in vivo*	β-actin	Partially responsible for cytochalasin resistance	2045
D154A/D157A	Yeast, *in vivo*	*ACT1*	Partial dominant lethal	1160
F169Y/A260S	Dros, *in vivo*	*Act88F*	(−) Structure	1095
F169Y/C257T	Dros, *in vivo*	*Act88F*	(+) Structure	1095
L176M	*in vitro*	*Act88F*	>Profilin; =ATP; =DNase; =NDG; (+) ADP-ribosylation	639, 1238, 2010
R177Q	*in vitro*	*Act88F*	>Profilin; =ATP; =DNase; =NDG; zero ADP-ribosylation	639, 1238, 2010
R177A/D179A	Yeast, *in vivo*	*ACT1*	Recessive heat sensitive; =cold; <profilin; <phalloidin	1160, 2233
R177A/D179A	Yeast, *in vivo*	*ACT1*	F-actin *in vivo*; unlabellable with rhodamine-phalloidin	637
R183A/D184A	Yeast, *in vivo*	*ACT1*	(+) Growth	1160, 2289
D187A/K191A	Yeast, *in vivo*	*ACT1*	(+) Growth	1160, 2289
K191M,K191M/C374A	Yeast, *in vivo*	*ACT1*	(+) Growth	983
E195A/R196A	Yeast, *in vivo*	*ACT1*	(+) Growth	1160, 2289

E205A/R206A/E207A	Yeast, *in vivo*	*ACT1*	Possibly dominant lethal	1160, 2289
R210A/D211A	Yeast, *in vivo*	*ACT1*	Recessive weak heat sensitive; =cold	1160, 2289
K213A/E214A/K215A	Yeast, *in vivo*	*ACT1*	Recessive cold; heat sensitive; low spore viability; <profilin	1160, 2233, 2289
D222A/E224A/E226A	Yeast, *in vivo*	*ACT1*	Recessive heat sensitive; =cold; <profilin	1160, 2233
T234S	Dros, *in vivo*	*Act88F*	(+) Structure	1095
E237A/K238A	Yeast, *in vivo*	*ACT1*	Partial dominant lethal	1160
E241A/D244A	Yeast, *in vivo*	*ACT1*	Partial dominant lethal	1160
G245D	Cells	Hs β	<Polymerization	903, 1024, 1025
G245D	Dros, *in vivo*	*Act88F*	Antimorphic	1075, 1105
G245D	Cells	Hs β	Stable; (−) morphology at high expression levels	1024, 1025
G245D	Yeast, pure	Ch β	<Polymerization; <velocity; =S-1; >K_m	875
G245K	Yeast, pure	Ch β	<Polymerization; =S-1; =velocity	875
E253A/R254A	Yeast, *in vivo*	*ACT1*	Partial dominant lethal	1160
R256A/E259A	Yeast, *in vivo*	*ACT1*	Recessive cold; heat sensitive	1160
L266D	Yeast, *in vivo*	*ACT1*	Recessive cold sensitive viability and polymerization	895
L266D	Yeast, pure	*ACT1*	>cold sensitivity in polymerization	895
del269-272	Yeast, *in vivo*	*ACT1*	(+) Growth	1327
E270A/D275A	Yeast, *in vivo*	*ACT1*	Recessive lethal	1160, 2289
V278T	Dros, *in vivo*	*Act88F*	(+) Structure	1095
D286A/D288A	Yeast, *in vivo*	*ACT1*	Recessive lethal	1160
I289F	Dros, *in vivo*	*Act88F*	Hypomorphic; myofibrillar degeneration	1104
R290A/K291A/E292A	Yeast, *in vivo*	*ACT1*	Recessive lethal	1160
A295D	Cells, *in vivo*	β-actin	Partially responsible for cytochalasin resistance	2045
T304G/M305T/Y307Term	Dros, *in vivo*	*Act88F*	Unstable; hsp inducing	1082
E311A/R312A	Yeast, *in vivo*	*ACT1*	Recessive cold, heat sensitive; <spore viability	1160, 2289
K315A/K316A	Yeast, *in vivo*	*ACT1*	(+) Growth	1160, 2289
E316K	Dros, *in vivo*	*Act88F*	Flightless; almost =muscles; (−)mechanics; weak hsp induction	919, 1160
E316K	*in vitro*	*Act88F*	=Polymerization; =NDG; <thermodynamic stability	919, 920
E316K	*in vitro*	*Act88F*	>Profilin; =ATP; =DNase	639
K326A/K328A	Yeast, *in vivo*	*ACT1*	Possibly dominant lethal	1160
E334K	Dros, *in vivo*	*Act88F*	(−) Structure; weak hsp induction	919
E334K	*in vitro*	*Act88F*	=Polymerization; =NDG; =thermodynamic stability	919, 920
E334K	*in vitro*	*Act88F*	>Profilin; =ATP; =DNase; =ADP-ribosylation	639, 2010
E334A/R335A/K336A	Yeast, *in vivo*	*ACT1*	Recessive lethal	1160, 2233, 2289
K336M,Q	Yeast, *in vivo*	*ACT1*	(+) Growth	983
V339I	Dros, *in vivo*	*Act88F*	No actin accumulation	919
V339I	*in vitro*	*Act88F*	=Polymerization; =NDG; <thermodynamic stability	919, 920
V339I	*in vitro*	*Act88F*	>Profilin; =ATP; =DNase; =ADP-ribosylation	639, 2010
Y356Term	Dros, *in vivo*	*Act88F*	Antimorph, some actin (?); strong hsp; nuclear swelling	993, 1082
W356Y,H	Yeast, *in vivo*	*ACT1*	(+) Growth	983
K359A/E361A	Yeast, *in vivo*	*ACT1*	(+) Growth; <profilin	1160
Q360E	Dros, *in vivo*	*Act88F*	(+) Structure	1095
E360H/E361H	Dicty, pure	*ACT15*	=Motility	984, 1126
E361A/D363N	*in vitro*	*Act88F*	>Profilin; =ATP; =DNase; =NDG	639
Y362C,I	*in vitro*	*Act88F*	>Profilin; <ATP; <NDG; =DNase	639
Y362V,L	*in vitro*	*Act88F*	<Profilin; <ATP; <NDG; =DNase	639

TABLE 4 Continued

Position	Expression	Gene	Results	References
Y362F	*in vitro*	*Act88F*	=Profilin; =ATP; =NDG; =DNase	639
D363H/E364H	Dicty, pure	*ACT15*	=Motility	984, 1126
D363Y	*in vitro*	*Act88F*	=Profilin; =ATP; =NDG; =DNase	639
D363H	*in vitro*	*Act88F*	>Profilin; =ATP; =NDG; <DNase	639
D363A/E364A	Yeast, *in vivo*	*ACT1*	Recessive heat sensitive; =cold; <profilin	1160, 2233, 2289
E364K	Dros, *in vivo*	*Act88F*	(−) Structure; strong hsp induction	919
E364K	*in vitro*	*Act88F*	=Polymerization; =thermodynamic stability; <NDG	919, 920
E364K	*in vitro*	*Act88F*	>Profilin; <ATP; <DNase; =ADP-ribosylation	639, 2010
G366S	Dros, *in vivo*	*Act88F*	Antimorphic; strong hsp induction	976, 1082
G366D	Dros, *in vivo*	*Act88F*	(−) Structure; strong hsp induction	919
G366D	*in vitro*	*Act88F*	=Polymerization; =thermodynamic stability; <NDG	919, 920
G366D,P,A	*in vitro*	*Act88F*	>Profilin; <ATP; =DNase	639
G368S	Dros, *in vivo*	*Act88F*	(+) Structure	1095
G368E	Dros, *in vivo*	*Act88F*	Flightless; almost (+) muscles; (−) mechanics	922
G368E	*in vitro*	*Act88F*	=Polymerization; =NDG; =thermodynamic stability	919, 920
G368E	*in vitro*	*Act88F*	=Profilin; <ATP; =DNase; =ADP-ribosylation	639, 2010
G368T,S,Q	*in vitro*	*Act88F*	=Profilin; =ATP; =DNase; =NDG	639
G368K	*in vitro*	*Act88F*	>Profilin; =ATP; =DNase; =NDG	639
G368Term	*in vitro*	*Act88F*	>Profilin; <ATP; =DNase; =NDG	639
R372H	*in vitro*	*Act88F*	=Polymerization; =NDG; =thermodynamic stability	919, 920
R372H	*in vitro*	*Act88F*	>Profilin; >ATP; =DNase; =ADP-ribosylation	639, 2010
K373Term	Yeast, *in vivo*	*ACT1*	Lethal	983
C374A	Yeast, *in vivo*	*ACT1*	(+) Growth	983
C375S	Yeast, pure	Ch β	<Polymerization; <Motility; No Profilin binding	878
C374Term	Yeast, *in vivo*	*ACT1*	Temperature sensitive	983
6P375Term	Yeast, *in vivo*	*ACT1*	Temperature sensitive	983

Site of mutation: mutants are named by using the one letter amino acid code with the 'wild type' amino acid first, followed by its position in the sequence (using skeletal muscle numbering) and the new inserted amino acid. Deletions are denoted by 'del', combinations of mutations in the same protein are separated by a solidus (/) and alternative mutations at the same site separated by commas but no spaces. Mode of expression: *in vitro*, expression in rabbit reticulocyte lysate; *pure*, purified from named source; *in vivo*, expressed in the host. Gene names: see glossary and Hs α and Hs β, human α skeletal and β-non-muscle actin genes; and Ch β, chicken β-non-muscle actin gene. Organism names: Dicty, *Dictyostelium*; Dros, *Drosophila*. Properties of mutant actin: NDG, mobility on non-denaturing gels; thermodynamic stability, as determined on urea gradient gels; =, the same as wild-type actin; >, better than wild-type actin; <, less than wild-type actin; (+), normal-; (−), abnormal- structure, muscle appearance by light or electron miscroscopy; V_{max}, for myosin ATPase, K_m, for the binding of actin to myosin; ATPase, myosin ATPase; Motility, velocity in the *in vitro* motility assay; Mechanics, mechanical response of dissected myofibrils; DNase- (DNaseI), ATP-, S1- (S1 fragment of myosin), profilin- or tropomyosin- binding to each ligand relative to wild-type actin; F-actin, production of actin filaments; Processing, *N*-terminal processing; Heat- or cold-sensitive, inability of yeast to grow at high or low temperatures; Antimorphic, mutant actin preventing wild-type actin from functioning normally *in vivo*; Hypomorphic, produces normal actin but at a reduced level; hsp, induces heatshock protein synthesis; ADP-ribosylation, by *Clostridium* toxin ADP-ribosylating activity in only one actin sequence. For most positions amino acids are identical in the majority of actin sequences; in 248 positions at least 95% of the actins have the same amino acid. Invariant amino acids (34) occur at positions:

D11	N12	K18	G20	Y69	G74	M82
T106	E107	N115	E117	E125	L140	G146
T148	G150	V152	G156	D157	G158	V163
G168	H173	G182	G245	R290	S300	G301
M305	P322	K328	R335	W340	G342	

The number of invariant amino acids will of course decrease as more sequences are published. In 1993, with 103 unique actin sequences there were 57 unique residues; in 1994, with 123 unique sequences, the number of invariant residues was 49; in this edition with 141 unique sequences there are 34 invariant sequences. Some of the single amino acid changes among the total of 14.1% may be due to sequencing errors. Estimates of sequencing errors in the databases vary but this class of substitution is the most likely repository of such errors. It is intriguing that in the sequence of osrac1 (rice actin 1) the previously invariant residue G366 is changed to D, a change engendering mutant effects in act6_drome (see Table 4) although in the rice gene this is presumably compensated for by substitutions at other residues.

Evolutionary relationships within the actin family

A phylogenetic tree produced from the alignment data (Figure 4a) is in broad agreement with previous trees constructed using similar methods but with fewer sequences [66, 91, 96, 972, 1062] and with the taxonomic arrangement of sequences in both the alignment and actin sequence table. As a consequence of the high conservation of most actins, the tree-forming computer programs produce relatively few groupings which are reproducible and therefore considered significant [91, 1062] The major branches appear to correspond to conventional taxonomic classifications. The plant actins form a distinct group, as do the yeasts, myxomycetes, filamentous fungi, echinoderms, arthropods and chordates. The protozoa seem to fall into two distinct groups, with the slime moulds forming a separate group from the others. There are a small number of actins which seem to occur within unusual groups or branches. Such instances probably serve to emphasize the small amount of variation that occurs in actin sequences and the lack of significance in some branches and branchpoints. The position of the *Pneumocystis carinii* actin within the fungal group was recently used by Fletcher *et al.* [943a] to classify this taxonomically difficult species.

Conserved amino acids that may define different groups have been proposed for vertebrates [96], insect muscle [1062] and plants [91]. Chordate muscle and non-muscle actins form two distinct groups, which has been taken to indicate that the emergence of muscle and cytoplasmic specific actins predated chordate speciation [96]. Indeed, vertebrate cytoplasmic actins have greater similarity to arthropod actins than to vertebrate muscle actins [66] indicating a very ancient common origin for actin genes [1062]. Intriguingly, the amino acid differences occur in clusters in the atomic structure [396], a feature that cannot be appreciated from the primary amino acid sequence. The largest cluster includes amino acids 1–6, 103, 129, 354 and 357, which are included in the myosin head binding region and may reflect the different myosin isoforms encountered by these two actins. The other clusters include residues 153, 162, 176 and 298 in the central β-sheet of domain IIA and residues 225, 259, 266 and 271, which occur in the lower portion of domain IIB, close to domain IIA.

Insect muscle actins from *Drosophila melanogaster* (Diptera) and *Bombyx mori* (Lepidoptera) form a separate group from their cytoplasmic actins [1062, 1064]. Artemia (crustacean) actins divide into two groups, two of them (act1_artsx and act2_artsx) are close to the insect muscle actins. The other two act4_artsx and the newly completed act3_artsx sequence appear close together and nearer to, but not within, the branch containing the insect non-muscle actins. This may not be significant. Nine amino acid substitutions define the insect muscle group of actins. They do not cluster in the atomic structure. Four of the substitutions occur at positions included in those which define the mammalian skeletal muscle actins [1062] and only Ile-76 is conserved in both groups.

It seems very likely that muscle actin genes arose from cytoplasmic genes at least twice, and independently during animal evolution [1062, 1064]. Vertebrate muscle actin genes would have appeared in the deuterostome lineage [14, 154] whilst insect muscle actin genes would have emerged in the protostome lineage, after the separation of molluscs and arthropods, by duplication of ancestral genes. The ancestral genes of these insect muscle actins, which include subgroups for larval and adult musculatures, were clearly present before the separation of the Diptera and Lepidoptera [1062, 1064].

The generally high conservation of actin sequences means that, unlike the situation in many other protein families, it is difficult to identify important sites on the protein by simply seeking those regions with the most conserved sequences. Conversely, regions of sequence variation may

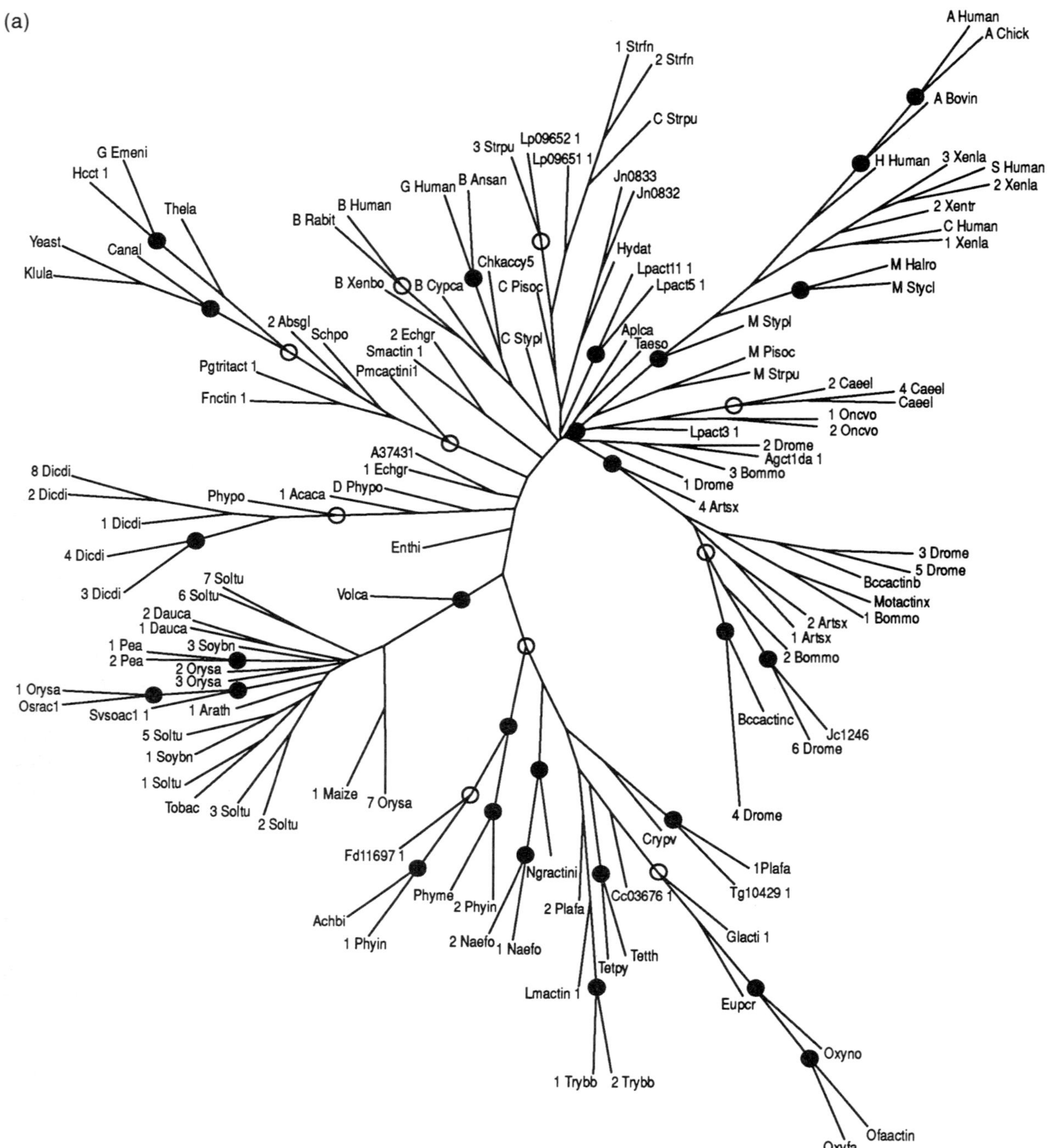

FIGURE 4 Phylogenetic tree of available actin and actin-related sequences. *(a) Phylogenetic tree of conventional actin sequences. The actin alignment data were used to produce a phylogenetic tree of all the actin sequences except for the following incomplete sequences: Owl:A60945, Owl:A61043, humactgaa, praactin, act_pinco, act_oxyfa, act4_soltu, acts_plewa, act_coscs, and act1_absgl. The tree was produced on Seqnet (SERC, Daresbury) from the edited PILEUP sequence alignment using routines (SEQBOOT, PROTDIST, NEIGHBOR, CONSENSE and DRAWTREE) of the PHYLLIP computer package (Phylogeny Inference Package version 3.5c, available from Dr J. Felsenstein, Department of Genetics, University of Washington, Seattle, USA). Similarity between sequences in PROTDIST was estimated using Kimura's approximation (Kimura M. (1983)* The Neutral Theory of Molecular Evolution. *Cambridge University Press, Cambridge). A thousand bootstraps were used. Figures in cartouches show the percentage of replicate trees which contained these branchpoints (branches occurring in >95% of replicates shown as solid circles and >80%, >95% as open circles, respectively). (b) Phylogenetic tree of actin-related sequences. The actin-like sequence alignment data was used to produce a phylogenetic tree of all the actin-related sequences except for the actr_yeast (yeast actin-related) sequence. The tree was produced on Seqnet (SERC, Daresbury) from the edited PILEUP sequence alignment as described in (a). A thousand bootstraps were used.*

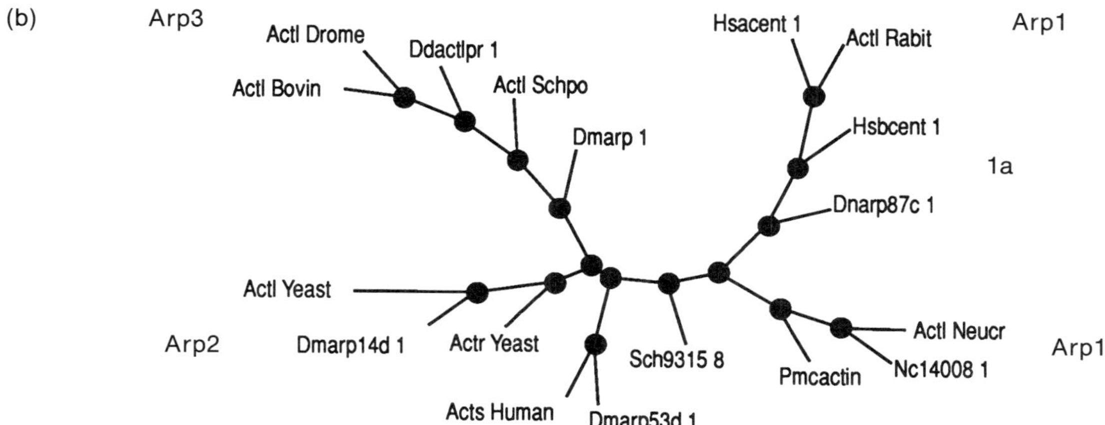

FIGURE 4 *Continued.*

identify either sequences at which the constraints are reduced or sequences which are important for the functions of specific isoforms. With actin there has been considerable interest in variable sequence regions and the potential for such regions to provide functional isoform differences.

Intron positions

Intron positions in the actins are much more characteristic of different phylogenetic groupings than are the sequence differences (Figure 5a). In the chordate line species, introns at position 41/42 are conserved whilst those at position 121/122 are conserved in all but vertebrate striated muscle actins, and those at 204 by all but vertebrate cytoplasmic actins. Introns at position 84/85 are characteristic of smooth muscle but are also found in coelenterates. All vertebrate actins sequenced to date have introns at positions 41/42, 267 and 327/328.

Lower invertebrates and arthropods have very few introns; those which are found occur at variable positions and do not echo those found in the chordate line. Genes from protozoans thus far sequenced have no introns whilst those of some fungi have an intron between codons 3 and 4. All angiosperm plants thus far sequenced have introns at positions at 19/20, 150/151 and 355/356.

A recent review of intron positions [158] examines the relationship of intron positions and secondary structure domains in actin (using the vertebrate actin structure). This review concluded that the position of introns is not correlated with the ends of such domains. However, if a slightly different analysis is performed in which an estimate of the probability of an intron occurring at an amino acid in either loop structure, β-sheet, α-helix or at or within 1.5 amino acids of a junction between any of these domains, then a rather different conclusion can be reached. There are seven intron positions within the approximate 190 amino acids in loop structure, five introns within the 126 amino acids in α-helix structure, three introns in the 57 amino acids in β-sheet structure, but 18 introns at the junctions. These give approximate probabilities that introns occur at amino acid residues in the various structural positions of 1:27, 1:26, 1:19 or 1:3 respectively, suggesting a high probability that introns occur at domain junctions. These positions have been marked on the diagram of the actin structure (Figure 5b) from which estimates of the number of amino acids in domains have been obtained. It may also be interesting to note in this context that some of the insertions/deletions in the actin-related proteins or actin 2s occur at sites close to intron positions; notably those around residues 41, 94, 198 and 320 (see Figure 6). Note that there may be small differences between published positions and those used in this review since all positions quoted are at those specified by the alignment with skeletal muscle actin and are therefore the actin 'consensus' position rather than the actual amino acid number of that particular actin. This seemed a logical way to proceed.

Comparison of actin isoform function

Differences in the function of actin isoforms have been sought both in intact cells and by more direct biochemical analysis [for review see 65, 123]. Actin isoforms segregate into different structures within cells [279, 1575, 1576, 1636, 1659, 1662, 1782, 5420] suggesting that they do have specialized functions and that there are mechanisms that give rise to a differential localization. In addition to mechanisms involving differential affinity to other proteins, it has been observed that mRNAs for different isoforms are also differen-tially distributed in cells in a distribution similar to that of their protein products [1601], suggesting the possibility that isoforms are delivered on site. Purified actin isoforms and actins from different species also differ in their intrinsic properties such as thermodynamic stability [920, 1349] and polymerization characteristics [51, 252, 316, 334, 971]. Actins from different species also differ in their affinity for profilin [727, 776], DNase I [1779, 1853, 2007], α-actinin and tropomyosin [2007]. In binding assays for non-muscle actins [51, 972, 1928, 1944] or rabbit muscle actin isoforms [972] with various rabbit myosin isoforms, significant differences in affinities were found.

The importance of isoform differences *in vivo* are not clear, but the presence of multiple actins with slight changes in biochemical properties both within an organism and especially within single cells would appear to indicate a need for different isoforms with specific functions. Such considerations probably explain the occurrence in so many species of multigene actin families. The physiological significance of isoform differences can be tested by substituting one actin isoform for another in a whole cell, plant or animal, although few studies of this type have been made to date. No difference was found in the incorporation of cardiac and cytoplasmic actin isoforms into the cytoskeleton of cultured non-muscle cells expressing a transformed cardiac actin gene. However, transformation of adult rat cardiomyocytes with different actin isoforms showed isoform-specific effects on both their incorporation into the myofibrils and cell morphology. Despite only 90% similarity the chicken β-actin (same amino acid sequence as actb_human) is able to substitute for the yeast *Saccharomyces cerevisiae ACT1* gene (act_yeast) without loss of viability, although the cells had altered morphology, slowed growth and temperature-sensitive lethality [1616]. Nine mutations encoding isoform-specific amino acid changes were introduced into the *Drosophila* flight muscle-specific actin gene *Act88F* (act6_-drome) and replaced in flies by P-element transformation. Only two of the mutations affected muscle structure. Given the large number of cloned actin genes and the availability of transformation systems for a wide variety of organisms and cell types the questions about isoform-specific functions can be expected to receive increased attention in the near future.

Actin-related sequences

Although most actins are highly conserved in sequence and length (374–376 amino acids), more

5(a)

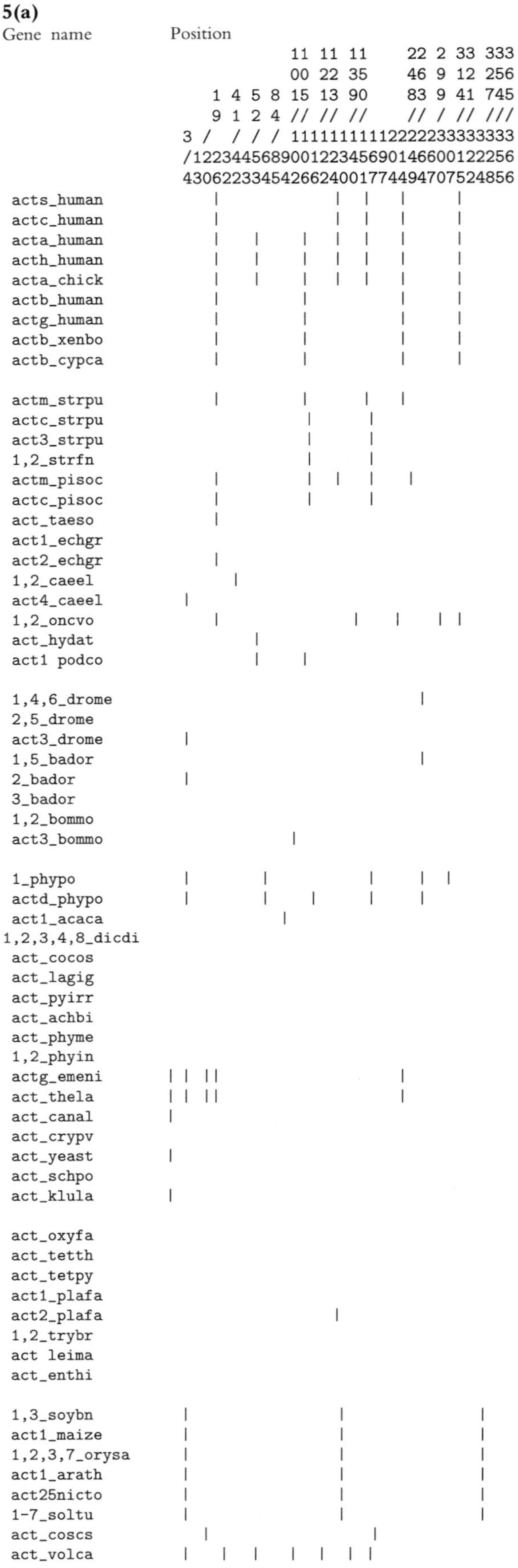

```
Gene name         Position
                             11 11 11     22 2 33 333
                             00 22 35     46 9 12 256
                     1 4 5 8 15 13 90     83 9 41 745
                     9 1 2 4 // // //     // / // ///
                  3 /  / / / 111111111122222233333333
                  /1223445568900122345690146600122256
                  43062233454266624001774494707524856
acts_human           |            |  |   |     |
actc_human           |            |  |   |     |
acta_human           |   |    |   |  |   |     |
acth_human           |   |    |   |  |   |     |
acta_chick           |   |    |   |  |   |     |
actb_human           |        |          |     |
actg_human           |        |          |     |
actb_xenbo           |        |          |     |
actb_cypca           |        |          |     |

actm_strpu           |        |      |   |
actc_strpu                     |     |
act3_strpu                     |     |
1,2_strfn                      |     |
actm_pisoc           |         |  |  |    |
actc_pisoc           |         |     |
act_taeso            |
act1_echgr
act2_echgr           |
1,2_caeel              |
act4_caeel        |
1,2_oncvo            |              |   |   |  |
act_hydat              |
act1 podco             |    |

1,4,6_drome                                |
2,5_drome
act3_drome        |
1,5_bador                                  |
2_bador           |
3_bador
1,2_bommo
act3_bommo                   |

1_phypo           |       |          |     |  |
actd_phypo        |       |    |     |     |
act1_acaca                  |
1,2,3,4,8_dicdi
act_cocos
act_lagig
act_pyirr
act_achbi
act_phyme
1,2_phyin
actg_emeni      | | ||                   |
act_thela       | | ||                   |
act_canal       |
act_crypv
act_yeast       |
act_schpo
act_klula       |

act_oxyfa
act_tetth
act_tetpy
act1_plafa
act2_plafa                        |
1,2_trybr
act leima
act_enthi

1,3_soybn         |               |              |
act1_maize        |               |              |
1,2,3,7_orysa     |               |              |
act1_arath        |               |              |
act25nicto        |               |              |
1-7_soltu         |               |              |
act_coscs           |                 |
act_volca         |   |  |   |  |  | |
```

5(b)

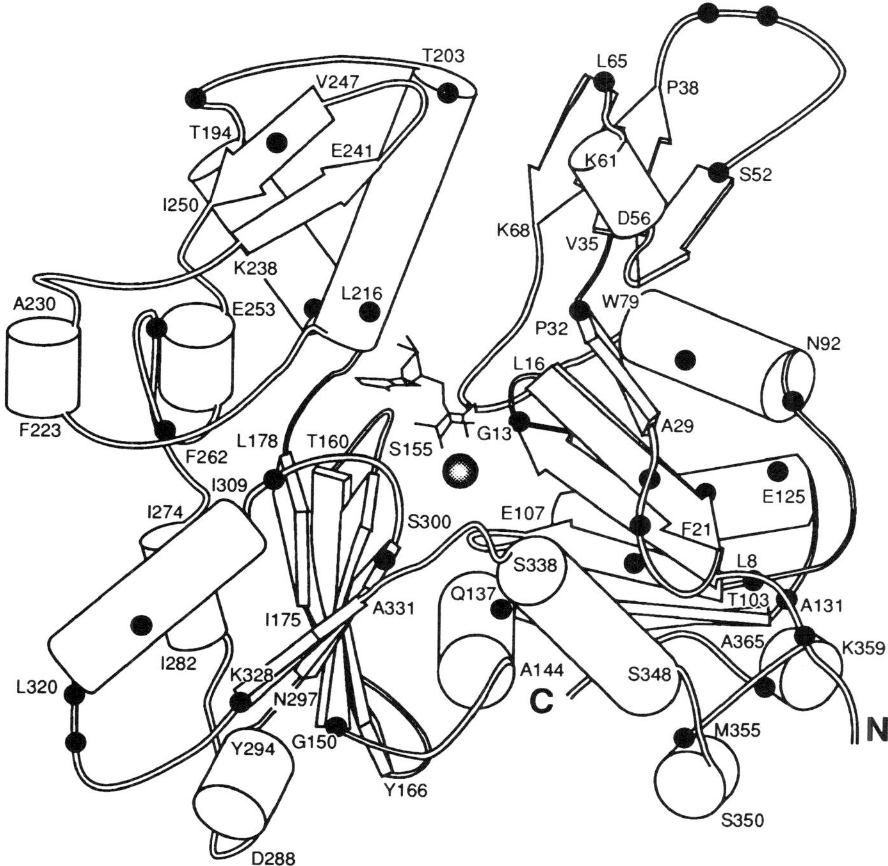

FIGURE 5 Intron positions in actin. *Intron positions in genes where data are available are shown in (a) as vertical bars under the position indicated on the header and in (b) as filled circles at their approximate positions on a three-dimensional cartoon of skeletal muscle actin. References to the data are available from Table 3.*

divergent actin-related sequences were recognized in several species a few years ago, and their number has grown quickly (Table 3). These sequences are found in very divergent species (see Table 3), from yeast to mammals, and there is every reason to expect that they are ubiquitous. All these actin-related proteins are similar in size to conventional actin, ranging from 376 amino acids (centractin or RPV-actin) to 418 amino acids (bovine and *Drosophila* actin related). In general they are much more divergent in sequence than any other conventional actins. Casual comparison of the alignment of current actin-related sequences in various databases (OWL, TREMBL, SWISSPROT and Genbank) (see fold-out section) quickly confirms this. The phylogenetic tree (Figure 4b) of actin-like sequences indicates this. When first discovered these sequences were given a variety of names: centractin or RPV-actin [730a, 898, 1128], actin-like, actin related. Recently, many of the major groups working on these proteins agreed [2153a], on the basis of sequence alignments and phylogenetic trees, to assign the proteins as members of actin-related protein subfamilies (Arp1, Arp2 and Arp3). Alignments of the sequences available in the current databases (Table 3; see fold-out section) and the phylogenetic tree (Figure 4) clearly show these groups. It seems likely that as more sequences become known that more Arp families will become apparent, though the emerging consensus [2153a] is that there are only a few such proteins and that they are highly conserved in all eukaryotes. The phylogenetic tree in Figure 4 strongly suggests that the sequences assigned to the Arp1 group most likely contain two subgroups, IA - Hsacent_1, Actl_rabit, Hsbcent_1, Dnarp87c_1 and 1B - Actl_neucr, Nc14008_1= and Pmcactin.

The observation that the same or closely related species of organisms contain more than one actin-related protein from different sequence families predicts that these proteins will have different functions within the cell. *Saccharomyces cerevisiae* contains the ACT2 gene (OWL:actl_yeast,= Arp1), which encodes a protein of 391 amino acids with 47% identity with the ACT1 gene [1112], and *Schizosaccharomyces pombe* has an ACT2 protein (427 amino acids) (OWL:actl_-pombe, Arp3), which shares 38% identity with its conventional (ACT1) counterpart [1026, 1027].

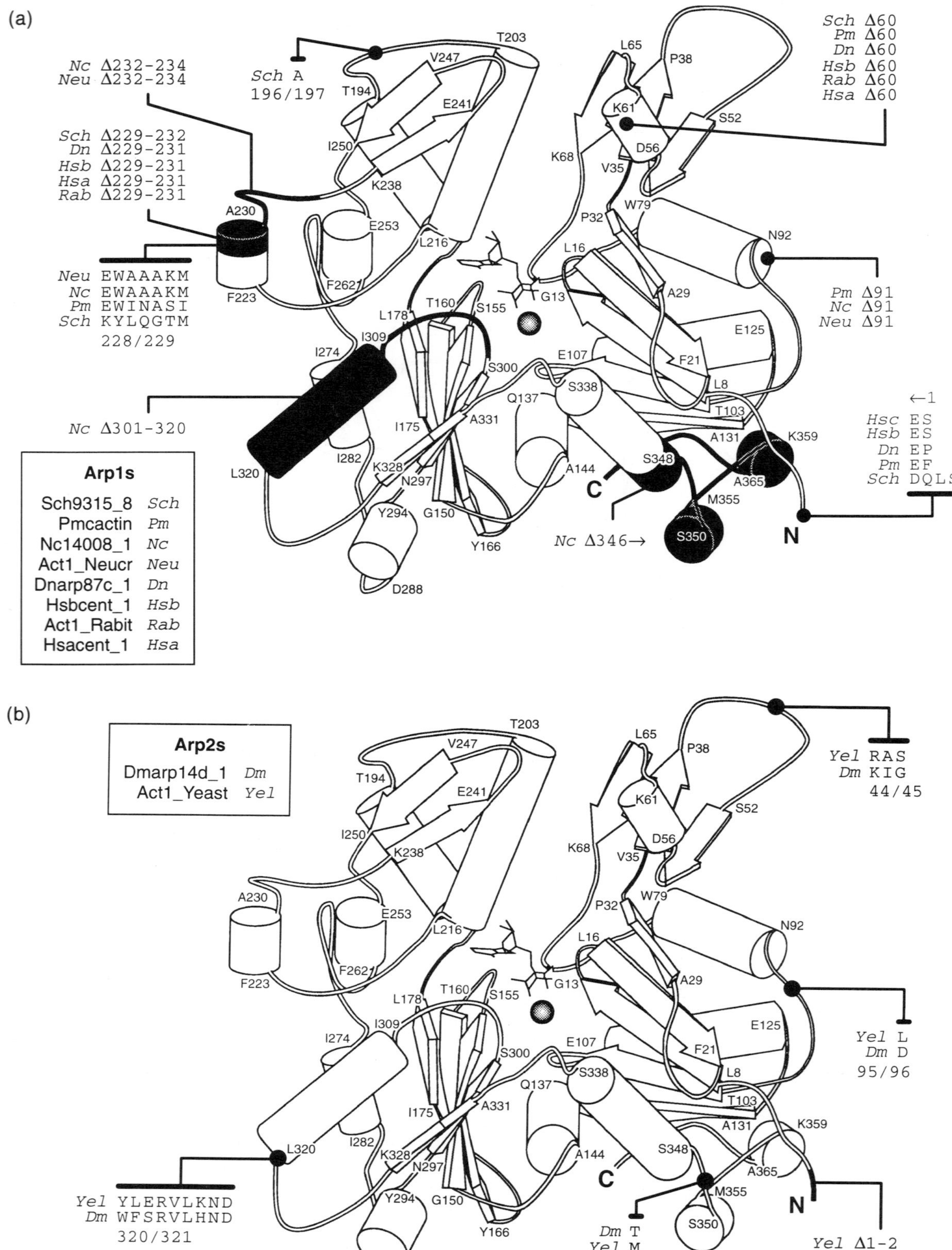

FIGURE 6 The positions of insertions and deletions of Arp1–3 superimposed on the three-dimensional map of conventional actin. *Insertions/deletions for ((a)–(c)) the Arp subfamilies Arp1–3 and (d) the 'unclassified' Arps. A representation of actin using cylinders for α-helix and arrows for β-sheet. Positions are numbered according to the skeletal muscle isoform. Insertions are indicated by a heavy line proportional to the length of insertion, deletions by shading of the structure in the region of the deletion and a delta (Δ) symbol preceding the amino acid(s) deleted. Insertions/deletions of single amino acid residues are not included if a compensatory length change was found within five amino acids in the same sequence. The positions of deletions/insertions were obtained by aligning each subclass of Arp individually with Act5-Human.*

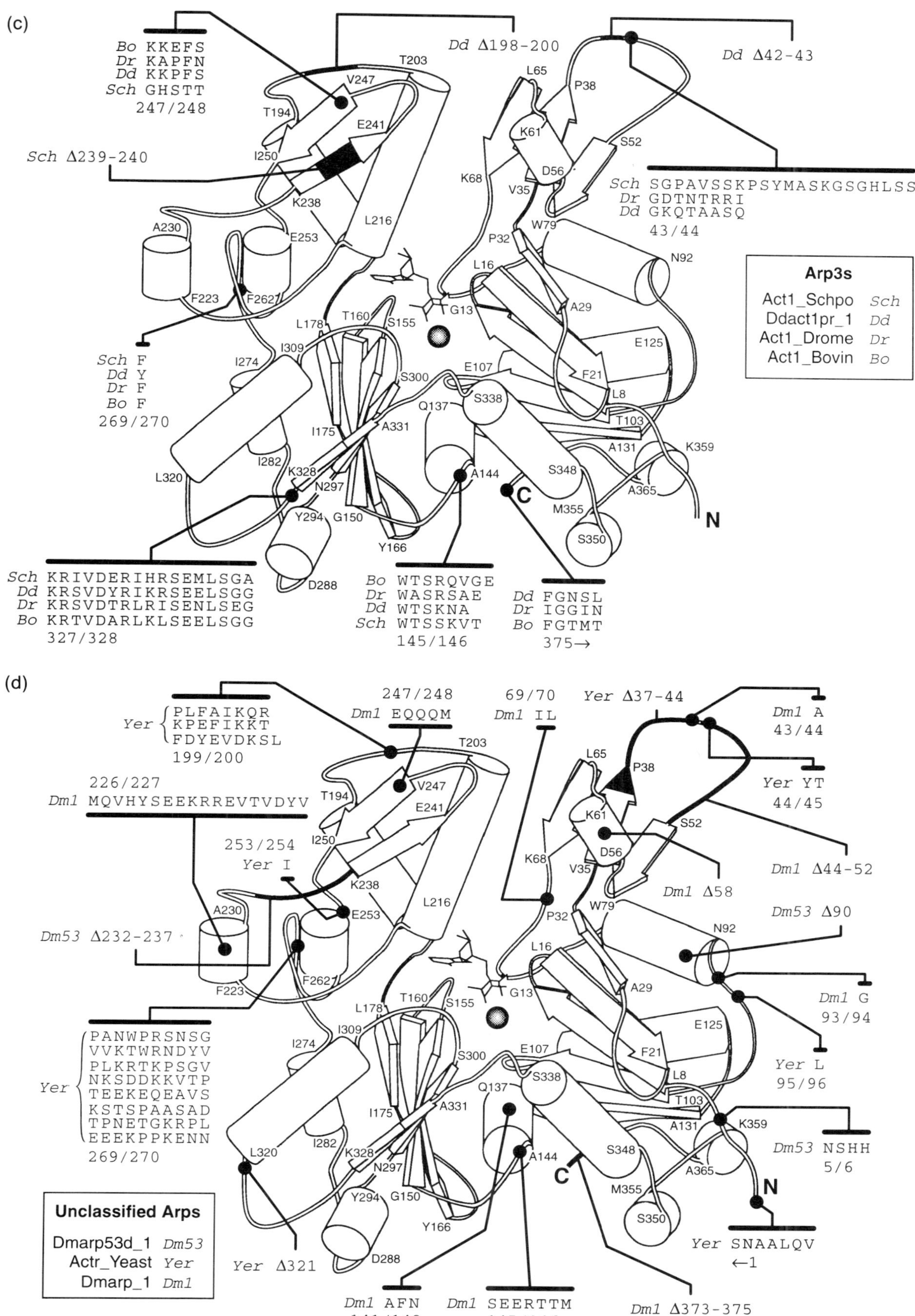
(c)
Bo KKEFS
Dr KAPFN
Dd KKPFS
Sch GHSTT
247/248
Dd Δ198-200
Dd Δ42-43
Sch Δ239-240
Sch SGPAVSSKPSYMASKGSGHLSS
Dr GDTNTRRI
Dd GKQTAASQ
43/44
Arp3s
Act1_Schpo Sch
Ddact1pr_1 Dd
Act1_Drome Dr
Act1_Bovin Bo
Sch F
Dd Y
Dr F
Bo F
269/270
Sch KRIVDERIHRSEMLSGA
Dd KRSVDYRIKRSEELSGG
Dr KRSVDTRLRISENLSEG
Bo KRTVDARLKLSEELSGG
327/328
Bo WTSRQVGE
Dr WASRSAE
Dd WTSKNA
Sch WTSSKVT
145/146
Dd FGNSL
Dr IGGIN
Bo FGTMT
375→
(d)
Yer PLFAIKQR KPEFIKKT FDYEVDKSL
199/200
247/248
Dm1 EQQQM
69/70
Dm1 IL
Yer Δ37-44
Dm1 A
43/44
Yer YT
44/45
226/227
Dm1 MQVHYSEEKRREVTVDYV
253/254
Yer I
Dm1 Δ44-52
Dm1 Δ58
Dm53 Δ90
Dm53 Δ232-237
Dm1 G
93/94
Yer L
95/96
Yer PANWPRSNSG VVKTWRNDYV PLKRTKPSGV NKSDDKKVTP TEEKEQEAVS KSTSPAASAD TPNETGKRPL EEEKPPKENN
269/270
Dm53 NSHH
5/6
Unclassified Arps
Dmarp53d_1 Dm53
Actr_Yeast Yer
Dmarp_1 Dm1
Yer Δ321
Yer SNAALQV
←1
Dm1 AFN
141/142
Dm1 SEERTTM
145/146
Dm1 Δ373-375

These two ACT2 proteins (an Arp1 and an Arp3) share 35% identity. Genetic complementation experiments show that both ACT1 and ACT2 genes are equally essential for cell survival and have different functions [1026]. Functional studies of actin-related proteins are now appearing. These proteins are much less abundant than actin, which has hindered their analysis. The best characterized group functionally are the Arp1s. In vertebrates, members of this subfamily form a short filament that is associated with dynein, and functions in the movement of vesicles [780, 898, 2153], whilst in lower eukaryotes (yeast and filamentous fungi) they appear to be involved in the function of the mitotic spindle [1026, 2152, 2295]. Of four polypeptides found in a complex from *Acanthamoeba* which bind to profilin–Sepharose [2145], two are actin-related proteins; one (47 kDa) is a homologue of the *S. pombe* ACT2, the other (44 kDa) of the ACT2 from *S. cerevisiae*. Antibodies locate the 47 kDa protein and smaller members of the complex to the cortex and filopodia of the amoebae. The anti-47 kDa antibody cross-reacts with a 47 kDa protein in *Dictyostelium* and rabbit muscle, but not with conventional actin. Evidence is accumulating that the actin-related proteins are ubiquitous in a wide range of eukaryotic cells with specific functions which differ from those of conventional actins.

In proteins which share a conserved core of residues, the crystal structure of one protein may provide a reasonable model for the main structural features of other members of the group. With this expectation it is clear from Figure 6 that the increased length of most ARP sequences derives from insertions in positions corresponding to surface loops in the skeletal muscle actin structure. This contrasts with amino acid substitutions in other conventional actins (see fold-out section) which occur with equal frequency in surface loops and other parts of the molecule. From the fact that the Arp insertions (and deletions) occur largely at the periphery of the molecule it seems likely that the core structure of the Arps will be similar to that of conventional actin. The ability of RPV (Actl_rabit, Arp1) to copolymerize with conventional actin and the observation that this Arp and the β-actin compete for the same chaperonin [756] appears to support this notion. The specific functions of Arps is only now becoming clearer. Like all actins, the Arps retain a conserved region corresponding to the ATP-binding site [396, 1026] and probably have an ATPase activity like that of actin, although the large deletion of amino acids 301–320 in Nc14008_1 will remove a significant part of the adenine-binding site (see Figure 6), which seems likely severely to affect or destroy ATP binding in this Arp. The presence of the sequence insertions at key functional sites on actin makes it likely that various ligand-binding sites are affected in Arps, whilst conservation of the insert sequences between members of the same Arp subfamily suggests that these inserts may also function as novel ligand-binding sites. In Arp1s, the moderately conserved sequence inserts at 228/9 would seem likely to disrupt or alter actin–actin contacts, yet some of these actins are known to form short filaments [780, 898, 2153]. The large *C*-terminal deletion of Nc14008_1 can be expected remove a number of actin functions associated with this region of the molecule. The deletions and large deletions of Actr_yeast (Arp2) are apparently an unusual feature of the Arp3 subfamily but are clearly in positions which will severely disrupt axial actin–actin contacts as well as the hydrophobic 'plug' (by the insertion at 269/270) which is believed to stabilize the F-actin helix [1251]. The numbers and sizes of sequence inserts varies considerably between Arp subfamilies; compare, for example, Arp3 and the 'unclassified' Arps (Figure 6). The Arp3 subfamily has large insertions in a number of actin-binding sites, and from atomic models seems unlikely to polymerize or bind monomer-binding proteins such as DNase I, profilin or gelsolin. The correlation of these structural changes with changed function of these proteins as compared with conventional actin promises to yield much useful information.

MOLECULAR GENETIC MANIPULATION OF ACTIN

Actin mutants

A large number of actin mutants have now been described and have been produced in different actins genes and studied in a variety of systems. Table 4 summarizes the mutants residue by residue from *N*- to *C*-termini, listing their effects *in vivo* and *in vitro*. Mutations in actin genes have been produced either by traditional genetic methods of obtaining mutants using mutagenesis of whole organisms followed by the selection of specific phenotypic traits or by protein engineering techniques.

Actin mutants have been recovered in *Drosophila melanogaster* [908, 976, 993, 994, 1057, 1082, 1119] and in *Caenorhabditis elegans* [1156] by their effects on flight or uncoordinated movement, in a mouse melanoma [1129] and in some transformed vertebrate cultured cell lines [125]. Recently, mutants have been recovered in the yeast (*S. cerevisiae*) by selecting for their ability to interact with mutants in actin-binding protein genes to suppress their lethal effects [2290] or to generate synthetic lethality [1159, 2369]. This approach is not only effective in identifying and investigating the function of unknown ABPs, but by sequencing the mutant actin, amino acids or motifs involved in the binding of actin to those proteins can be identified.

In vitro mutagenesis of cloned actin genes has been used to look more systematically at the role of individual amino acids on actin function. Expression of mutated genes can be achieved either *in vitro* or *in vivo*. Actins can be expressed *in vivo* either to study the effects of the actin mutations on cell or muscle biology, or to produce mutant actins in sufficient quantities for biochemical analysis. The expression of actin for this latter purpose has proved difficult and, until recently, it has not been possible to produce mutant actins in any quantity. As reviewed below these problems have been overcome in yeast (*S. cerevisiae*), the slime mould (*D. discoideum*) and the fruitfly (*D. melanogaster*).

Expression *in vitro*

The expression of mutant actin genes by transcription and translation *in vitro* in rabbit reticulocyte lysate produces small quantities (picograms) of actin which can be radioactively labelled. This is enough material for affinity binding assays and provides a useful means of rapidly screening large numbers of mutations to pinpoint those which are of interest for more detailed study. Mutant actins have been expressed from the genes for human skeletal muscle actin [1117, 1118, 1483], rat brain cytoplasmic actin [1101] and *Drosophila Act88F* [64, 919, 920, 921, 971, 2010]. This technique is valuable but severely limits the types of assays that can be used.

Expression *in vivo*

The advantage of *in vivo* analysis is that the actin is in its normal environment where the ability of the mutant to assemble and interact with the whole range of actin-binding proteins can be tested. This complexity can also be a disadvantage since, if the actin does not behave normally, it is very difficult to pinpoint which actin interactions are affected. Conversely, although individual interactions can be studied *in vitro*, it can be difficult to extrapolate to possible *in vivo* effects. It is certain that *in vivo* expression and analysis of actin mutants will be important for examination of the myriad roles of actin in cell biology. Mutants often show interesting and unusual effects, such as the pleiotropic effects found with many actin mutants in yeast [637]. They can also be used, by recovering mutants in other genes which suppress their mutant effects, to investigate the roles of proteins interacting with actin in the cell (for a review, see [164]).

Drosophila melanogaster

The *D. melanogaster* genome contains six actin genes; two encode cytoplasmic actins and the remaining four muscle actin isoforms. Mutations have been recovered only in the indirect flight muscle-specific *Act88F* (act6_drome) actin gene. The *Act88F* gene is only expressed in these tissues [1586] and encodes all the muscle actin found in these tissues [see 1119, 1382 for reviews]. Mutations in this gene affect only flight ability and not the viability of the flies; because of this, the *Drosophila Act88F* gene provides a unique system in which the effects of actin mutations on muscle assembly and function can be studied *in vivo*. In addition to actin mutants generated by whole-fly mutagenesis, *Act88F* mutants can be generated *in vitro* from the cloned gene and inserted into flies using P-element transformation [976, 919, 922, 993, 1094, 1104]. Once transformed into flies, the effects of an actin mutation can be analysed by flight testing, examining the appearance of flight muscles and mechanical testing of dissected flight muscle fibres [922, 1880]. It is possible to purify *Act88F* gene actin in small quantities (5 μg from 10 flies) sufficient for *in vitro* motility and force measurements [2248]. Studies of mutant actins have also been performed on actin preparations from whole flies (up to 10 mg from 100 000 flies), where the major actin isoform is *Act88f*, although other actin isoforms are present.

Caenorhabditis elegans

The genome of *C. elegans* contains four actin genes [1012] but only regulatory mutants have been characterized to date. However, since the actin genes have been cloned and a gene transformation system has been developed for this organism, it should only be a matter of time before *C. elegans* is exploited as a model system with which to study the *in vivo* effects of actin mutations.

Dictyostelium discoideum

The genetics of myosin and other cytoskeletal proteins have been developed in *D. discoideum*. No actin mutants have been recovered by selecting for abnormal phenotypes and this is probably due to the fact that the *Dictyostelium* genome contains at least 17 actin genes [1144]. Recently, this system has been exploited for the expression and purification of large quantities of actin following mutagenesis *in vitro* of the *Dictyostelium Act15* gene [1127]. To enable purification of the expressed mutant actins away from the endogenous actins, charge changes were introduced into the *Act15* gene by site-directed mutagenesis. Thus the mutant actins contained either a single substitution causing a charge change or two mutations; one the charge change, the other the substitution made to answer specific questions about actin function. It is possible that the charge change may affect actin function in unknown ways. However, it binds myosin, activates Mg.ATPAse activity of rabbit myosin and forms F-actin filaments that move in the *in vitro* motility assays. To date, the *Dictyostelium* expression system is one of the most easy to use and readily produces mutant actins in quantity. Its major limitation is that so far it has been developed to produce only *Dictyostelium* actin.

Saccharomyces cerevisiae

The budding yeast *S. cerevisiae* is one of the primary model organisms for studying eukaryotic cell biology. Its genome contains one conventional actin gene *ACT1* [953, 1073] essential for viability of cells. A large number of *ACT1* mutants are now available (see Table 4), including temperature-sensitive lethal alleles, which manifest a variety of cellular effects including slowing of growth, disruption of the normal actin filament network, partial inhibition of vesicle secretion and osmotic sensitivity [1078, 1116].

ACT1 I mutants have been recovered either by selecting *ACT1* alleles which interact with mutations in other genes, especially those of ABPs, to produce altered phenotypes, or have been made by *in vitro* mutagenesis of the *ACT1* gene and transformation of yeast cells. Three types of selection procedure have been used to obtain *ACT1* mutants: (1) by selecting mutants which suppress the phenotypes of other mutants (e.g. four *ACT1* mutants were selected which act as suppressors of *SAC6* mutants [2290] (*SAC6* encodes a protein homologue of fimbrin and the selected mutants presumably define its binding site on actin); (2) by generating synthetic lethality in combination with null alleles of ABPs which are not themselves lethal [866]; and (3) by extragenic non-complementation in which mutants uncover the normally recessive phenotypic effects of temperature-sensitive *ACT1* alleles [1159, 2369] leading to the identification of four new ABP-encoding genes and 11 new *ACT1* alleles.

The *GAL4 in vivo* assay, in which *GAL4* transcriptional activity is reconstituted *in vivo* only in response to protein–protein interactions has been used to identify *ACT1* affecting yeast ABP–actin interactions (6785). This elegant and simple approach together with knowledge of where in the actin molecule the mutation resides is likely to allow the rapid identification of several other ABP-binding sites.

S. cerevisiae is much used for heterologous protein expression and provides one of the few systems in which mutant actins can be obtained in sufficient quantities for biochemical analysis. Expression of heterologous actins is achieved by transformation with yeast episomal vectors [876, 995] whereas transformation of the yeast genome with mutant *ACT1* genes involve homologous recombination [1116] where the vector and gene or gene alone insert into the genome by recombination between homologous regions of the mutant and wild-type alleles. The level of heterologous actin expression from DNA vectors of wild-type or mutant chicken β-actin in yeast containing a functional *ACT1* gene is similar to that of the endogenous actin [585, 995]; whilst over-expression is lethal. Proteins produced by mutations in the genomic *ACT1* gene have also been expressed and purified [901, 904]. The absence of *N*-terminal processing of actin in yeast [904] may also be a disadvantage for heterologous actin expression. Whilst this system can easily produce 10–100 μg quantities from flask cultures, larger quantities require fermentors and careful control of the growth conditions.

Protein engineering techniques have been used on *ACT1* to study the role of individual amino acids in yeast actin. The ability to recover, express and study *in vivo* actin mutations in the yeast system has allowed the sequential replacement of charged residues, presumed to lie on the surface of the molecule, with alanines [165, 1161], and replacement of other amino acids previously implicated by other techniques [903, 904, 983, 1116, 1327].

Cultured cells

Several studies have looked at the effects of mutant actin genes on the growth and morphology of cell lines such as human fibroblasts [1025, 1103, 1132] and simian COS-1 cells [1117]. Mutant actins can cause cellular transformation *in vitro* [1025, 1103, 1132] and are responsible for a melanoma phenotype in mice [1129]. It has been proposed that the actin cytoskeleton may have an important role in cancer. Natural actin mutants are uncommon in neoplasia but effects on actin expression and on the cytoskeleton are usual following tumorigenic transformation [975].

Escherichia coli

Escherichia coli is usually the first-choice organism for the expression of recombinant proteins, and has no endogenous actin. The *Dictyostelium Act8* gene has been expressed in *E. coli* [947, 1047] but the actin forms inclusion bodies and only a very small amount can be extracted in its native state. Similar problems have hindered all other attempts to obtain native actin by expression in *E. coli* and low expression levels do not improve the situation. A cytoplasmic chaperone protein isolated from rabbit reticulocyte lysate catalyses the folding of denatured β-actins expressed in *E. coli* [1238]. This may provide a solution to the problems of expressing native actin in *E. coli* for biochemical studies.

ACTIN STRUCTURE

Overview of the three-dimensional structure of the actin molecule

From examination of the three-dimensional structure in Figure 7, it can be seen that the shape of the molecule approximates to a thick slice of a cube of side about 5–6 nm and thickness about 3.5 nm. The molecule is effectively divided into two domains of roughly equivalent size by a cleft containing the bound nucleotide and cation. These two domains are covalently connected by only two strands of the polypeptide chain which are close together at the base of the molecule, allowing the potential for significant relative movements of the domains. This region is consequently termed the 'hinge' region of the molecule. The domains were, from negatively stained images or low-resolution maps and prior to the atomic resolution of the structure, termed the large and small domains, respectively, despite their being approximately similar in mass. The atomic resolution model also shows that the two major domains can be themselves divided into smaller subdomains. Those in the small domain (on the right in the 'conventional' view of actin) numbered 1 and 2 and those in the large domain, 2 and 3. The hinge is close to the *N*- and *C*-termini that come close together in subdomain 1. With the molecule orientated as it is in Figure 7, the lower face of the molecule (subdomains 1 and 3) represents the barbed end of the polymer, as determined by the pattern of myosin binding to actin filaments. A further structural nomenclature has derived from comparison of actin with (in particular) hexokinase and the 70 kDa-ATPase heat-shock protein. These proteins have remarkably similar detailed three-dimensional structures, despite only minimal sequence relationships [10]. There is an existing nomenclature for hsp70 in which the small domain (subdomains 1 and 2) and the large domain (subdomains 3 and 4) are termed domains I and II respectively. Subdomains 1 and 2, and 3 and 4, are termed IA and IB, and IIA and IIB, respectively. As this nomenclature has precedence, we shall use it throughout this compilation. Comparison of the DNase I data from Kabsch and colleagues [396], the information available from the gelsolin segment 1 crystals (containing the same skeletal muscle isoform [1284]) and the profilin crystals (containing the β-non-muscle isoform [1337]) show that all three available structures of actin are very similar, although there are differences in the position of the chain in subdomains IB and IIB and in the position of the *C*-terminus (Figure 7).

The F-actin structure model

The availability of a three-dimensional map at atomic resolution has allowed the modelling of a detailed structure for actin polymers by fitting the detailed monomer structure into the constraints of a consensus structure [1253, 1219] gained by computer-aided manipulation of images from negatively stained, rapid-frozen and shadowed material and from cryo-electron microscopy [1172, 1296, 1299, 1363]. From these data and, more recently, from better-preserved images of rapidly frozen material, an overall diameter of about 10 nm was obtained [1296] and a prediction that the long axes of monomers were approximately parallel to the long axis of the filament. All but positively-stained sectioned material also shows, with varying degrees of resolution, the helical organization of monomers in the filament. The pitch of the major helix visible by electron microscopy (the long pitch helix) is about 70–80 nm. However, most descriptions use the distance

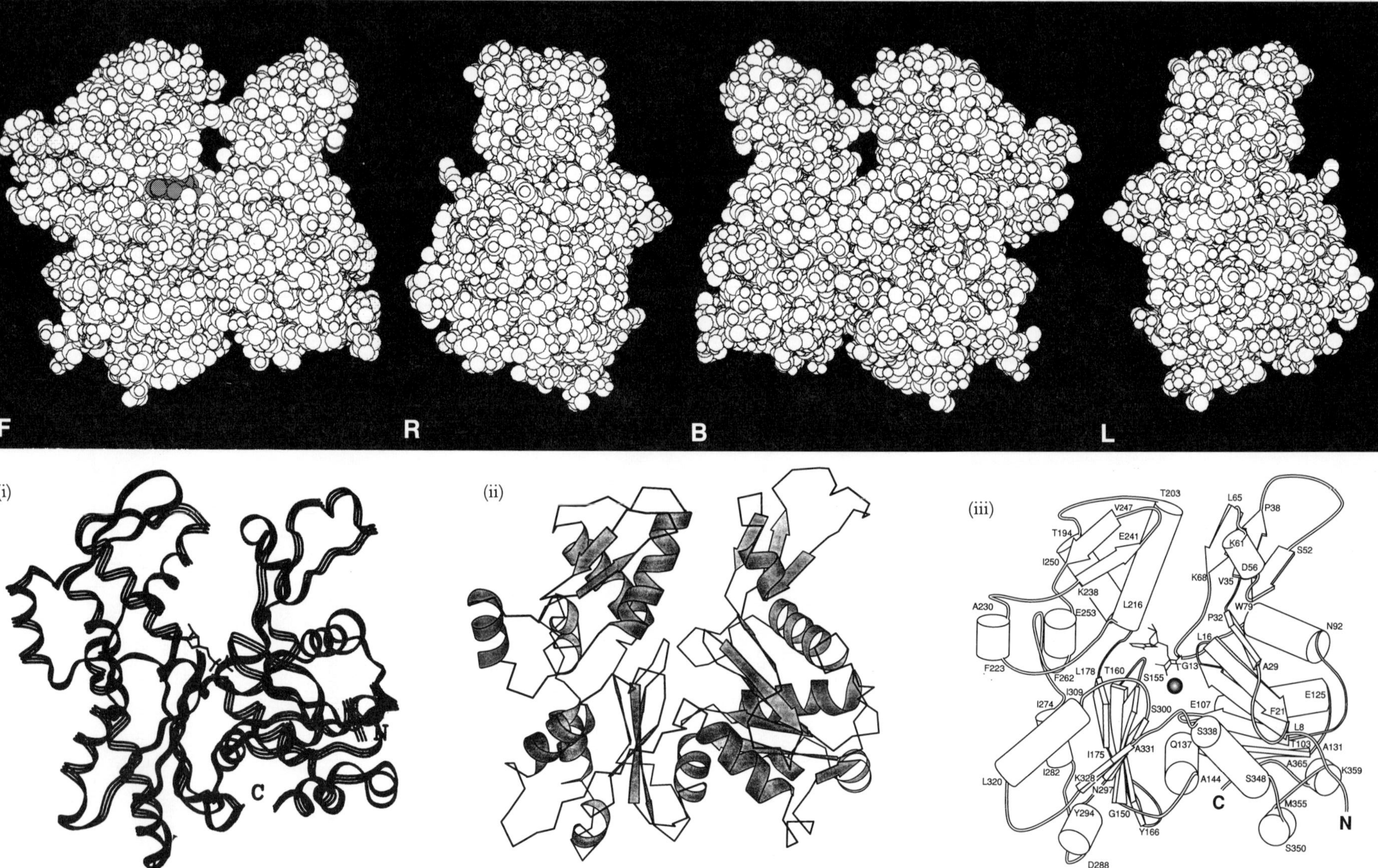

FIGURE 7 Different structural representations of skeletal and β-non-muscle actin. *Van der Waals representations are formed into a montage along the top to show all four views of actin. The rotation is as if the molecule were being turned clockwise from the perspective of the pointed end (top). Underneath are (i) a ribbon diagram of β-non-muscle actin from the profilactin crystal (Courtesy of Professors Lindberg and Schutt); (ii) a ribbon diagram of the skeletal muscle isoform (Courtesy of Dr Alfonso Valencia) and (iii) the cylinder and arrow diagram for comparison. The position of the ATP is shown in (i) and (iii) and the position of the cation in (iii).*

of half-pitch (half-turn) of the helix since this is the shortest distance between segments of the filament where monomers have equivalent positions. The long-pitch helix consists of two linear strands of actin monomers wrapped around each other. In some negatively stained images it is possible to see partial separation between the strands confirming that the structure is indeed two-stranded or, more formally, that the bonds between monomers along the axis of the filament are somewhat stronger than those between monomers across the filament [69]. The relative affinities of lateral and longitudinal bonds is also dependent on the nucleotide bound [533]. The structure of the filament predicted by Holmes *et al.* [69] is similar to that observed in the ADP.Pi state and changes on release of phosphate [1311]. The pitch of the helix does not seem to be tightly proscribed [1223, 1224]. The half-pitch varies between about 34 nm and 38 nm and contains between 12 and 14 monomers per half-pitch. The variability in pitch may result from compensatory twists of the filament or, as more recently asserted, from random differences in rotation of monomers. The degree of freedom for each monomer is about 5–6 degrees [1222]. The rise per monomer on each of the strands is then about 5.5 nm.

This helical pattern can be enhanced by binding myosin fragments (S1 or HMM) to filaments *in vitro* in the absence of ATP (rigor complexes). Negatively stained decorated filaments show an arrowhead appearance because of the projection of the three-dimensional filament on to two-dimensional film. These images confirm that each actin monomer must be oriented the same way round with respect to the long axis of the filament and that the myosin-binding regions of the molecule must be on the outside. The filament is thus polarized in the sense that both ends of the filament expose opposite ends of the actin molecule and that the same edge of each actin molecule is exposed on the outside of the filament. The overall polarity of the filament is defined by the binding pattern of myosin as the barbed or pointed ends. The barbed end appears to dominate assembly and disassembly events in cells.

The orientation of monomers within the filament has been studied in a number of ways; by measuring the distances (by FRET) between equivalent amino acids on adjacent monomers [92], by visualizing a particular amino acid (Cys 374) using undeca-gold labels [94], by localizing the positions of binding sites on the actin molecule of ligands that discriminate the barbed [764] or pointed [1261] ends of the filament or bind laterally [583, 822]. These data have confirmed that monomers are oriented such that the 'A' subdomains all point towards (and in the terminal subunits form) the barbed end whilst the 'B' subdomains are similarly related to the pointed ends. These interpretations are consistent with the reconstructions of the myosin S-1:F-actin structure [1320, 1334]. Domain I is on the outside of the filament whilst the domains 'II' of each strand associate at the centre of the filament to hold the strands together. Further refinement, derived from computer-fitting using the atomic structure of the monomer, suggests that essentially the association between subdomains A on one monomer and subdomains B on a neighbouring monomer along a strand form the basic long-axis intrastrand contacts, whilst monomers in the two strands are out of register along the long axis of the filament by approximately half a monomer so that binding between subdomains IIA in one strand and IIB in the other constitute the major interstrand contacts. Amino acids involved in the contact sites will be discussed in the section on binding sites (below).

The actin helix can be described in two ways; firstly as a single-start left-handed helix where a rise of 2.75 nm per monomer gives a pitch of 36 nm for 13 monomers and six turns or a two-start right-handed double helix with a half pitch of 36 nm (Figure 8). The diameter of the filament in these models is 9–10 nm. The fitting of the atomic structure into a filament has used predictions gained from X-ray diffraction patterns of oriented gels of F-actin which diffract to a resolution of better than 1 nm [69] and from frozen-hydrated images of F-actin [1320, 1334]. Calculated diffraction patterns for monomers oriented in all possible combinations were compared with the measured pattern of oriented filaments. A best-fit solution involved only small adjustments of chain position to avoid steric effects [69] and some rotation of subdomains. The predicted model has domains II close to the filament axis and the outer edge of subdomain IA at highest radius. In this model, the axial strands of the two-start helix are stabilized by interactions between residues in subdomain IA with residues in subdomain IIB; residues in IIA with residues in IIB and residues in IIA/IA with residues in IB. Some of these interactions are similar to those between DNase (mimicking the barbed end of an adjacent monomer) and the pointed end of actin. In contrast to these potentially extensive interactions along the filament, density at the centre of the model is low involving possible interactions between a hydrophobic loop between subdomains IIA and IIB on one strand with a hydrophobic pocket formed between two monomers on the adjacent strand. The existence of this hydrophobic loop is contradicted in the profilin:actin crystal where the loop appears to be folded back within the molecule

[1337] (see Figure 7) suggesting that interactions between the two strands of the polymer are even weaker than predicted by the F-actin model and rely predominantly on interactions between residues at the back of subdomain IA and a helix in subdomain IIA. Slightly different F-actin structures have been observed in scruin-cross-linked F-actin where subdomain IB is tilted relative to the conventional model [800, 1312, 1314].

The characteristics of this model can be compared with models constructed from FRET and morphological data [92, 94]. Fluorescence resonance energy transfer data map the distances between fluorescent groups and can, depending on the number of groups studied, yield a skeleton three-dimensional map of residue positions. Whilst the relative positions of residues 41, 374 and nucleotide are similar, their radii from the filament axis are larger. The models from diffraction data and FRET do, however, agree that domain I is on the outside of the filament and II close to the axis. Successive refinement of F-actin structure from myosin-decorated filaments has led to a model that is not inconsistent with the diffraction model [31] but that uses undecagold attached to Cys374 to show that this residue is close to the filament surface [1296]. The ultrastructural and diffraction models are also consistent in that they put the binding site of myosin in subdomain IA on the outer surface of the filament, a prediction that is consistent with the three-dimensional modelling of the myosin S-1 and F-actin interaction [1320, 1334]. Strong longitudinal interactions and weak lateral interactions between strands has been observed directly in negatively stained preparations of F-actin where the two strands appear to be able to dissociate partially [12]; this property appears also to relate to the bound species in the nucleotide site (see below). The variability in the number of subunits per half-turn of the actin long-pitch helix has been explained by relative slipping of the two actin strands [12, 1201], and by random angular disorder in the subunits [1222]. Whatever the mechanism, the magnitude of the angular variation is of the order of 5–6 degrees [1222]. Assembly is sensitive to hydrostatic pressure, implying an increase in volume during assembly [1239].

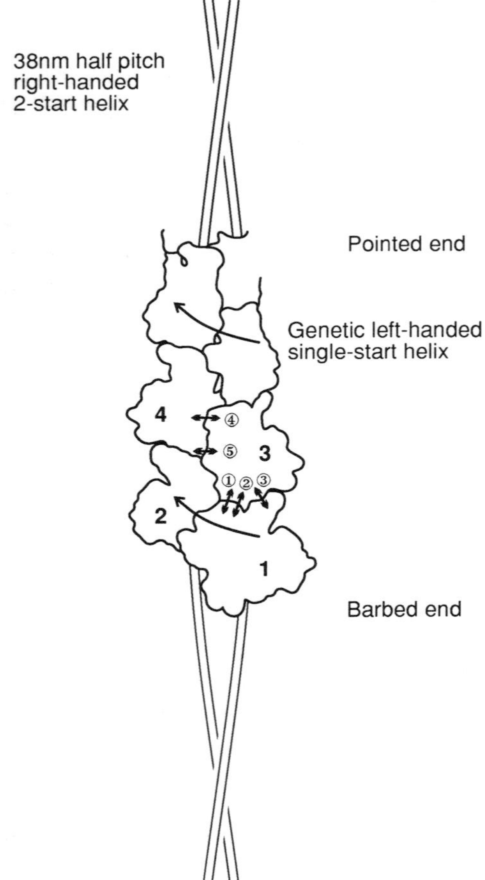

FIGURE 8 A representation of the overall structure of F-actin. *Seven monomers of F-actin are schematized to show the 1-start genetic helix (curved arrows) and the 2-start double helix (indicated by the twisting rods above and below) relative to barbed and pointed end. The arrows indicated 1–5 are the proposed monomer interactions detailed in Figure 11 and the text.*

The ribbon structure

The ribbon structure is an independent feature within profilin:actin crystals where the orientation and interaction between monomers in a chain of actin monomers is directly accessible to crystallographic analysis [1337]. The ribbon structure reveals alternate front and back views of actin monomers with the back views on the right with the putative barbed end towards the bottom (Figure 9). Subunits are oriented essentially diagonally with, alternately, subdomains IB and IIB and IA and IIA towards the filament axis. Subdomain IIA is at highest radius. The interactions between monomers are different from those in the F-actin model; subdomains IB and IIB interact with IA and IB, respectively, on the $n+1$ monomer towards the putative pointed end. This set of interactions is repeated along the ribbon. In this structure, there are not two separate strands, all interactions are diagonal with adjacent monomers and the DNase loop of subdomain IB crosses the ribbon axis. Whilst this structure cannot be reconciled easily with the other reconstructions based on images of filaments assembled *in vitro*, or with the peptide inhibition data (see below) or with the observation of strand separation in F-actin, it is not inconsistent with the derivatization data. Comparison of the effects of derivatization of amino acids involved in the ribbon

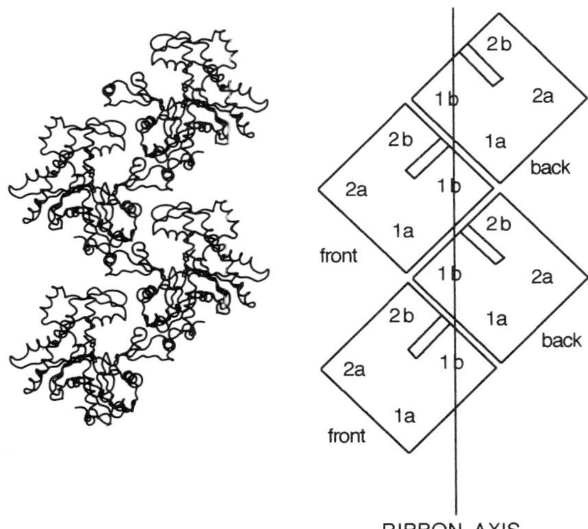

FIGURE 9 The profilactin crystal ribbon structure. *A representation of the shape of the α-carbon chains in four actin monomers taken from the profilactin crystal (Lindberg and Schutt, personal communication). Alongside is a diagram showing the orientation of monomers and sites of contact. These are discussed more fully in the text.*

structure on polymerization (Table 5) show that derivatization at a number of interacting residues disrupts assembly. The interactions between monomers in the ribbon are extensive, both in terms of the area buried and in the high potential for weak bonding, suggesting the possibility of a biological role for this interaction. It has been proposed that monomers can reorientate from the ribbon structure to the F-actin structure and that this conformational reorganization might contribute to the mechanism of muscle contraction [1336]; this remains to be demonstrated experimentally. It is conceivable that events at filament ends or at sites of filament anchorage utilize an alternative ribbon-like interaction that can isomerize. The observed [12] and predicted [69] weak interactions between adjacent strands in the F-actin model, which could lead to filament unravelling at the ends, may be strengthened at ends by either the nucleotide cap and or by an alternative interaction of monomers. That an alternative orientation of monomers may occur at the ends of filaments is suggested by the observation that monomers bound to the gelsolin GA2 complex are oppositely polarized as they are in the sheet structure [684].

Comparison of actin structure with other proteins

From a more detailed examination of the three-dimensional structure, the pattern of a five-stranded antiparallel β-sheet pattern enclosed by α-helices on both surfaces is repeated in domains IA and IIA on either side of the ATP-binding cleft (Figure 10) suggesting gene duplication at some stage prior to the evolution of eukaryotes (all of which contain actin). These structural motifs are also present in sugar kinases (including hexokinase), exopolyphosphate phosphatases, uncoating ATPases [1237] and in the 60 and 70 kDa heat-shock proteins [10]. The amino acid sequences are highly divergent except around the ATP-binding site [10] leaving open the possibility of convergent evolution, but more likely suggesting that the ancestral structure evolved as an individual domain in prokaryotes. Although the primary biological functions of these three classes of protein are significantly different they all bind and can hydrolyse ATP in the cleft between the domains; this underlying function may be the property that has been exploited and diversified during evolution.

The similarity of structure of these actin structural homologues is further suggested by the observation that two actin capping proteins, aginactin and cap 32/34, bind to F-actin via an hsp70 molecule [641, 678] although hsp70 alone does not appear to bind.

There is also a rather intriguing sequence similarity between a *C*-terminal region of collagen type VI α3 chain and the region 126–196 which extends from subdomain 1A, forms much of 2A and extends into 2B. There is a perfect match with the sequence 183–194 that forms a helix at the back of subdomain 2B [1204], the extreme *C*-terminal end of which is involved in actin self-assembly (see Figure 11).

COVALENT DERIVATIZATION SITES

Actin has been extensively derivatized by using a range of covalent derivatization agents to study almost all aspects of actin structure and function (Table 5). The most reactive side chains of amino acids available for derivatization are the primary amines, sulphydryls and carboxyls, although many other amino acids have been derivatized. Of the sulphydryls, Cys374 is the most reactive and is preferentially labelled in solution at physiological pH by iodoacetamide or maleimide derivatives. Cys10 can also be selectively labelled after blocking Cys374 with NEM either by partially denaturing actin with urea or by using large hydrophobic derivatizing agents [1458]. Of the 19 or so lysine residues in the actin molecule, Lys61 is the most reactive [274]. If it is labelled with a large group then derivatization impairs assembly. Lys373 can

TABLE 5 Covalent derivatization of actin

AA derivatized	Labelling reagent	Functional parameter measured*	Actin isoform /purpose of labelling	References
AA-1	Acetylation	**Polymerization**	All actins?/*N*-terminal processing	1387, 904, 1415, 1091, 1092, 1477, 1450, 1099, 1481, 1483, 1416, 1479
Acidics 1–7	EDANS/EDC	**Polymerization** **Myosin binding**, Tropomyosin binding	RSMA/	1388
Cys10	BFP	Polymerization Myosin binding	RSMA/^{19}F nmr	1192
	DABM	Polymerization	RSMA/FRET	519, 1289, 1384, 1184
	DACM	Monomer activation *Polymerization*	RSMA/G-actin	1339, 1445
	DDPM	Polymerization	RSMA/FRET	519
	DTNB		RSMA/	1218
	DTNBC		RSMA/	1218
	IAEDANS		RSMA/FRET	412, 1289
	IAM	Polymerization Nucleotide binding Cation binding	RSMA/	1391
	IAZB	Polymerization Myosin binding	RSMA/	1447
	NBDN	**Polymerization** **Phalloidin binding**	RSMA/internal cross-link with C374	1410
	nmr probes	Polymerization Myosin binding	RSMA/^{2}H and ^{19}F nmr	1316
	Phalloidin	**<Polymerization** **<Phalloidin binding**	RSMA/	1989
	PIA		RSMA/	1443
	PYM		RSMA/	1443
Lys18	AA	*Cation binding* Polymerization	RSMA/	238
	IA	*DNase binding*	RSMA/	699, 887
Arg28	CHD	Tropomyosin binding Myosin activation Polymerization	RSMA/	1428
	APG		RSMA/photoactivated cross-linker	598
His40	DEPC	***Polymerization*** Phalloidin binding Myosin activation Tropomyosin binding	RSMA/	1423, 1434, 1467
Gln41	Cadaverines (via TG)	Polymerization Myosin activation Myosin binding Antibody binding	RSMA/FRET	1265, 1490, 2306
	Putrescine (via TG)	Polymerization	RSMA/	1405, 1422
Lys50	AA	*Polymerization* Cation binding	RSMA	238
	IA	*DNase binding*	RSMA/	699, 887
Tyr53	DZT	***Polymerization*** **Phalloidin binding** **Nucleotide binding** Myosin binding		1386, 2034
Lys61	AA	*Polymerization*	RSMA/	238
	FDAP	Polymerization Myosin binding Tropomyosin binding	RSMA/correlation times	1372
	FITC	Myosin binding Myosin activation **Polymerization** DNase binding Phalloidin binding Tropomyosin binding Actin ATPase	Various/FRET; CD-spectra; FRAP	310, 590, 308, 1394, 1461, 276, 1455, 2033, 1508, 1291, 1456

TABLE 5 Continued

AA derivatized	Labelling reagent	Functional parameter measured*	Actin isoform /purpose of labelling	References
	IA	*DNase binding* *Polymerization*	RSMA/	274, 699, 887
	nmr probes	Polymerization Myosin binding	RSMA/ ^{2}H and ^{19}F nmr	1316
	PFP-ITC	Polymerization Tropomyosin binding Myosin binding	RSMA/19F nmr	590
	PLP	**Polymerization** **DNase binding** Myosin binding	RSMA/	1403
	RITC	Microinjection		
	TNBS	Polymerization	RSMA/Ab production	1741
Lys68	AA	*Polymerization* *Cation binding*	RSMA/	238
	IA	*DNase binding* *Polymerization*	RSMA/	699, 887
Tyr69	DABSYL-Cl		RSMA/FRET	1290, 1400
	DNS-Cl	**Polymerization** **Nucleotide binding** DNase binding Myosin binding Tropomyosin binding Phalloidin binding	RSMA/FRET	1185, 1184, 2034, 1290, 1178, 1288, 1400, 1357, 2321
	FDNB	***Polymerization***	RSMA/	1417, 1491
	MNS-Cl	**Polymerization** **DNase binding**	RSMA/	1395
	TNM	**Polymerization** **Myosin activation** Phalloidin binding Myosin binding	RSMA/	1400, 1464, 1189, 1441, 2034
His73	N^{τ}-Methylation		All actins except *N. gruberi*/normal post-translational modification	1497, 1381, 1398, 930, 1780, 1421, 1429, 1430, 1499, 1432, 1138, 1489, 1416, 1439, 1117
Arg95	CHD	***Tropomyosin binding*** Myosin activation Polymerization	RSMA/	1428
	APG		RSMA/photoactivated cross-linker	598
	ADP-ribosylation (turkey erythrocyte ADP-ribosyltrans-ferase A)	Cyt-D G-actin ATPase DNase I binding Gelsolin binding Polymerization **Nucleation**	G- and F-actin	2332
Lys113	AA	*Polymerization* Cation binding	RSMA/	238
	IA	*Polymerization*	RSMA/	274, 887
Glu117	PAL	Phalloidin binding	RSMA/	2072
Met119	PEAL	Phalloidin binding	RSMA/	2072
Arg147	CHD	Tropomyosin binding Myosin activation Polymerization	RSMA/	1428
Arg177	ADP-ribosylation (*C. botulinum* C2 toxin)	**Polymerization** **Actin ATPase** **Gelsolin binding** DNase binding EF-2 binding	β, γ and smooth muscle G-actins	1951, 1952, 2032, 2042, 2043, 2044, 2073, 1953, 2052, 1498, 5, 2010, 2014
	ADP-ribosylation (*C. perfringens*, *C. iota*, *C. spiriforme* and *C. difficile* toxins)	**Polymerization** **Nucleotide exchange** **Actin ATPase** **Gelsolin binding**	All G-actin isoforms	1977, 2032, 310, 2050, 2073, 2049, 477, 1434, 2059, 1497, 5, 2010, 2014
	Endogenous ADP-ribosyl-transferase	**Polymerization**	Non-muscle	1401, 1435, 2067

TABLE 5 Continued

AA derivatized	Labelling reagent	Functional parameter measured*	Actin isoform /purpose of labelling	References
Lys191	AA	*Cation binding* Polymerization	RSMA/	238
	PBM		RSMA/cross-linking to C374 in F-actin	1409
Thr203	Phosphorylation		*Physarum* actin/fragmin complex substrate for actin kinase	1414, 1415, 1418
Cys217	DACM	*Monomer activation* *Polymerization*	RSMA/	1445, 1339
Lys238	2,4PD	**Tropomyosin binding** Polymerization Myosin activation	RSMA/	644
	AA	Cation binding Polymerization	RSMA/	238
	IA	*DNase binding*	RSMA/	699, 887
Cys257	DACM	*Polymerization* Monomer activation	RSMA/	1339, 1445
	IAF	*Cation binding*	RSMA/	1269
Lys284	IA	*DNase binding* *Polymerization*	RSMA/	274, 699, 887
Cys285	BFP	Polymerization Myosin binding	RSMA/nmr	1192
	IAM	Polymerization Nucleotide binding Cation binding	RSMA/	1391, 1431
Lys291	AA	*Cation binding* Polymerization	RSMA/	238
	IA	*DNase binding*	RSMA/	699, 887
Tyr306	2-Azido-ATP	**Myosin activation**	RSMA/affinity label	510
Lys315	AA	*Polymerization* Cation binding	RSMA/	238
Lys326	AA	Cation binding Polymerization	RSMA/	238
	N^{ϵ}-Di- or tri-methyllysine		*Acanthamoeba* actin only/normal post-translational modification	1936, 1500, 216, 1158, 1143
Lys328	AA	Cation binding Polymerization	RSMA/	238
Lys336	8-Azido-ATP		RSMA/photo-affinity label	487
	IA	*Polymerization*	RSMA/	274, 887
Met355	PEAL	Phalloidin binding	RSMA/	2072
Trp356	8-Azido-ATP	*Polymerization*	RSMA/photo-affinity label	487
Lys359	AA	*Cation binding* Polymerization	RSMA/	238
Arg372	ADP-ribosylation (turkey erythrocyte ADP-ribosyltransferase A)	Cyt-D G-actin ATPase DNase I binding Gelsolin binding Polymerization **Nucleation**	G- and F-actin	2332
Lys373	NBD-Cl	Polymerization Myosin activation Myosin binding DNase binding Tropomyosin binding	RSMA/	464, 199, 340, 965, 186
Cys374	5MF	Polymerization	RSMA; *Dictyostelium* actin/	308
	BFP	Polymerization	RSMA/nmr	1192
	BPM	Gelsolin binding	RSMA/photoactivated cross-linker	636, 1359, 612

TABLE 5 Continued

AA derivatized	Labelling reagent	Functional parameter measured*	Actin isoform /purpose of labelling	References
	CRIA	Polymerization	RSMA/caged fluorophore for microinjection	346, 347
	DABM	Polymerization DNase binding Myosin binding Tropomyosin binding	RSMA/FRET	1184, 1383, 1288
	DACM	Caldesmon binding *Monomer activation* *Polymerization*	RSMA/	591, 1445, 1339, 525
	DBF	Polymerization	RSMA/	1501
	DDPM	Polymerization DNase binding Myosin binding Tropomyosin binding	RSMA/FRET	1288, 1292, 1451
	Disulphide formation with low MW thiols	**Polymerization** **ATPase** Phalloidin binding	RSMA/	201
	DNPG	**Polymerization** *Phalloidin binding*	RSMA/	1960, 473
	DNS-A	**Myosin activation** Polymerization Myosin binding	RSMA/	1398, 1442
	DNS-cysteine	Polymerization Tropomyosin binding Myosin binding	RSMA/	1202, 1368
	DTNB	Caldesmon binding	RSMA/	671, 1218, 1985, 2114
	DTNBC		RSMA/	1215
	DTNBG	**Polymerization** **Actin ATPase** **Nucleotide binding** Phalloidin binding	RSMA/	1406, 1488
	EIA	Polymerization ***In vitro* motility** Cytochalasin binding Phalloidin binding Tropomyosin binding Spectrin binding	RSMA/ phosphorescence anisotropy	341, 1330, 1446
	FM		Caged G-actin	2334
	IA	Polymerization Nucleotide binding Cation binding	RSMA/	931, 1391, 1397, 1451
	IAE	*Polymerization* Phalloidin binding	RSMA/FRET	341, 1375
	IAEDANS	Phalloidin binding Myosin binding Cation binding Tropomyosin binding Polymerization	RSMA; *Dictyostelium* actin/ FRET; FP	1292, 416, 387, 1258, 2033, 1294, 308, 1357, 493, 435, 377, 385, 393, 413, 436, 1398, 1178, 1390
	IAF	**Profilin binding** Polymerization Myosin activation	RSMA; amoeba/ microinjection; FRET; myosin binding	782, 226a, 1379, 271, 341, 342, 1495, 1496, 342a
	IAM	Polymerization	RSMA/	1451, 1431
	IANBD	*Actobindin binding* Polymerization	RSMA/FRET	203, 602
	IATR	**Polymerization**	RSMA/Diffusion coefficients	1483, 339, 340, 341, 338, 2374
	IAZB	**Polymerization** *Myosin binding*	RSMA/	1447
	MBB		RSMA/	1265
	NBD-Cl	Polymerization	RSMA/FRET	284, 340

TABLE 5 Continued

AA derivatized	Labelling reagent	Functional parameter measured*	Actin isoform /purpose of labelling	References
	NBS	**Polymerization** **Myosin activation** Nucleotide binding	RSMA/	1451
	NEM	**Polymerization** **Profilin binding**	RSMA/nmr	1192, 199, 744, 340, 407, 1451, 441
	nmr probes	Polymerization Myosin binding	RSMA/ ^{2}H and ^{19}F nmr	1316
	Phalloidin	Phalloidin binding Polymerization	RSMA/	1989
	PBM	**Polymerization**	RSMA/cross-linking of F-actin	1409
	PCMB	**Polymerization** **Spectrin binding** **Cytochalasin binding**	RSMA/	407, 1402, 2019
	PIA	**Myosin binding** **Profilin binding** *Actobindin binding* Phalloidin binding Polymerization BeF_3 binding DNase binding	RSMA/ assembly probe	464, 196, 618, 218, 718, 1443, 744, 1461, 781, 602, 625, 1767, 460, 608, 186, 2328
	PIP		RSMA/anisotropy	2328
	PRODAN	Polymerization ADF binding Cytochalasin binding Myosin binding Profilin binding DNase binding Phalloidin binding	RSMA/FRET	862, 1450
	PYM	Polymerization	RSMA; *Physarum* actin/	1438, 1443
	Spin-labels maleimide/IA	Polymerization Myosin binding Cation binding Phalloidin binding DNase binding Cytochalasin binding Monomer orientation	RSMA/EPR spectra	486, 833, 1196, 1197, 1246, 1872, 1862, 1302
	TMB	Thymosin binding	RSMA/	681
	Tb-DTPA-maleimide		RSMA/FET	1380
	Undeca-gold-maleimide	Polymerization Myosin binding	RSMA/	1296, 1328
Unknown				
K residues	^{125}I-HBE	Polymerization ABP binding	Cardiac; platelet/ radiolabelling	1392, 1910, 814, 1741
K residues	Acetaldehyde		RSMA/	1502
K residues	Methylisocyanate	**Polymerization**	RSMA/	267
K residues	BNHS	Polymerization	Chick SMA/micro-injection	298, 263
K residues	NVOC-Cl (caged actin)	**Polymerization** Polymerization (uncaged) *In vitro* motility (uncaged)	G-actin	2334
K residues	RNHS		RSMA/micro-injection	251a
K residues	TNBS	**Polymerization** **Myosin activation** **Myosin binding** Tropomyosin binding α-Actinin binding	RSMA/Ab production	1741, 1466, 1465
K residues	Ubiquitin	Polymerization Myosin activation Myosin binding	Insect skeletal/ natural actin derivative	1382, 1393, 1412

TABLE 5 Continued

AA derivatized	Labelling reagent	Functional parameter measured*	Actin isoform /purpose of labelling	References
?	Cytochalasin-B		Erythrocyte actin/	2063
H residues	DEPC	**Polymerization** **Myosin activation** Myosin binding	RSMA/	1423, 623
	Sulphation (non-enzymic		Non-muscle/*in vivo*	1482
?	Glycosylation	Polymerization	Non-muscle/*in vitro* and *in vivo*	1472a
E residues	17-β-Oestradiol dehydrogenase		Non-muscle/*in vivo* modification	576
C residues	Glutathione (non-enzymic)		Non-muscle/*in vivo*	1399, 1478, 2330
C residues	ADP-ribosylation (non-enzymic)		Non-muscle/*in vivo*	1435
C residues	IAM	Nucleotide binding		1345, 546
?	Nitroxide-radical esters		RSMA/EPR	1187
C, M and W residues	Performic acid oxidation		RSMA/Ab production	1741
D and E residues	Carbodiimide	**Myosin interaction**	RSMA/	1510
S, T and Y residues	Phosphorylation	**Polymerization** Myosin activation	Various/†on tyr;	1494, 1396, 1471, 1486, 1407, 1411, 1413, 1414, 1487, 1484, 1464, †1448, †1449, †1426, †1419, †1433, 2331
H, W and C residues	Photo-oxidation	**Polymerization** **Myosin activation** α-Actinin binding	RSMA/	1464
?	Reserpine	**Polymerization**	Various/photolinks to G-actin	1470, 2039
K & C residues	SA	**Polymerization** **Myosin activation** **α-Actinin binding**	RSMA/	1464
<u>Cation site</u>	Co^{II}		RSMA/FRET	412, 1185, 1184, 1456, 1460
	Fe^{II}		RSMA/generate free-radical cleavage 159/160 and 301/2	432
	Gd^{III}	**Polymerization** **Myosin activation**	RSMA/nmr	1182, 382
	La^{III}		RSMA/nmr	1182
	Mn^{II}		RSMA/EPR	1182, 1372
	Ni^{II}		RSMA/FRET	1460
	Tb^{III}	**Polymerization**	RSMA/FRET	1182, 1383
<u>Nucleotide site</u>	2-azido-ADP	**Myosin activation**	RSMA/affinity label	510
	8-azido-ATP		RSMA; β/γ actin/ affinity label	487, 496
	8-Br-ATP	**Myosin activation**	RSMA/affinity label	486
	AMPPLP		RSMA/affinity label	1403
	DNP-ATP	Polymerization Myosin activation	RSMA/affinity label	474, 441
	ϵ-ADP		RSMA/FRET	520
	ϵ-ATP	Polymerization Phalloidin binding DNase binding	RSMA/FRET	519, 1291, 284, 1292, 1459, 1460, 55, 2321
	FTP		RSMA/FRET	1185, 1291, 1184, 1383, 1182, 1178
	TNP-ADP		RSMA/FRET	520
	TNP-ATP		Binds F-actin only	2313
<u>Actin cross-linking</u>				
Gln41–Lys113	AZP (transglutaminase)	Polymerization	RSMA/intermolecular cross-link	1422

TABLE 5 Continued

AA derivatized	Labelling reagent	Functional parameter measured*	Actin isoform /purpose of labelling	References
Lys 326 or 328–?	ANB-NOS	Polymerization **Myosin ATPase** ***In vitro* motility**	RSMA F-actin	2333
Cys374–?	BPM	Polymerization	RSMA/intramolecular (G-actin) or intermolecular (F-actin)	1359
?–?	DMS (plus other imidoesters)	**Polymerization** **Tropomyosin binding** DNase binding Nucleotide exchange Myosin binding Myosin activation	RSMA/intramolecular cross-link	774, 1468, 1469, 786
?–?	EDC	**Polymerization** **Myosin activation** **Myosin binding**	RSMA/intermolecular cross-link	786
?–?	Glutaraldehyde	**Polymerization** **Myosin activation** **Myosin binding**	RSMA/intermolecular cross-link	1474, 786
Cys374–?	MBS	**Polymerization** **Myosin binding** **Myosin activation** **Tropomyosin binding** Phalloidin binding DNase binding	RSMA/intramolecular (G-actin) or intermolecular (F-actin)	1389, 596, 1427, 1437, 823, 584
Cys374–Cys374 Cys374–Cys10	NBDN	**Polymerization** **Phalloidin binding**	RSMA/intra- and intermolecular cross-link	1410
Cysteine link	Oxidation	**Actin ATPase**	RSMA/intermolecular cross-link	1259
Cys374–Cys374	PBM (paracrystals)	Gelsolin binding	RSMA/	1462
Cys374–Lys373	PBM (F-actin)	**Nucleotide exchange** **Actin ATPase** Polymerization DNAase binding Myosin activation	RSMA/intermolecular cross-link	1409, 1420, 1440, 522, 1463, 584, 380
?–?	TEA	**Self-nucleation** **Nucleotide exchange** **Phalloidin binding** Co-polymerization Myosin activation DNase binding	RSMA/Intermolecular cross-link	1736

*In this column: plain text indicates that a functional parameter of the derivatized actin was measured and that it was not significantly altered; bold text indicates that derivatization significantly disrupted that functional parameter, underlining denotes that the function caused changes in the fluorescence or other signal from the derivatization group and italics indicates that the function caused changes to the accessibility of that amino acid to that derivatizing agent. For example, ***myosin binding*** indicates (i) that myosin binding to the derivatized actin was measured, (ii) that binding was affected by derivatization and (iii) that the accessibility of that amino acid to the derivatizing agent was altered when myosin was bound. See the glossary for abbreviations used in the table

also be labelled after blocking of Cys374 [199]. Of the other groups, tyrosines 53 and 69 can be selectively labelled with DZT [1386]. At the simplest level, derivatization with small reagent molecules can give information about which amino acid side chains are exposed to solvent. Whilst this information is available in detail from the crystal structure, the crystal data are essentially static data and will not reveal transient conformational changes which change the exposure of residues. Such changes in conformation have been detected during cation exchange and nucleotide exchange and for changes caused by ABP binding [473]. Derivatization has also been used extensively to map the relative spatial relationships of particular amino acids and with the nucleotide and cation binding site by FRET [1458]. The data obtained with this approach have, with one or two exceptions, been consistent with structure models obtained by other techniques.

Derivatization of actin with fluorescent groups has also been used to study binding of ligands, notably cations and nucleotides, by monitoring the effect of ligands on fluorescence [387]. Much of the best kinetic data have been obtained in this

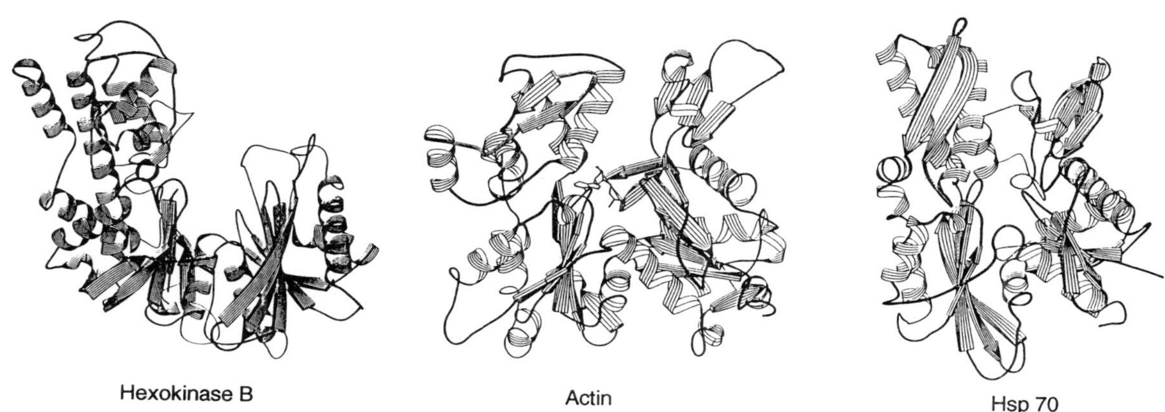

FIGURE 10. Comparison of the three-dimensional structures of hsp70 and hexokinase with actin. *Ribbon diagrams at the same scale of actin, hsp70 and hexokinase are set together. The major similarities are the ATP-binding cleft and the structure of sub-domains IA and IIA. Images courtesy of Dr Alfonso Valencia.*

way (see sections on cations and nucleotides). Fluorescent actin has also been used to monitor assembly by direct changes in fluorescence signal in polymerized actin [264], but also by energy transfer experiments between monomers carrying complementary fluorophores [341]. Actin derivatized at cysteine 374 has been the choice derivative for microinjection [342] since assembly is largely unaffected [264]. However, binding of some of the monomer-binding proteins is inhibited [744]. The general disadvantage of derivatization is that these covalent changes necessarily cause a change in the local or global structure of actin, so data obtained with derivatized actin may not transfer to the native molecule. The disruptive effects of specific derivatizations have been exploited in a large number of studies to implicate amino acids in various ligand binding or self-association sites (see Table 7).

POST-TRANSLATIONAL MODIFICATION SITES

N-terminal processing

Processing of the *N*-terminus occurs with all actins thus far examined and appears to proceed during or shortly after translation. The major consequence of processing is the addition of an acetyl group to the α-amino group of the terminal amino acid via an acetylase reaction. The functional significance of acetylation is not known, but mutated actins expressed in yeast that can no longer be acetylated do not appear to alter the phenotype of the organism, but blocking processing *in vitro* affects polymerization, probably nucleation [971]. The *N*-terminal acetylated amino acid of the mature actin is usually, but not always, different from that at the *N*-terminus of the product specified by mRNA. In these cases, the removal of one or two amino acids precedes acetylation. The precise mechanisms differ owing to the different sequences in class I and II actins. In class I actins, which have a methionine residue preceding an acidic residue, methionine is acetylated then removed prior to acetylation of the acidic residue to form the mature terminus [1098]. In Class II actins, which have a Met-Cys-acidic-sequence at the *N*-terminus, methionine is removed followed by acetylation and removal of cysteine prior to acetylation of the acidic residue to form the mature terminus [1098]. Yeast and fungi are class I by sequence but are processed by the acetylation of the terminal methionine to form the mature protein [904]. Plants and the divergent type 5 non-muscle actin in vertebrates are variants of class II having a Met-Ala-acidic-sequence, whilst *Acanthamoeba* actin has a Met-Gly-acidic-sequence. Both of these class II variants are otherwise processed like true class II actins to yield an acetylated acidic terminus. *Entamoeba histolytica* actin, a divergent actin in other ways, has an unusual terminal glycine that is acetylated [981]. The aminopeptidase involved in processing has recently been isolated from rat liver [1535]. A list of references relating to *N*-terminal processing is given in Table 3.

Methylation

Most actins (with the exception of *Naegleria gruberii*) contain a methylated histidine (*N*-τ-methylhistidine) at the equivalent of position 73 of vertebrate skeletal muscle actin. Some lower eukaryotic cells also contain a di- or tri-methylated lysine at position 326 (*N*-ϵ-di- or tri-methyllysine). Methylation occurs during or shortly after translation and has no known functional correlates. The endogenous methyltransferases utilize S-adenosyl methionine as methyl donor and can recognize the peptide YPIDHGIIT, residues 69–77 of actin [1475]. A list of references relating to methylation can be found in Tables 3 and 5.

Phosphorylation

There are a number of reports that actins can act as substrates for various protein kinases *in vitro* at serine, threonine or tyrosine residues. There are very few reports supporting a role for phosphorylation as a regulatory post-translational modification *in vivo*. A list of references is given in Tables 3 and 5. Of those which might be of potential significance in cells are reports of insulin-stimulated serine and tyrosine phosphorylation [1396, 1449] and of actin in an actin:fragmin complex in *Physarum polycephalum* by a selective kinase at Thr203 [1414, 1418]. Phosphorylation of this threonine abolishes the nucleation and capping activities of the complex. Phosphorylation of tyrosine in *Dictyostelium* regulates the assembly of actin *in vivo* and the behaviour of the cells [1426, 1433].

ADP-ribosylation

ADP-ribosylation of actin was first characterized *in vitro* using one of the *Clostridium* toxins (see Table 5). Derivatization occurs on arginine-177 using NADP as donor [1951]. Derivatization of actin at this site has been confirmed by site-directed mutagenesis [2010]. Ribosylated actin will not polymerize, but ribosylated monomers can cap the barbed ends of filaments [1951]. The ubiquity of actin and its sequence conservation in eukaryotic cells presumably makes it an attractive target for evolution of toxins. Only the actin from *Euplotes crassus* has a

TABLE 6 Proteolytic cleavage of actin

Cleaving agent	Sites cleaved/sizes of fragments	Properties of fragments	References
2NTC	Cys		1864
Arg C proteinase	F-actin at 39		1227
Arginine esterase			1514
Botulinum neurotoxin type E	Arg and Lys		1512
Calpain II	42 and 40 kDa fragments		1538
Carboxypeptidase A	374, 375		383, 670, 253
Carboxypeptidase B	G-actin at 372 F-actin at 373*	*<caldesmon-binding	717
Chymotrypsin	Ca-G-actin at 44, 57, 67/68	Need both parts for Ca-binding	691, 1513, 1269, 1227, 1509, 1520, 395, 498, 1297
Cyanogen bromide	At methionines		1504, 874, 1511, 931, 1526, 838
E. coli A2 protease	G-actin at 42	-polymerization	1347, 1524, 1523
Fe^{2+} free radicals	159, 301		432
HIV proteases			1503, 1536
Kallikrein			1514
Proteinase K	G-actin at 47	-polymerization, <motility, <ATPase	1518
S. aureus V8	16 kDa *C*-terminus	=Gc protein-binding	838, 1505, 1516, 1931
S. griseus pronase	Similar to trypsin Phalloidin-F-actin protected		1504, 2048
Subtilisin	G-actin at 39–52 39–51, 47/48	=hMM binding; =myosin activation; <motility; BeF_3 binding	253, 608, 1049, 1227, 1347, 1533, 1513
Thermolysin			2072, 2073
Thrombin	Ca.ATP-G-actin stable EDTA G-actin at 28, 39		1528, 1530
Trypsin	Ca-G-actin 26, 62, 68, 372, 373 Mg-G-actin protected Ca-F-actin at 372, 373* Mg-F-actin at 372 only§ Phalloidin-F-actin protected	*<caldesmon-binding; <TNI binding; altered filaments §>ATPase; >C_c; >exchange; BeF_3 binding	1347, 1514, 1519, 1227, 1509, 608, 201, 1520, 395, 1519, 498, 1431, 931, 722, 1527, 1307, 2048, 1347, 2073, 2072, 838, 1372, 1539, 1513, 525
Various			1527
Venom proteases	EDTA G-actin at 39		1530

Proteinases, sites of cleavage (where known) and the properties of cleavage or of the fragments are listed.

substitution in this position to a lysine, suggesting the importance of a basic residue at this site. More recently, ADP-ribosylases have been identified in eukaryotic cells [1435, 2067] and shown to derivatize actin selectively *in situ*. The functional significance of such ribosylation is not yet known.

Sulphation

Radiolabelled sulphate is incorporated into actin at several (unknown) sites in cultured cells [6168].

PROTEOLYTIC CLEAVAGE SITES

G-actin is moderately resistant to proteolytic cleavage and yields, under many incubation conditions in the presence of a number of proteinases, a resistant 33 kDa fragment often called the 'core'. The major sites of cleavage of trypsin and chymotrypsin are in the DNase-binding loops between residues 60–70 and 40–52 and at the extreme *C*-terminal at 372 and/or 373. DNase [1509] and phalloidin [1513] protect both cleavage sites in subdomain IB and myosin S-1 protects the 60–70 loop. F-actin is also substantially more resistant to proteolysis unless Mg.ADP is in the medium, when cleavage at 60–70 loop can occur [494]. Similarly, the sensitivity of F-actin to cleavage by thrombin only occurs if the cation has been removed by EDTA [1530]. There are some other differences in cleavage patterns caused by polymerization; whereas G-actin is cleaved at 372, F-actin is cleaved at 373 by carboxypeptidase B [717]. Chymotrypsin can also cleave G-actin to give 1–44 *N*-terminal fragment which remains attached to a 60–375 *C*-terminal fragment. The two fragments can be dissociated by removing the calcium.

The 33 kDa core structure does not polymerize nor activate S-1 myosin ATPase but does still retain the ATP binding site [498]. Surprisingly, the core will assemble after mild denaturation in urea. To date there are no reports of cleavage of actin into the two structural domains. A summary of proteolytic data is given in Table 6.

LIGAND-BINDING SITES

CATION-BINDING SITES

There is a single, exchangeable high-affinity divalent-cation-binding site located within the cleft of actin close to the nucleotide and three or more intermediate and lower affinity cation-binding sites elsewhere on the molecule (see Table 2 and [34] for review). The high-affinity cation-binding site has relative affinities for cations in the order Ca > Mn > Cd > Mg > Zn > Ni > Sr [425, 426, 427, 399, 1277], whilst the lower affinity sites do not greatly discriminate between calcium and magnesium and will bind monovalent cations [377]. The relative affinities for both classes of binding site and the free ion concentrations of magnesium (0.5 mM) and calcium ions (0.1 μM) in the cytoplasmic compartment of cells [405, 431, 507], suggest that the sites would all be occupied by magnesium *in vivo*. This has been confirmed by the use of electron-probe X-ray analysis in muscle [402]. Thus, the characteristics of assembly and properties of Mg-actin measured *in vitro* are more relevant to assembly of actin *in vivo* than are those of Ca-actin. This has some practical corollaries since actin is invariably prepared with calcium at the high-affinity site and suitable conditions for exchange of magnesium for calcium must be employed to ensure conversion to Mg-actin. The use of low concentrations of EGTA in the presence of actin can mediate exchange within minutes [389].

Measurement of the affinities and exchange rates of cations on actin has been complicated by several factors, not least that the stability of actin is compromised in the absence of cations, that nucleotide and cation binding are closely interrelated and that binding of ions at the lower affinity sites influences binding at the high-affinity site. However, whilst the latter problem has confused interpretation of binding studies, these low-affinity sites are unlikely to influence events *in vivo* since they have K_d values of the order of tens of millimolar, whilst the intermediate affinity sites will be saturated with magnesium or potassium. There is still some conflict with regard to whether or not structural isomerizations further complicate the interrelationships of binding at the various sites but the emerging consensus suggests that binding is essentially competitive with allosteric changes mediated by the lower affinity (and probably physiologically irrelevant) sites *in vitro*. Furthermore, the relationships between nucleotide and cation binding appears to resolve to independent exchange of either cation or nucleotide albeit with different relative affinities for different cation nucleotide combinations (see nucleotide section below).

Polymerization was thought to be enhanced by replacement of calcium by magnesium at the low-affinity sites [377, 408]. The situation is more complex in that the two cations have different effects on different aspects of polymerization. Nucleation of Mg-actin is many orders of magnitude more efficient than it is with Ca-actin [348] whilst the elongation rates of Ca- and Mg-actin are similar [211]. Self-nucleation in cells is presumably inhibited by the presence of stoichiometric amounts of monomer-binding proteins that may either inhibit nucleation in a catalytic manner or simply sequester monomer. The critical concentration for Ca-actin assembly is about 10-fold higher than that for Mg-actin [255]. The low-affinity sites may be involved in the potassium ion-induced assembly effects [309] or occupied by magnesium at higher magnesium concentrations [34]. Lithium can also reduce the off-rate of monomers thereby stabilizing F-actin by conferring a structure similar to that of ADP.Pi F-actin [860].

Earlier estimates for the affinity of the single high-affinity site measured by equilibrium dialysis were in the micromolar range [408]. Subsequent measurements using a range of fluorescent techniques have shown that the consensus affinities for magnesium and calcium are about 10 nM and 2 nM, respectively at neutral pH [34], the pH believed to prevail in the cytoplasmic compartment. The site would therefore be saturated with magnesium *in vivo*. The lower-affinity sites would also be mainly occupied by magnesium, although some may be occupied by potassium ions whose concentration is some 50–100-fold higher than magnesium, but whose K_d may only be 5–50-fold higher (see Table 2). Exchange of cation at the high-affinity site is the result of simple competition and is preceded and limited by the rate at which water molecules are

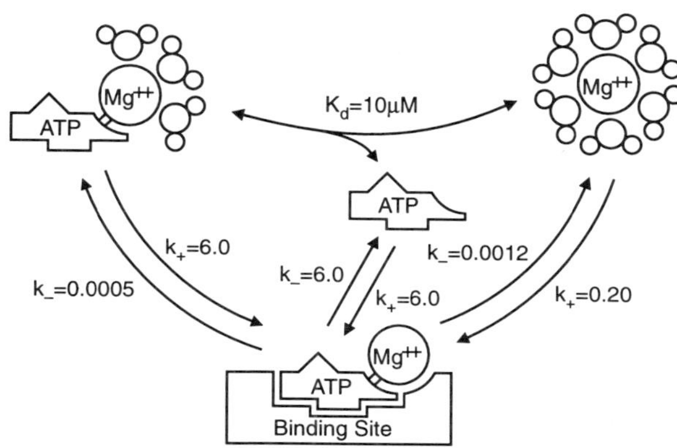

FIGURE 11 Kinetics of interactions between ATP, Mg and water at the nucleotide-binding site. *Figure 11 shows the rate constants for the interactions between Mg, ATP and water at the nucleotide-binding site of actin (taken from 2105 with permission). Rate constants are given in the following units:* k_-, s^{-1}; k_+, $M^{-1} s^{-1}$.

shed from the hydrated ion [34]. The rate of shedding of water molecules from magnesium is some 10^3-fold slower than from calcium [381] which is reflected in the rate-constants for exchange [385] (see Table 2). The rate constants for the interactions of Mg, ATP and water at the binding site are shown in Figure 11. The differences in the behaviour of Mg- and Ca-actin appear to result from a conformational change induced on the monomer by exchange of one cation for the other at the high-affinity site which can be detected by antibodies [411, 414]. It is likely that these changes involve the cleft between the two domains and may also explain the enhanced rate of ATP hydrolysis of Mg- but not Ca-G-actin in the presence of cytochalasin D [433] which binds at the barbed end of the cleft (see below).

The binding of cation and nucleotide are interrelated. Removal of the cation using chelators dramatically reduces the affinity of nucleotide binding [392, 404, 430, 565] and its dissociation may be a prerequisite for nucleotide exchange [413]. The cation at the high-affinity site appears to be closely associated with the nucleotide since both nucleotide and cation can be displaced by the stable Cr–ATP complex [434]. The location of the high-affinity site from the atomic model of the Ca-actin:DNase structure [396] shows that the cation is bound in the cleft to the barbed-end side of the nucleotide. The cation interacts with the β-phosphate oxygen of ADP in Ca-ADP-actin, whilst it is bound to the β- and γ-phosphate oxygens in Ca-ATP-actin. Whether magnesium would interact differently is not known. The cation also binds to Asp11, Glu137 and Asp154 [396]. These data agree with the location relative to the nucleotide site, Tyr69 and Cys374 previously estimated by FRET [1458] and nmr [1191] measurements. Monomer structure is influenced by bound cation as judged by the alteration of susceptibility to proteolysis of Arg26 and Lys68 by exchange of calcium for magnesium [1347]; in the presence of bound magnesium, trypsin cleaves only at Lys373 [288]. Cations also affect polymer structure; the standard volume change on assembly is significantly different depending on whether assembly was induced by divalent or monovalent (potassium) cations [1239].

NUCLEOTIDE-BINDING SITES

Nucleotide binding has been measured by using a number of different techniques. Binding studies are complicated, in the same way that cation binding is, by the observation that (i) removal of nucleotide causes a rapid denaturation of actin, (ii) that cation binding and nucleotide binding are interrelated, but also (iii) because measurements of nucleotide exchange on monomers under physiological conditions is difficult because of the induction of assembly under these conditions. In cells, a high concentration of unpolymerized actin is maintained by the interaction of actin with a number of monomer-binding proteins. These practical problems have been tackled, with greater or lesser degrees of success, by measuring exchange at low ionic strength (to prevent assembly), by exploiting denaturation as a measure of how much nucleotide has bound to actin and by using actin:ABP complexes which may both protect actin against denaturation and prevent assembly under physiological conditions.

These various data have resolved a single nucleotide-binding site which, from X-ray structure information, has been located in the cleft between domains I and II adjacent to the high-affinity cation-binding site. The nucleotide molecule is partially buried within the cleft (Figure 7) and would thus appear to require some conformational breathing of the two domains to allow exchange. Such a breathing motion has been predicted by normal mode analysis [1362]. A second, lower-affinity site has also been reported, but the lower affinity and lower selectivity would indicate that it is unlikely to have any physiological significance [711].

The major high-affinity binding site is highly selective for di- and tri-nucleotides although a number of nucleotide analogues will also bind (see Table 2). One molecule of ATP is hydrolysed during each assembly and disassembly cycle of an actin monomer and the hydrolysis occurs within the polymer. Thus, ATP monomers assemble to yield, eventually, ADP monomers at the end of the cycle. Nucleotide hydrolysis is not necessary for assembly *per se*, since ADP-G-actin and actin bearing non-hydrolysable nucleotide analogues will assemble at nearly equivalent rates [194]; rather, it provides a conformational switch that drives the cycle of assembly and disassembly (see the section on polymerization) by promoting disassembly. The interaction of G-actin with nucleotide is complicated by the association of nucleotide with the cation at the high-affinity site, by the rapid and irreversible denaturation of actin when cation- and nucleotide-free and affected by the nucleotide species already bound for exchange. Nucleotide will bind to cation-free actin with a dissociation constant of millimolar order [505], but this is assumed to be irrelevant in cells, with the possible exception of the binding of nucleotide by newly synthesized and folded actin. The physiologically important reaction appears to be the replacement of bound ADP by ATP on monomers released from polymers. The mechanism of this interaction involves cations and therefore values for the rates of the various steps are directly dependent on divalent cation concentrations, composition and pH. This complex dependency has led to the publication of many different values and to some uncertainty as to the mechanism of binding and exchange. The affinity of both ATP and ADP (or nucleotide analogues) is increased in the presence of either magnesium or calcium ions (see Table 2). The co-localization and contribution of the cation to nucleotide binding is confirmed by the atomic structure data [396] which show a direct interaction between them (see Figure 11). The dissociation constant, under physiological conditions, for ATP binding to ADP monomer is about 10^{-10} M and about 4–12-fold lower for ADP to ADP-G-actin [712]. The free cellular concentrations of ATP are usually estimated to be between 1 and 3 mM in non-muscle cells [503, 545] and nearer 10 mM in muscle [478, 544]. The total concentration of ADP in the cytoplasmic compartment of muscle and non-muscle cells has been estimated to be about 1 mM [544, 545], but in muscle, at least, with a free concentration some 10–30-fold lower at 30–100 μM [497, 543], the large discrepancy being due to the significant amount of ADP bound to F-actin in muscle. Under these equilibrium conditions the binding site would be occupied by ATP. The rate of exchange, which is governed by the dissociation rate constant (k_{-N}) for nucleotide release, is most sensitive to cation concentration (see Table 2). At roughly physiological conditions (low Ca, moderate Mg and pH 7.0), the exchange rate (k_{-ADP}) on monomer is about 0.003 s^{-1} [712], a relatively slow rate, which could become rate-limiting under the dynamic assembly–disassembly conditions in cells. It is not surprising, therefore, to find that this crucial exchange which refurbishes monomers for reassembly, is regulated by a number of monomer-binding proteins which are known to maintain the monomer pool available for assembly. The rate of nucleotide exchange (nucleotide dissociation) is inhibited by a number of monomer-binding proteins, including DNase I that bridges the cleft between domains I and II [491, 1883], myosin [400], by thymosin, ADF and cofilin which bind close to the *N*- and *C*-termini near the hinge and can compete with profilin for binding [482, 485, 771] and by gelsolin [511, 554] and phalloidin [589]. These effects appear to be modulated by differing affinities for the actin-binding proteins for the ADP- and ATP-G-actin species [305, 451, 485, 511] suggesting different conformations of G-actin with ADP or ATP in the binding site. However, whilst cofilin and thymosin inhibit exchange, profilin catalyses it, even in the presence of thymosin [481, 482]. The profilin-binding site sits right on the hinge (Figure 7) and may therefore influence the opening up of the cleft to allow exchange. The proportion of actin monomers in the cytoplasmic compartment of cells that carry ATP rather than ADP is not known. Thus the proportion of monomers charged with ATP and therefore capable of assembly under cellular conditions may be determined by the activities of catalysts like profilin that are themselves regulated via the inositide signal-transduction pathway [482].

Although the rate of exchange of nucleotide on the monomer is relatively slow, it is very much slower on the polymer [499, 506, 524, 531, 569]. Half-times for exchange appear to be of the order of hours or days rather than minutes. The increased accessibility of fluorescent ATP derivatives to solution quenchers in F-actin (rather than in G-actin) [540] suggests that the apparent opening of the nucleotide site in F-actin is countered by decreased mobility of domains to allow exchange. Exchange on the polymer is increased by myosin [524] but reduced by binding of phalloidin.

Hydrolysis of ATP on actin monomers occurs with an overall rate constant of about 0.05 s^{-1} after their incorporation into polymer [537] to yield an ADP.Pi intermediate that is quite stable [450]; the rate constant for release of the phosphate ion is about 0.006 s^{-1} [446]. The properties of G- and F-actin are dependent on the nucleotide occupying the binding site. ADP-actin has a critical concentration for assembly of about 1–2 μM at pH 7.0, whereas that for ATP-actin (N.B. ATP-F-actin really means F-actin where the monomer(s) at the rapidly assembling barbed end carries ATP; the rest of the polymer will be ADP.Pi or ADP) is about 0.1 μM [538]. The critical concentration for ADP.Pi actin prepared by adding back inorganic phosphate to ADP-F-actin is close to that of ATP-actin [538] suggesting that a conformational change accompanies release of Pi. However, compared with the slow post-hydrolysis release, exogenous inorganic phosphate can exchange rapidly at the γ-phosphate site of ADP-actin [184, 538], suggesting that any conformational changes are more complex. The increase in critical concentration is due to a much slower off-rate (k_-) for ADP.Pi monomers [539]. The affinity for inorganic phosphate is about 0.5 mM for the $H_2PO_4^-$ ion which is about 1 mM total phosphate at pH 7.0 [454]. The cytoplasmic concentration of phosphate is thought to be about 1–2 mM in non-muscle [418, 445], but nearer 10 mM in muscle [544] and so whilst phosphate is a potential regulator of actin disassembly in non-muscle cells where filaments are turned over rapidly (as would pH due to its effects on $H_2PO_4^-$ ion concentration), disassembly in muscle cells would be maximally inhibited, a prediction consistent with (but maybe irrelevant to) their slower turnover. The phosphate analogues, AlF_4 and BeF_3, also bind with micromolar affinities and cause similar changes in assembly behaviour and structure as those caused by exogenous phosphate [1311].

The effects of hydrolysis of nucleotide and phosphate release on critical concentration [538] support the proposition of a conformational change associated with the status of the nucleotide-binding site. The structure or stability of actin filaments may also be different, as judged by electron microscopy of negatively stained filaments [533] and the rheology of filament solutions [468], depending on whether subunits carry bound ADP, ADP.Pi or ATP, although these differences have been disputed [5837]. Accessibility of thiol groups is also different depending on which nucleotide is bound [546]. These conformational changes might be recognized by actin-binding proteins [570]. Normal mode analysis of the monomer structure shows that domains I and II might twist relative to one another, suggesting that the proposed hinge might affect both nucleotide exchange and conversely that nucleotide might affect motion [1362]. Reconstructions from images of ADP-F-actin and ADP.BeF_3-F-actin suggest that the structure of subdomain IB becomes more disordered on release of phosphate [1311]. The conformational differences of actin with different bound nucleotide species may be due to the location of the binding site in the cleft where bound nucleotide provides one of the major bridges between the two domains of actin and may therefore determine the relative orientation of the domains. The differences in orientation of equivalent domains in the structurally homologous sugar kinases and heat-shock proteins [1252] lend credence to this possibility.

The adenine ring of the nucleotide fits into a pocket made by residues 214, 213, 303, 305, 306 and 336 in the DNase crystal [396]. These data are supported by affinity labelling studies using 2-azido- and 8-azido-ATP where residues Tyr306 and Lys336 are labelled respectively [487, 510]. The 2′- and 3′-hydroxyl groups of the ribose sugar appear to hydrogen bond with Glu214 and Asp157, respectively [396]. The phosphate groups of the nucleotide contribute to the linkage of domains I and II via hydrogen bonds with amide nitrogens on the loops around residues 302 and 158 on subdomain IIA and around residue 15 on subdomain IA. The γ-phosphate interacts predominantly with subdomain IIA whilst the β-phosphate interacts exclusively with subdomain IA. The α-phosphate interacts with subdomain IIA. Hydrolysis of the β/γ-bond may thus significantly affect the interdomain interactions and give rise to the conformational changes that accompany hydrolysis and phosphate release. The α-phosphate interacts with Gln302 and Asp157, the β-phosphate with Gln15, Leu16 and the cation, whilst the γ-phosphate interacts with Ser14, Asp157, Gln158 and Val159 [396]. The structural similarities of actin, sugar kinases and chaperonin-type proteins is centred around position and function of this nucleotide-binding site. These unexpected structural similarities do not correlate obviously with overall sequence homology, but have prompted a more detailed examination in the region of the ATP-binding regions in the cleft. Some conserved amino acids have been found in these regions; particularly the two phosphate-binding loops (residues 7–29 and 150–170), the adenosine-binding loop (residues 297–314) and two connecting pieces (residues 129–148 and 332–352). These sequence motifs have since been used to search the existing sequence database and have shown that a number of other proteins including heat-shock and heat-stabilizing proteins and sugar kinases from prokaryotes and eukaryotes share its ATP-binding structure and might be structurally related to actin [10, 396].

SELF-ASSOCIATION SITES

Actin can form a number of three-dimensional supramolecular structures, the helical filament [31], three-dimensional para-crystals [1231], two-dimensional para-crystalline sheets [1171, 1173, 1341, 1363], tubular structures [1233] in the presence of metal ions or lipid surfaces, and filaments can organize into liquid crystalline arrays in solution [1206, 1236]. Monomers can also form conventional crystals with DNase [396], with segment 1 of gelsolin [746] and a ribbon structure that is found in profilin:actin crystals [1336]. Actin can also be induced to crystallize in the absence of a binding protein in the presence of polyethylene glycol [1310]. The relationships between the various structures are not well understood, but the structures of F-actin have been

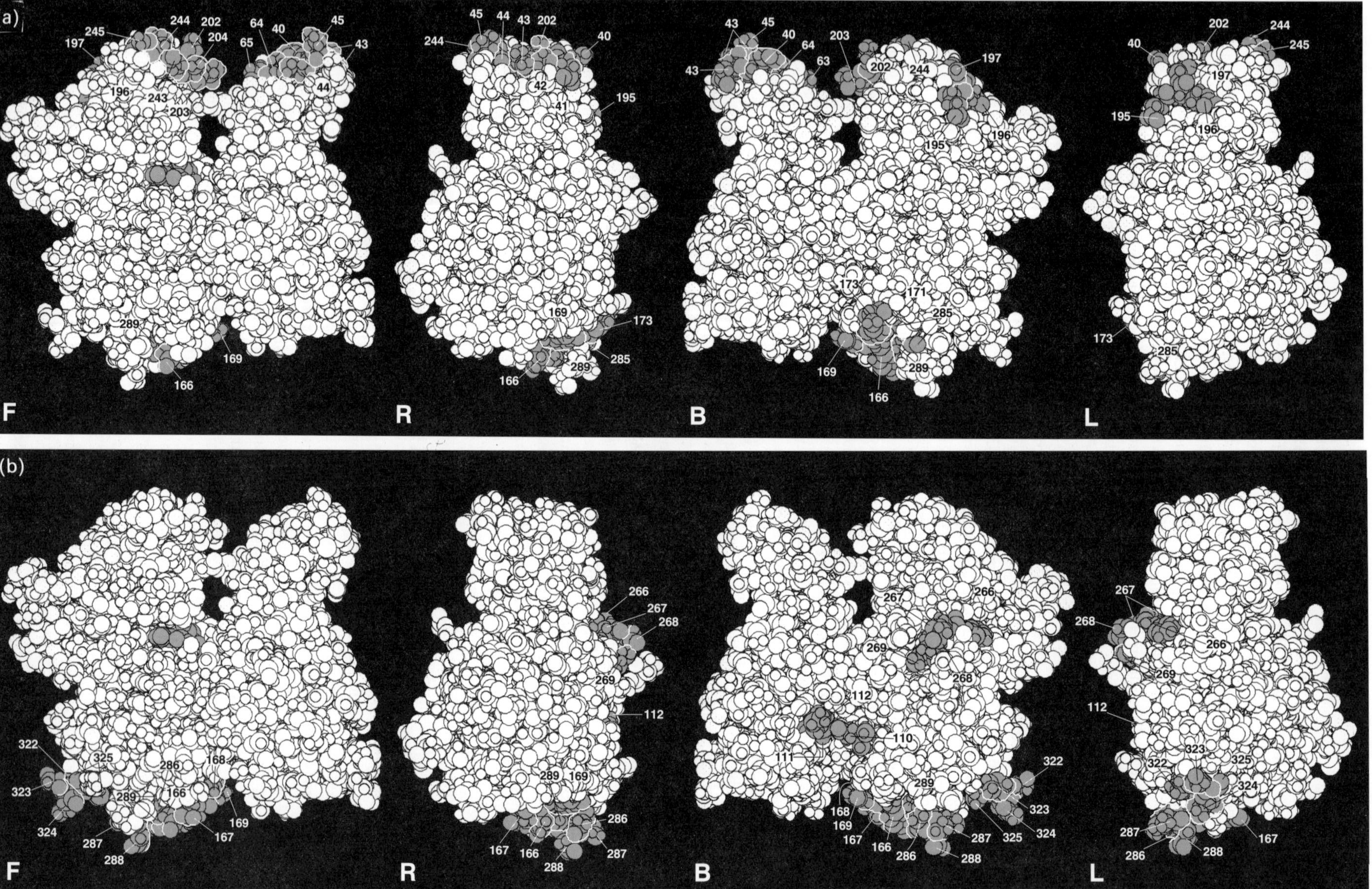

FIGURE 12 Surface binding sites for self-assembly. *Van der Waals surface images of skeletal muscle actin (lacking 373–375) with amino acids presumed to be involved in stabilization of F-actin [69] (see Table 7). Use the legend to Figure 7 for orientation of the actin monomer. (a) shows the 'donor' sites from monomer 3 in Figure 8 to monomer 1 (barbed end side) labelled 1–3 in Figure 8 and those across the filament to monomer 4 labelled 4 and 5 in Figure 8. (b) shows the complementary 'acceptor' sites on monomers 1 and 4, respectively.*

modelled at atomic resolution from available ultrastructural data and the monomer structure [69], whilst the ribbon structure has been solved directly from profilin:actin crystals. The available evidence suggests that the modelled filament provides the best description of the filaments assembled in the presence of magnesium ions and high salt and observed *in vitro* and *in situ* by a number of different morphological techniques. The ribbon structure is very different from the modelled filament in terms of the actin:actin interactions involved and at present is not known to exist outside the profilin:actin crystal. Whether the conformation of the biological filaments can shift between these two structures is not known.

In the F-actin model, the axial strands of the two-start helix are stabilized by interactions between residues 322 and 325 in subdomain IA with residues 243–245 in subdomain IIB; residues 286–289 in IIA with 202–204 in IIB and 166–169 and 375 in IIA/IA with 41–45 in IB (see Figures 8 and 12). These interactions are similar to those between DNase (mimicking the barbed end of an adjacent monomer) and the pointed end of actin. In contrast to these potentially extensive interactions along the filament, density at the centre of the model is low involving possible interactions between a hydrophobic loop consisting of Phe266, Ile267, Gln268 and Met269 on one strand with a hydrophobic pocket formed by Tyr166, Ala169, Leu171, His173, Cys285, Ile289, Gln63, Ile64 and residues 40–45 between the pair of monomers on the opposite 'strand' of the polymer. However, the recent structure of the profilin:actin crystal suggests that this loop, modelled in the original DNase structure, may not be a loop, but is folded back into subdomain IIB. However, mutations in the plug at position 266 increase the critical concentration for assembly [895] suggesting the possibility of an isoform difference. The only significant interstrand contacts in non-muscle actins may therefore be between 110 and 112 and 195–197 [69].

Derivatization of several amino acid residues either blocks or modifies assembly or conversely, assembly affects derivatization or signals from derivatives (see Table 5). In particular, derivatization of His40, Glu41, Tyr53, Lys61 (with bulky groups), Tyr69 and Arg177 generally affects polymerization. Furthermore, peptides 177–198 and 199–279 (essentially the whole of subdomain IIB) compete with monomers for assembly [691]. Surprisingly, the smaller segment inhibits polymerization more effectively. Many of these groups are at, or close to, proposed interaction sites in the filament. Inconsistent with a simple view of the model is the common effects on polymerization of derivatization of Cys374. This apparent discrepancy may be due to the uncertainty of the *C*-terminal structure in the DNase crystal where the *C*-terminal residues are missing [396] but where in the profilin crystal, the *C*-terminus forms a helix.

In the ribbon structure there are extensive close interactions between monomers, predominantly between residues 40–42, 44–46, 60, 61, 63 and 64 in subdomain IB with residues 2, 4, 5, 99, 100, 130, 131, 353, 358 and 360 on subdomain IA and between residues 202, 204, 205 and 243–247 on subdomain IIB with residues 48, 50, 87, 91, 92, 95 and 97 in subdomain IB of the $n+1$ monomer. This pattern of interaction continues along the ribbon (Figure 9).

ACTIN-BINDING PROTEIN SITES

Introduction

There are several different approaches that have been used to study the binding of actin-binding proteins to actin. Definitive information comes only from the atomic models derived from actin crystallized with an actin-binding protein. At present variable amounts of information are available for the binding sites on actin for DNase I, gelsolin (segment 1) and for profilin. Although these data are potentially definitive in terms of identification of amino acids involved in weak bond formation, they may provide only limited or no information about conformational changes during functional interactions and may represent only one of several possible structural interactions between those proteins. The general spatial location of protein-binding sites on F-actin has been examined by direct microscopy of negatively stained or rapidly frozen samples [1296]. Additional information has also been obtained by identification of proteolytic fragments of actin that bind by analysing the effects of protein binding on the accessibility of particular amino acids to covalent derivatization or, conversely, by the effects of derivatization on protein binding and by covalently cross-linking actin and the binding protein using 'zero-length' cross-linkers. In addition, it is possible to ask whether pairs of proteins (or protein fragments) compete for the same binding region on actin (see Table 8) or whether actin sequence-selective antibodies can inhibit binding. Some circumstantial evidence can also be gained by looking for common actin-binding sequences on different actin-binding proteins. Data on actin-binding sites have been reviewed periodically [160, 840, 1296]. A list of references relating to the identification of binding sites on actin is given in Table 7. *Protein Profile* issue 7 is devoted to the spectrin family.

TABLE 7 Summary of amino acids implicated in ligand-binding sites

Ligand bound	Amino acids involved	Evidence	References
Myosin	1, 2, 3, 4, 11	Cross-linking	583, 822
	Ionic site: 1, 2, 3, 4, 24, 25 Hydrophobic site: 144, 341, 345, 349, 352 2nd subunit site: 40–42/79–92 332–3	Modelling X-ray structures	1321
	1, 2, 3, 4	Mutagenesis	902, 1127, 901, 585, 877
	1–6	Deletion	619
	1–7	Antibody competition and derivatization	604, 722, 623, 625, 624, 1313
	1–28	Cross-linking	595, 724
	1–44	Peptide binding	716
	1–113	Cross-linking	595
	1–226	Peptide binding	724
	11 and 12	Mutagenesis	1118
	18–28 and 40–113	Antibody competition, peptide studies	716, 837, 752, 753, 724, 722, 575
	18–29	Antibody competition	575
	20–41	nmr	732
	24, 25, 99, 100	Mutagenesis	984
	28	Cross-linking	598, 599
	39–40	Proteolysis enhancement	692
	40–130	Cross-linking	595
	50	Cross-linking	599, 2201
	61	Derivatization	590, 1456
	61–69	Proteolysis protection	608
	95	Cross-linking	598, 599
	95–113	Antibody competition	724
	96–103	Antibody competition, peptide studies	721
	112–125	Antibody competition, peptide studies	721
	245	Mutagenesis	875
	305, 325, 355, 374	Antibody competition	722
	338–348	Peptide binding	723
	348–358	Antibody competition	723
	360–372	Peptide-binding	722
	360, 361, 363, 364	Mutagenesis	2233
	360, 362, 363, 364 and	Cross-linking	612, 822, 833
	362, 364, 366, 368, 372, 368-term	Mutagenesis	921
	370–375	Proteolysis	620
	374	Cross-linking	612, 652
	C-terminus	Antibody competition	2211
	Ionic site: 1–4, 24, 25 Hydrophobic site: 144, 341, 345, 349, 352– 2nd subunit site: loop332–333, 99, 100	Used in Figure 13	
Self-assembly sites	322–325 → 243–245 286–289 → 202–204 166–169 + 375 → 41–45 266–269 → 166, 169, 171, 173, 285, 289, 63, 64, 40-45 110–112 → 195–197	F-actin prediction from actin : DNase structure and image reconstruction (used in Figure 12)	69, 82, 396, 1253
	40, 41, 42, 44, 45, 46, 60, 61, 63, 64 → 2, 4, 5, 99, 100, 130, 131, 353, 358, 360 202, 204, 205, 243, 244, 245, 246, 247 → 48, 50, 87, 91, 92, 95, 97	Actin ribbon from actin : profilin structure	804, 805, 1335, 1338
	N-terminal acidics	Derivatization	1388
	11	Mutagenesis	1118
	18–29	Antibody competition	2306
	40, 53, 61, 177, 374	Derivatization	2033, 1403, 1454, 1406, 1497

TABLE 7 Continued

Ligand bound	Amino acids involved	Evidence	References
	42, 43	Proteolysis	253
	61	Derivatization	1456
	73	Mutagenesis	1117
	177	Ribosylation	1997, 2032, 310, 2050, 2073, 477, 1434, 2039, 2049, 1497, 5, 1951, 1952, 2042, 2044, 2073, 1953, 2052, 1498, 1955
	177–198	Peptide competition	2167
	191, 213 or 215–374	Cross-linking	823
	199–279	Peptide competition	2167
	245	Mutagenesis	1025, 1024, 903, 875
	266	Mutagenesis	895
	373–375	Proteolysis	288, 1307
DNase	1–83	Chimaeric *Tetrahymena*	686
	11, 12	Mutagenesis	1118
	39, 41, 43, 45, 61, 63, 203, 207	X-ray crystal structure	396
	42/43	Proteolytic protection	1524
	One of: 50, 53, 61, 68, or 69	Cross-linking	823
	50, 61, 68	Derivatization	699
	62, 68, 69	Protects against proteolysis	1178, 1509
	61, 69	Derivatization	1395, 1403
	363 and 364	Mutagenesis	921
Actobindin	1, 2, or 3 and 100	Cross-linking	838, 602
Tropomyosin	*N*-terminus	Antibody competition	753
	N-terminus	Mutagenesis	877
	61	Derivatization	1456
	95	Derivatization	1428
	215, 307	Modelling of structure	94, 74
	222–233	Deletion	237, 977
	238, 326, 328, 336 (on)	Derivatization	644, 829
	336 (off)	Derivatization	644, 829
	374-375	nmr	764
	374-375	Deletion	1307
Profilin	113, 166, 167, 169, 171, 172, 173, 284, 286, 287, 288, 290, 354, 355, 361, 364, 369, 371, 372, 373, 375	X-ray crystal	804, 805, 1335, 1338
	del1-12	Mutagenesis	921, 1238
	176, 177, 316, 339, 361, 362, 363, 364, 366, 368, 372, 368-term	Mutagenesis	921
	364	Cross-linking	840
	374	Derivatization	744, 745, 782
	375	Proteolysis	745
Cofilin family	1–12	Cross-linking	858, 767
	1, 2, 3, 4, 11, 361, 363, 364	Cross-linking (depactin)	826, 827
	357–375	Cross-linking	826, 827
	374	Cross-linking and derivatization (ADF)	622, 862
	1, 2, 3, 4; 361, 363, 364, 374–5	Used in Figure 21	
Vitamin-D-binding protein	360–372	Antibody competition	693
Thymosin	374	Cross-linking	681
α-Actinin family	1–12, 86–123	Cross-linking	762
	83–117, 350–375	nmr (dystrophin)	733, 734
	86–117, 350–375	Peptide binding (dystrophin)	647
	86–117, 350–375 on adjacent long-pitch monomer (α-actinin)	Cryoelectronmicroscopy	2244
	99, 100	Mutagenesis (fimbrin)	2288
	103, 356-375	Cross-linking	728
	105–120, 360–372	Peptide competition (filamin)	721, 754
	112–125, 360–372	Peptide and antibody competition	729, 730
	112–117, 360–372	Used in Figure 18	
	?	Mutagenesis (fimbrin)	2290
Troponin-I	373–375	Deletion	743

TABLE 7 Continued

Ligand bound	Amino acids involved	Evidence	References
Calponin	326–355	Cross-linking	757
	373–375	Proteolysis	2202
Caldesmon	1–7	Antibody competition	574
	1–12	Cross-linking	591
	1–44	Peptide competition	2249
	370–375	Deletion	620
	373–375	Deletion	743
	374	Cross-linking	670, 671
	374	Derivatization	619
Gelsolin family	1, 2, 3, 4, 11	Cross-linking (fragmin ≡ segs 1–3)	825
	1–12	Cross-linking (whole molecule)	634
	1–18	Cross-linking (segs 1–3)	828
	112–117, 360–372	Segs 2–3 equiv. to actinin site	846
	285–375	Antibody competition (whole molecule)	600
	350, 374, 375, 144	Crystal structure (seg 1)	661
	356–375	Cross-linking (segs 1, 2–3 and 4–6)	828
	360–372	Peptide binding (Gelsolin seg. 1)	651
	374	Cross-linking (whole molecule)	635, 636, 1359
	345, 350, 374–375, 144 seg. 1 112–117, 360–372 segs 2–3	Used in Figures 16 and 17	
Scruin	Subdomain 1/junctions 3 and 4	Electronmicroscopy	2326
Cytochalasins	10	Derivatization	2019
	139, 295	Mutagenesis	2045
Phallotoxins	10, 117, 119, 355	Cross-linking	442, 1410, 2072
	79, 198, 279	Three-dimensional modelling	82
	121?	Changes in *Tetrahymena* actin sequence	977
	374	Cross-linking	1989
	374	Derivatization	1967
Nucleotide	213, 214, 303, 305, 306, 336 adenine ring 157, 214 ribose 14, 15, 16, 18, 157, 158, 159, 302	X-ray structure	396
	114–118	Peptide-binding	584
	306	Derivatization	510
	336, 356	Derivatization	487
Cation	11, 137, 154, β-*O* of nucleotide	X-ray structure	396
	18, 68, 191, 291, 359	Derivatization	238
	26, 68, 372, 373, 39-51	Proteolysis	1347

Evidence relating to the identification of binding sites on actin are listed for ligands where information is available.

Myosin-binding site

Actin and myosin form one of the major motor protein complexes in all eukaryotic cells. An understanding of how myosin and actin produce force and movement has been an important goal in muscle research for many years and, with the realization that non-muscle myosins are universally responsible for many intracellular movements and cell motility, has become a key area of study within cell biology (see the *Protein Profile* issue on myosin [41]).

The S-1/F-actin reconstructions show that each myosin head interacts with two actin monomers, giving rise to a binding site which spans the two actin subunits (n and $n-2$ in the genetic 1-start helix). This is consistent with experimental evidence that myosin binds two actin monomers [580, 582] and may account for the enhancement of actin assembly by S-1 [186]. The myosin-binding site on the actin has been divided, based upon structural studies of the myosin molecule, into so-called primary and secondary sites (for a recent review, see [14]). The primary site contains contacts with both the IA and IIA domains which involve both ionic and stereospecific interactions. Four contacts on actin have been predicted within the primary site: (1) the *N*-terminus and the acidic residues 1–4 and 24–25; (2) the loop consisting of P332, P333 and E334; (3) a hydrophobic site containing residues 144, 341, 345, 349 and 352; and (4) a contact with residues 40 and 42 in the actin subunit $(n-2)$ below. A lysine-rich loop in the myosin (not resolved in the X-ray structure) is close enough to interact with the cluster of negatively-charged

TABLE 8 Competition for binding sites on actin

Simultaneous binding of this actin-binding ligand	With this one	References
α-Actinin	Villin tail, talin, destrin	783, 766, 2228
Actophorin	Profilin	741
ADF (actin depolymerizing factor)	Gc, DNaseI, villin head, cofilin	783, 532, 622
c-Abl	Gelsolin	836, 2266
Caldesmon	Tropomyosin, myosin, dystrophin	591, 671, 647
Cytochalasin	Villin, tropomyosin	1977, 2065
DNase I	Gc, ADF, myosin	664, 622, 608
Drebrin	Tropomyosin, α-actinin	2228
Dystrophin	Caldesmon, tropomyosin	647
EF-1α	EF-2	2199
EF-2	EF-1α	2199
Filamin	Myosin	721
Gc (vitamin D-binding protein)	Profilin, ADF, DNase I	1516, 532, 664
Gelsolin	c-Abl	836
Segment 1		
Segments 2–3	Villin tail	783
Jasplakinolide	Phalloidin	2380
MAP1A	MAP2	2251
MAP2	MAP1A	2251
Myosin	DNase I, caldesmon, filamin, phalloidin	608, 671, 721, 2385
Phalloidin	Jasplakinolide, myosin, tropomysosin, gelsolin	2380, 2385
Profilin	Gc, Actophorin	1516, 741
Talin	α-actinin	766
Tropomyosin	Caldesmon, cytochalasin, dystrophin, drebrin, phalloidin	591, 2065, 647, 2370, 2385
Villin	Cytochalasin	1977
Tail	Segments 2–3, α-actinin, villin head	783
Headpiece	ADF, α-actinin, segment 2–3, villin tail	783

One actin-binding protein might compete with, allosterically regulate, or bind independently of another actin-binding protein to G- or F-actin. Instances where any of these interactions have been observed are listed.

amino acids at the *N*-terminus, which may constitute an ionic site (see the *Protein Profile* issue on myosin [41]). This interaction should be sensitive to ionic strength.

Actin mutants (see Table 4) have been used to examine the importance of the acidic groups at the *N*-terminus for the myosin interaction. These residues are particularly variable in different actin isoforms (see the actin alignment). Mutation of the *N*-terminal acidic residues 1–4 abolishes (D3K/D4K) or reduces S1 binding (D3Δ/D4Δ and D3A/D4A, where Δ denotes a deleted residue) [876], but not all do so. Mutations D1H, D1H.D4H do not affect K_{app} for actin activation of myosin ATPase [1127]. All of these mutants (D2N/E4Q, D2Δ/S3Δ/E4Δ, D3K/D4K, D3Δ/D4Δ, D3A/D4A, D1H and D1H/D4H) reduce actin activation of myosin ATPase activity [877, 901, 902, 1127]. D1H, D1H/D4H, D3Δ/D4Δ and D3A/D4A reduce *in vitro* motility [877, 1127]. Overall, these results indicate that the *N*-terminal negative charge cluster is essential for the ATP-dependent actin–myosin interaction and the development of movement. Two further acidic residues mutated simultaneously to histidines (D24H/ D25H) confirm the importance of these amino acids for myosin binding and movement [984]. The double mutant allows the F-actin to bind under rigor conditions, but there is a loss of sliding activity and the filaments are gradually lost from the surface on addition of ATP. D24H/D25H actin does not activate myosin ATPase activity.

The secondary site occurs on the actin monomer ($n-2$) immediately below the primary site. The myosin structure in this region is not seen in the X-ray data, but the proposed secondary site includes actin residues in the *C*-terminus of the 79–92 α-helix and residues 93, 95, 99 and 100 in the loop that extends barbed-endwards from it. The extent of this binding region along the loop probably depends on the myosin isoform since the unseen myosin polypeptide sequence differs in length between chicken and *Dictyostelium* S1. The importance of residues E99 and E100 has also been implied by the E99H/E100H double mutation in the *Dictyostelium act15* gene [984]. This mutant actin has a tendency to bundle, but single filaments moved, albeit with a much reduced velocity *in vitro*.

The myosin family of proteins can be divided into a number of subfamilies that have similar actin-binding ATPase domains (myosin head) but different tail regions that mediate self-assembly and other ligand interactions [831]. In slime mould myosin I, an additional ATP-independent actin-binding site exists in the glycine- and proline-rich region at the *C*-terminus of the tail [709, 791]. The majority of studies on myosin-binding and interaction of myosin with actin during force generation involve the use of the mammalian fast muscle myosin II isoform, or its functional proteolytic fragments S1 and HMM, with the skeletal muscle isoform of mammalian actin. Each myosin head is an independent force generator and may bind to two adjacent molecules of actin in the filament during the cross-bridge cycle [579, 581]. The cycle involves transitions between a number of conformational intermediates of actin bound to myosin which may involve shifts in emphasis from one binding site to another or changes in structure within a binding site. The most stable interactional conformation is the rigor conformation which exists at the end of the cross-bridge cycle after the force-generating change and can be frozen experimentally by the removal of ATP necessary for release of the myosin head. Myosin S-1 can also bind to one or two actin monomers in solution in the presence of ATP. Because it is not easy either to identify or to maintain intermediate conformations for biochemical or structural studies, the data on myosin binding are biased by the experimental convenience of rigor and for the interaction of ATP.S-1 with monomer. The precise sequence of interaction of myosin with actin sites during the contractile cycle remains to be clarified. Several recent reports review myosin binding to actin [74, 94, 119, 396, 787, 788, 803].

The strong rigor binding site(s) on actin have been studied by computer-aided modelling of ultrastructural data and lead to a clear indication that the major binding site for myosin centres on subdomain IA on the outer surface of the filament [94, 1002, 1364]. Cross-linking studies on the bound complex between myosin S1 and G- or F-actin show some differences [612] but implicate amino acids at both the *N*- and *C*-termini. Acidic residues 1, 2, 3, 4 and 11 can be cross-linked to S1 and 360, 362, 363 and 374 to LC1 [612, 822, 833]. In addition, sequences 1–28 and 40–113 can be cross-linked to myosin using a different cross-linker [595] probably via arginines 28 and 95 respectively [598]. However, antibodies that bind the seven *N*-terminal residues of actin only marginally affect binding and in some cases are able to bind simultaneously with S1 [624, 625, 760, 1313]. Site-directed mutagenesis of the *N*-terminal 1–4 sequence of actin shows that charge abolition of the acidic residues involved in cross-linking has little effect on S1-binding [585, 922] and that even charge reversal only moderately inhibits binding [586]. Removal of *C*-terminal residues 373–375 or 370–375 also only reduces binding [619, 620] but inhibits the myosin ATPase and abolishes *in vitro* motility [902] whilst antibodies to 18–29 block the weak binding of myosin in the presence of ATP and also inhibit the ATPase [575]. Furthermore, addition of extra negative charges at the *N*-terminus increased the V_{max} but did not affect either binding or motility [902] suggesting an endo-allosteric role for the *N*-terminus. Perturbation of signals from or the effects of derivatization of acidics at the *N*-terminus and lysines at the *C*-terminus on binding of S1 suggest also that these regions are close to the binding site (see Table 7). In addition, complementary studies using actin peptides show that residues 1–28 and 1–44 (via activation) can bind to S1 whilst 1–18 and 82–119 are inactive [716, 837].

The observations listed above suggest that whilst the actin terminal regions of these peptides might be close to and available for covalent cross-linking, they may play only a supporting role in rigor binding. Further evidence using similar approaches implicates residues beyond 18 in binding to myosin; antibodies to sequences 18–28 and 40–113 disrupt rigor binding [724, 752, 753] and nmr studies also implicate sequence 20–41 [732]. Charge reversal of residues 99, 100 and 24, 25 disrupt *in vitro* motility [984]. Furthermore, S1 bound to actin both blocks proteolytic cleavage of the 61–69 loop in subdomain IB [608] and perturbs signals from derivatized Lys61 [590], suggesting that these residues are also close to the binding site. Binding of two S-1 molecules to G-actin or S-1 to F-actin exposes a new proteolytic site at His40 in subdomain IB suggesting both that at least two distinct binding sites exist on actin for myosin and on myosin for actin and that the S-1 ternary complex mimics S-1-F-actin [1227].

Activation of myosin presumably involves both binding (at different or additional sites?) and allostery. Sequences 1–28 and 1–24, but not 1–18 activate myosin ATPase [716, 837]; whilst mutation of the *N*-terminal acidic residues or removal of two *C*-terminal residues reduces, but does not abolish, activation [901, 922, 1307]. Nitration of tyrosine-69 inhibits myosin activation (see Table 5) suggesting that sites other than the termini are close to or involved in activation.

Whilst the ultrastructural evidence that the myosin head binds on the outer surface of subdomain IA and overlaps on to subdomain IB is essentially irrefutable, evidence identifying sequences involved in binding and activation, whilst implicating the *N*- and *C*-terminal regions, suggests that binding involves a sequence somewhere between 18 and 44 on a loop that emerges from subdomain IA and perhaps extends over into subdomain IB where it can perturb Lys61.

The spatial position of the myosin-binding sites has been predicted by fitting the X-ray crystal structure of both *Dictyostelium* [1334] and vertebrate

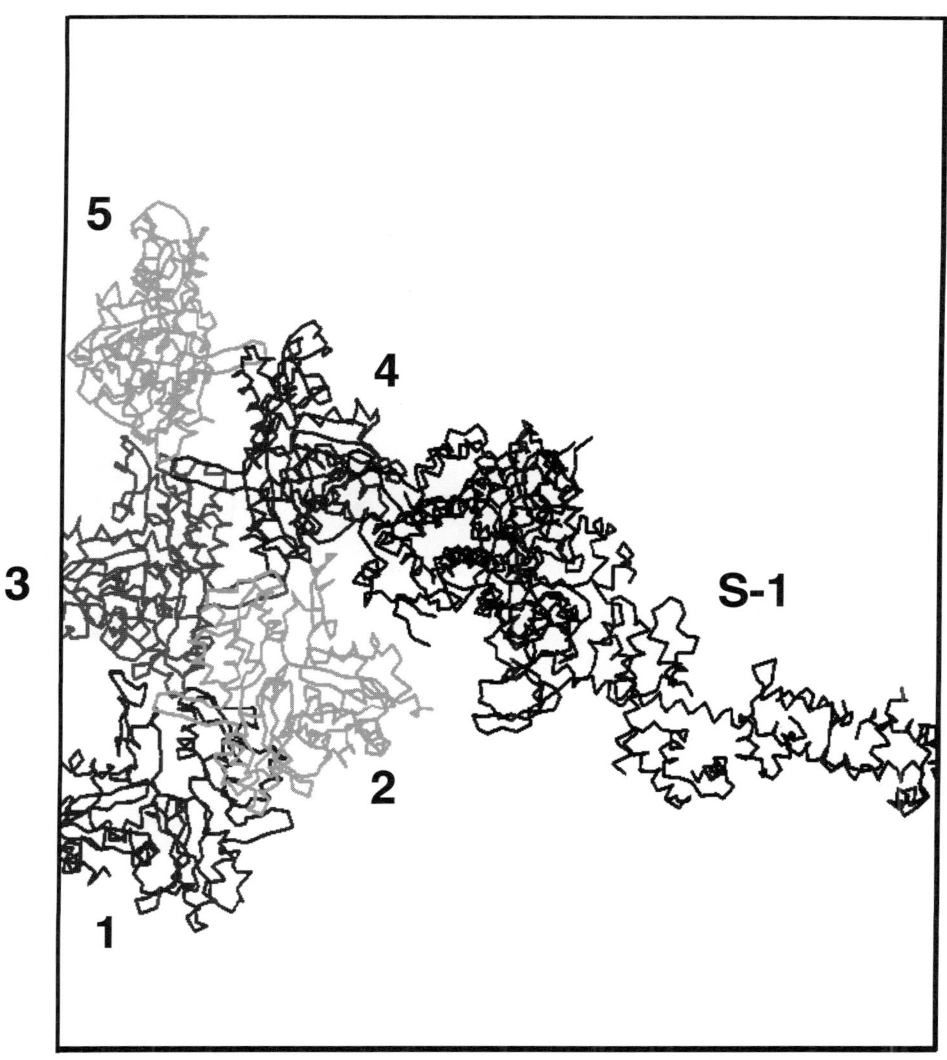

FIGURE 13 Structure of S-1 : F-actin complex. *Diagram showing the proposed relationship between myosin S-1 and five subunits of a modelled actin filament [1320]. The primary interaction is with monomer 4, although additional interactions may occur with subdomain IA of monomer 2 (see text). Only the α-carbons and peptide bond atoms are shown. The orientation of the filament is such that the barbed end is at the bottom of the Figure. Monomers 2, 3 and 4 in this Figure are equivalent to monomers 1, 2 and 3 in Figure 8.*

skeletal muscle [787, 788] myosin S-1 and a predicted structure for F-actin into an envelope of a shape reconstructed from transmission electronmicrographs of S-1-decorated F-actin preserved in vitreous ice (Figure 13). The model is consistent with EM mapping of the positions of the SH1 cysteine and the nucleotide-binding site relative to the actin interface and regions of myosin sequence protected from proteolysis whilst bound to actin [788].

These sites are all shown on a single actin monomer in Figure 14.

The effects of myosin binding on the structure of F-actin have also been studied; whilst myosin binding does not appear to alter the orientation of a spin probe attached to Cys374, suggesting no movement of subdomain IA [1302], S-1 binding does alter the accessibility of nucleotide to the surrounding water [575] suggesting a change in the relative orientation of the two domains.

Tropomyosin-binding site

Tropomyosin is a linear dimeric molecule about 40 nm long consisting of an α-helical monomer wrapped about a partner in a coiled-coil arrangement [1199]. The molecule consists of six or seven (non-muscle and muscle isoforms, respectively) linearly repeated actin-binding sites (each consisting of two subsites) that bind the sequential seven (or six) actin subunits along the major groove on each side of the actin filament helix in each half-turn of the helix. The identification of tropomyosin-binding sites is complicated by the apparent

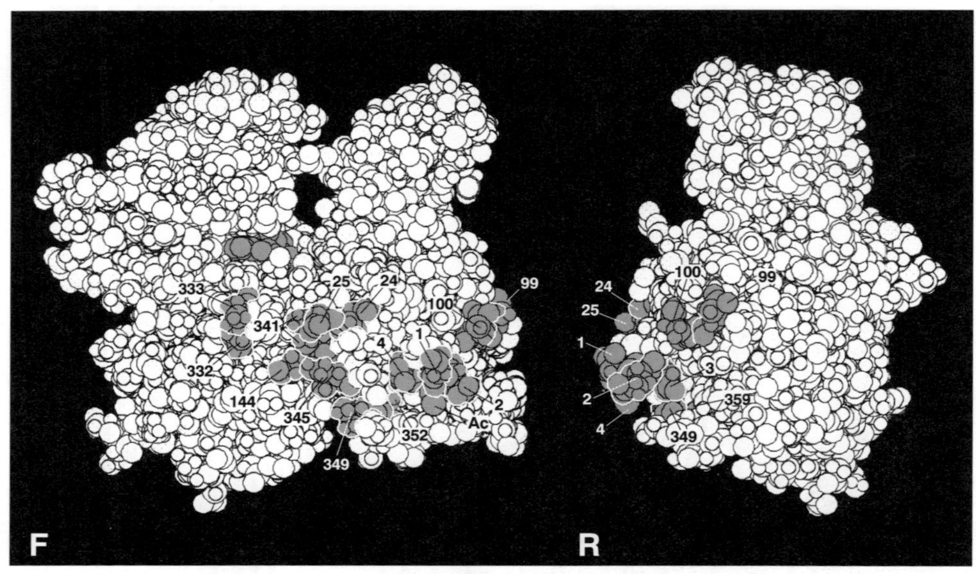

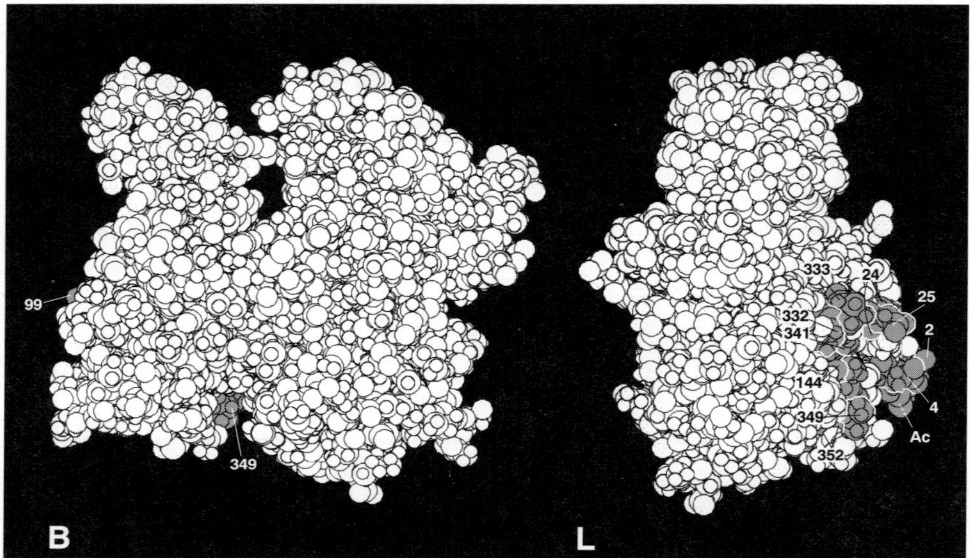

FIGURE 14 Surface binding sites for myosin. *Van der Waals surface images of skeletal muscle actin (lacking 373–375) with amino acids presumed to be involved in binding of myosin. See Table 7 and text for discussion. Front (F), right (R), back (B) and left (L) views are shown (see Figure 7).*

shift in position of the molecule in the filament groove from a position that inhibits myosin interaction with the filament (the 'off' site) to one that permits interaction on activation via the thin or thick filament (the 'on' site). Image reconstructions of ultrastructural data suggest that tropomyosin moves towards the axis of the filament from the 'off' to 'on' position to expose the myosin-binding sites on subdomain IA and IB during activation [731]. Tropomyosin is at a radial distance of about 3.8 nm from the axis when occupying the 'on' sites that approximately diagonally bisect the two subdomains IIA and IIB [1296, 1363, 1364]. Several recent reports review tropomyosin binding to actin [74, 396, 1296, 1322, 2224].

Biochemical evidence for the interaction of tropomyosin with the *N*- or *C*-terminal regions of actin is contradictory. Neither antibodies to *N*-terminal regions [753] nor charge-shifting of the *N*-terminal acidic residues [877] affect binding of tropomyosin. Nmr indicates that the *C*-terminal peptide 355–375 binds [764] and that removal of residues 374 and 375 inhibits binding [1307]. Derivatization studies have shown that modification of lysines 238, 326, 328 and 336 are reduced in the 'on' position, whilst lysine-336 is protected in the 'off' position [644, 829]. Tropomyosin-binding is blocked by derivatization of Lys238 [644]. Similarly, derivatization of Arg95 blocked tropomyosin binding whilst conversely, derivatization was blocked by binding [1428]. The most convincing data come from the observation that *Tetrahymena pyriformis* actin does not bind tropomyosin at all [237] and it lacks eight amino acids

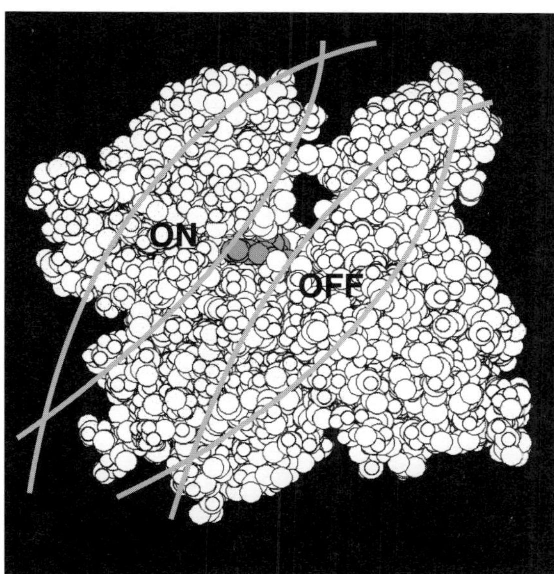

FIGURE 15 Surface binding sites for tropomyosin on the front of actin. *Van der Waals surface image of the front view of skeletal muscle actin indicating the approximate position of tropomyosin from the structural and biochemical data (see Table 7).*

in the region 222–233 [977]. Interestingly, in hsp70 the sequence 217–236 is replaced by just three amino acids suggesting an actin-specific function at this site.

The morphological evidence lays tropomyosin in the 'on' position over subdomain IIB where it could interact with the helix 223–230 and cause labelling changes of lysines 238, 326 and 328 [94]. The exposure of Lys336 in the 'off' position [829] is consistent with a distal shift and may account for the *C*- and *N*-terminal effects. The effects of Arg95 labelling are difficult to reconcile with either putative location as shown in Figure 15.

Caldesmon-binding site

Caldesmons are a family of elongated molecules with a globular head which contains the actin-binding and calcium regulatory sites. Caldesmons play a major role in the regulation of contraction in smooth muscle, probably by competing with myosin heads for binding sites on actin or by some related steric mechanism [619]. There is evidence for interaction of caldesmon with both the *C*- and *N*-termini of actin. Antibodies to sequences 1–7 significantly block binding, whilst those directed at 18–28 inhibit less so [574] and a zero-length cross-linker covalently couples caldesmon to the sequence 1–12 [591]. At the *C*-terminus, removal of 374 and 375 weakly inhibits, whilst removal of 373–375 or 370–375 strongly inhibits binding of caldesmon [620, 743]. Caldesmon can be cross-linked to form a disulphide bridge onto Cys374 [670, 671] and derivatization of this amino acid inhibits binding [619]. There have been no reports of any effect of calcium on the extent of the caldesmon-binding sites.

Binding site of the gelsolin family

Gelsolin is the vertebrate homologue of a family of proteins that contain varying numbers (1–6) of a repeat domain that together confer nucleating, capping and severing activities to members of the family [56, 145, 784]. Most members of the gelsolin family are regulated by both calcium ions and phosphatidylinositol-4,5-bisphosphate [56]. Gelsolin contains six repeat segments that appear from sequence comparisons of the segments to have a 1-2-3-1-2-3 organization, but where segments are numbered 1 through 6 such that segment 1 is similar to 4 and 2 to 5, etc. [784]. Members of the family from lower eukaryotes and invertebrates include the fragmins and severins which consist of segments 1–3 by comparison with gelsolin, suggesting that gelsolin arose from a gene duplication event. Proteins expressing a single segment exist in yeast and some other lower eukaryotes, suggesting that the whole family arose as series of gene duplications [56, 145]. Each segment has a molecular mass of about 15 kDa (120–130 amino acids). The capping and severing functions of gelsolin appear to be mediated through three functional binding sites. Segment 1 functions as a monomer-binding segment and is the site occupied in the GA1 binary complex; segment 1 binding is calcium independent but dissociated by inositide. Segments 2–3 specify a calcium-dependent polymer-binding site whilst segments 4–6 specify a second calcium-dependent monomer-binding site [784] used in the GA2 complex. It is interesting to note that the two actin monomers in the GA2 complex appear to be oppositely polarized [723]. It is now known from the atomic structure of segment 1:actin crystals that segment 1 binds to the barbed end of a monomer between subdomains IA and IIA (Figure 16) [750]. It is likely that segments 2–3 bind on the outer surface of domain I of the same monomer since the replacement of segments 2–3 in gelsolin with the actin-binding site of α-actinin confers a normal actin-binding stoichiometry and functionality [846] (see Figure 17). Segments 4–6 are predicted to bind to a similar region to segment 1, but on an adjacent monomer on the opposite strand of the polymer [643, 784]. These data both imply some structural homology of the two

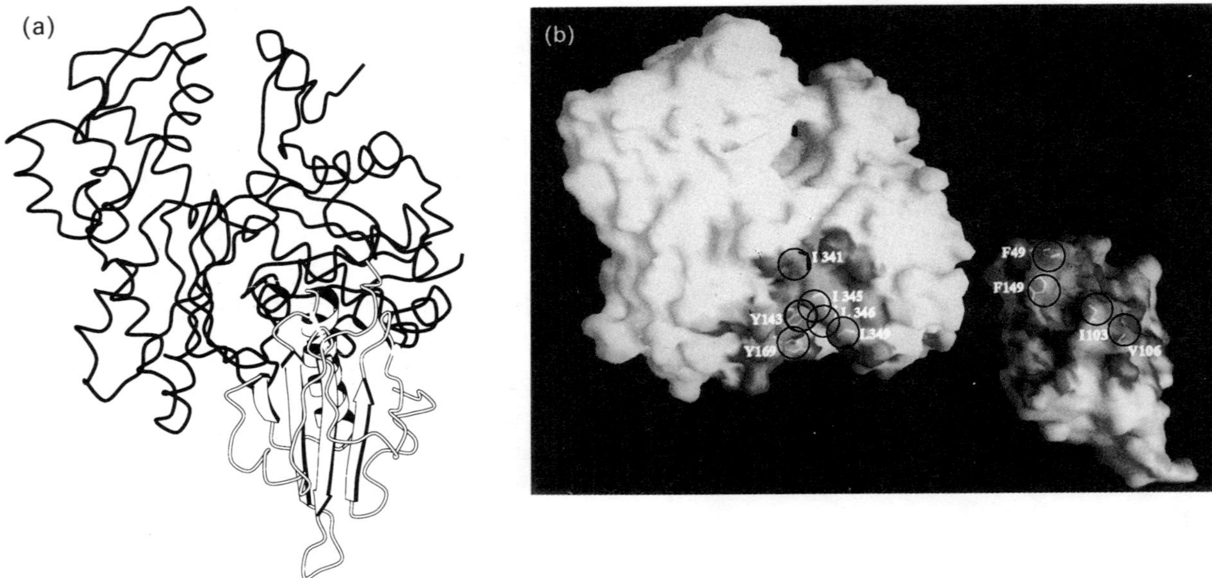

FIGURE 16 Molecular structure of the gelsolin segment 1:actin crystal. *Figure (a) shows the actin alpha carbon chain (shaded) and gelsolin segment 1 (central helix and pleated sheet as a ribbon diagram) [1284]; (b) shows a three-dimensional rendering of the hydrophobic contact surfaces of the two molecules (Courtesy of Dr P. McLaughlin).*

domains and that domains 2–3 of gelsolin bind at a similar site to α-actinin on the outer surface of the filament at the back of domain IA.

Crystallographic data from segment 1:actin crystals (Figure 15) show that segment 1 is bound closely to the barbed end of actin between subdomains IA and IIA at the hinge region of the molecule. This binding site is consistent with both the barbed-end binding characteristics of gelsolin and with the inhibitory effects of gelsolin on nucleotide exchange [554] that appear to require relative movements between domains I and II. The structure predicts that amino acids in the hydrophobic patch around residues 345, 346 and 352 can interact with a similar hydrophobic region on gelsolin and that residues 144, 146, 148, 167, 169, 334, 350, 351 and 354 hydrogen bond to residues on segment 1 [750] (Figure 17). Whole gelsolin can be cross-linked to acidic residues in sequence 1–12 [634] and both whole gelsolin and segment 1 to Cys374 [635, 636]. Both the *N*- and *C*-termini are close to this region. Fragments of gelsolin have been expressed in and isolated from *E. coli* [784]. Both segment 1 and segments 2–3 can be cross-linked to acidics in sequence 1–18 [828], whilst segments 4–6 can be cross-linked to residues at the *C*-terminus in sequence 356–375 [828]. Site-specific antibodies against sequence 285–375 inhibit gelsolin binding to F-actin [600] and a peptide of actin, 360–372, binds to segment 1 of gelsolin [651]. These data do not help to discriminate between the F-actin site of segments 2–3 and the G-actin sites of segment 1 and segment 4–6. Fragmin, related in sequence to segments 1–3 of gelsolin, can also be cross-linked to *N*-terminal acidic residues 1, 2, 3, 4 and 11 [825].

Binding site of the α-actinin family

The α-actinin family of F-actin cross-linking proteins includes a range of molecules that share homologous actin-binding domains [647] allowing them to cross-link actin filaments by the formation of dimers or tetramers. The family includes the spectrins, fimbrinplastins, dystrophins, gelation factor from *Dictyostelium*, and filamin subfamilies (see the *Protein Profile* issue on the spectrin superfamily). The globular actin-binding domain is contained in the first 250 or so amino acids at the *N*-terminus of all members of the family [9, 756], and may be composed of more than one subdomain. Whole α-actinin binds near Thr103 and to the *C*-terminal 356–375 sequence [728]. The isolated 27 kDa actin-binding domain can be cross-linked to *N*-terminal sequence 1–12 and to sequence 86–123, a loop on the outer surface of subdomain IA [762]. These data are partly supported by nmr measurements implicating sequences 83–117 and 350–375 in the binding of the homologous actin-binding region of dystrophin [733, 734] and by the observation that peptides 112–125 and 360–372 both bind and compete in binding directly to α-actinin [730], and peptides 105–120 and 360–372 both bind to filamin and compete with actin for filamin. These data locate the binding site of this family of proteins on the outer (with respect to the filament axis) edge of the filament behind subdomain IA, and involving the α-helix 360–372 and the sequence 95–125 (see Figure 8). *Tetrahymena* actin only binds α-actinin weakly [17]. This divergent actin has, among other changes, an A to Q change

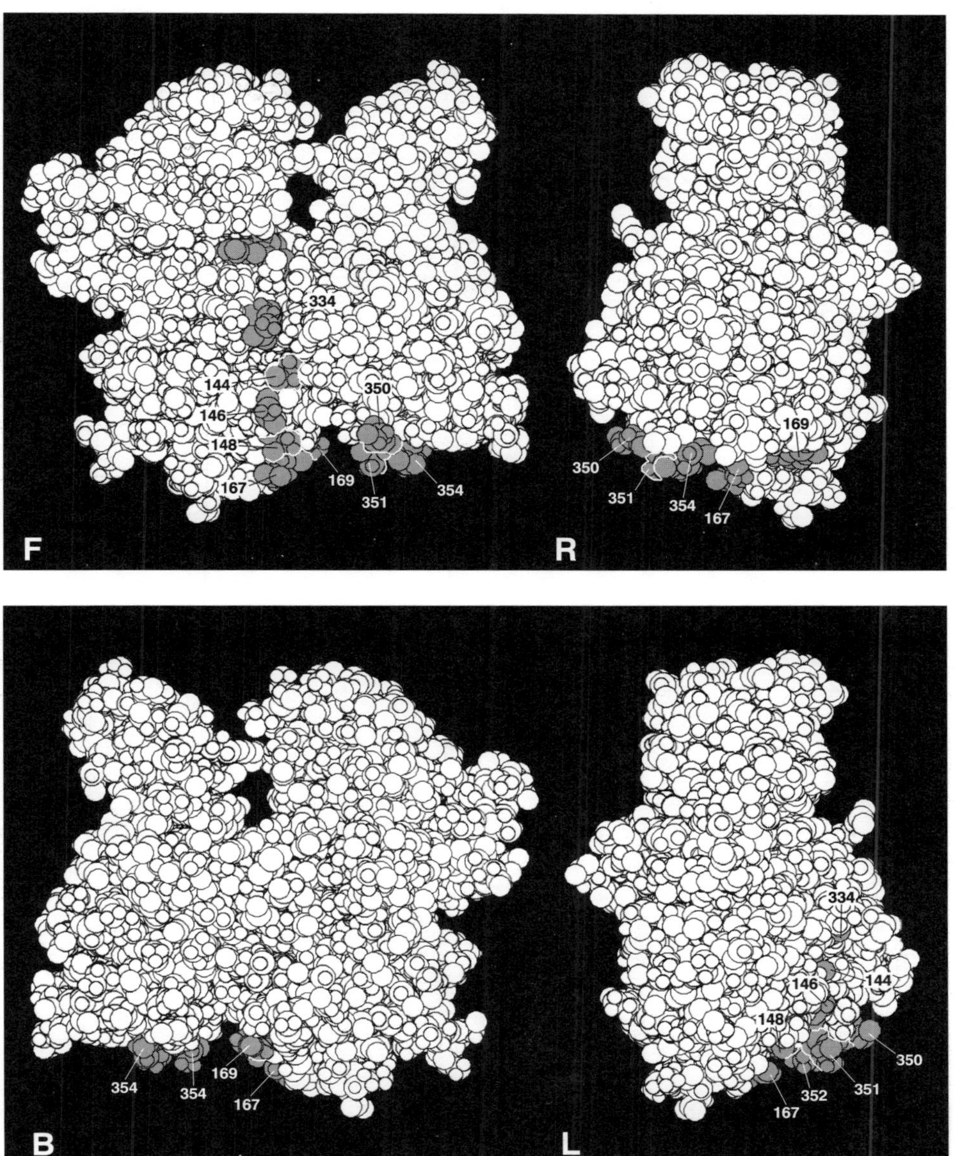

FIGURE 17 Surface binding sites for gelsolin. *Van der Waals surface images of skeletal muscle actin (lacking 373–375) with amino acids implicated in binding of gelsolin in addition to those in Figure 16. Front (F), right (R), back (B) and left (L) views are shown. See Table 7 and text for discussion. Note that the additional gelsolin site may be similar to that of α-actinin family (see Figure 17).*

at position 114 and a Q to A change at 360, both in areas of interest. Recently, image processing of electron micrographs of actin filaments decorated with the *N*-terminal domain of α-actinin suggests that the binding of this domain occurs at two sites on the α-actinin *N*-terminal domain and sites on two neighbouring monomers along the long-pitch helical strands [2244]. The contacts are both in subdomain IA, but at least one of them is poorly resolved. Residues 86–117 and 350–375 are clearly involved. This biochemical localization is consistent with the ultrastructural evidence that α-actinin can cross-link or bundle oppositely polarized actin filaments in ladder-like arrays with a periodicity along the filament of about one half-turn (or multiples thereof) where the actin monomers are in equivalent positions relative to the filament [758].

Profilin

Profilins are a group of proteins that can form a binary complex with actin monomers. Profilins catalyse the exchange of nucleotide on actin monomers [482] and can be dissociated from the complex by phosphatidylinositol-4,5-bisphosphate [666]. The *N*-terminal sequence of the profilins is conserved within the group, but the *C*-terminal portion of the molecules differ. The *C*-terminal region in lower

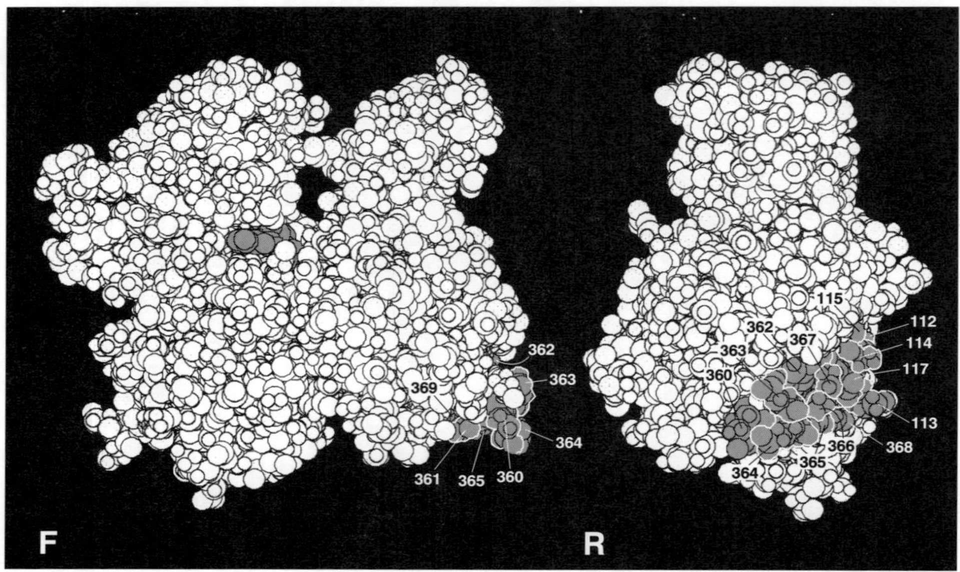

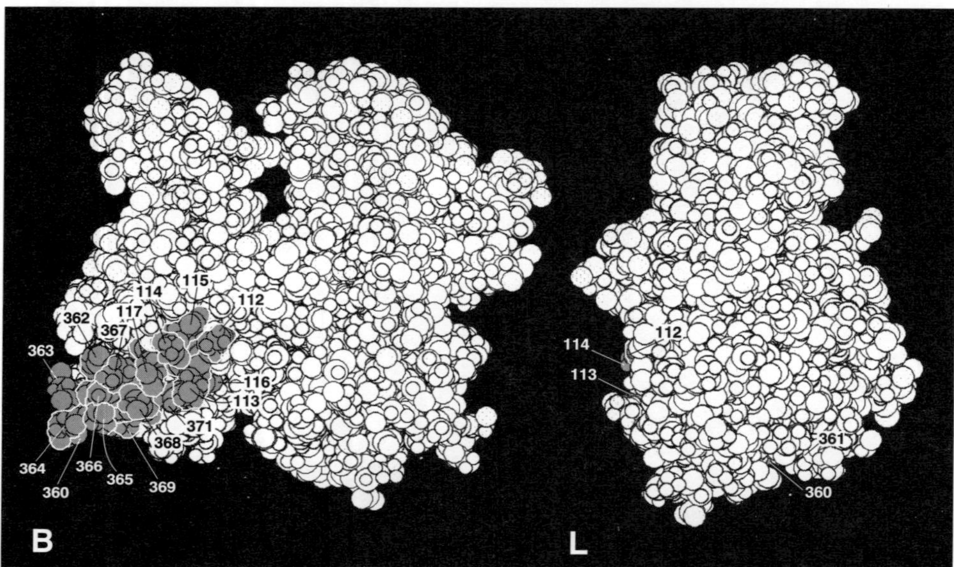

FIGURE 18 Surface binding sites for α-actinin. *Van der Waals surface images of skeletal muscle actin (lacking 373–375) with amino acids presumed to be involved in binding of α-actinin. Front (F), right (R), back (B) and left (L) views are shown. See Table 7 and text for discussion.*

eukaryote and yeast profilins shares sequence similarity with part of the repeat domain in the gelsolin family [145].

The atomic model of the structure of profilin:actin crystals indicates that profilin interacts with the barbed end of the actin molecule between subdomains IA and IIA, in a similar position to that at which segment 1 of gelsolin binds (Figure 19) [805, 1336]. Profilin appears to make close contact with the two loops containing residues 166 and 288. Close contacts are made with profilin by residues 113, 354, 355, 361, 364, 369, 371, 372, 373 and 375 in subdomain IA. Close contacts are also made with residues 166, 167, 169, 171, 172, 173, 284, 286, 287, 288 and 290 in subdomain IIA (Figure 20). The biochemical data indicate that profilins bind close to the *C*-terminus and require that terminus for binding. Derivatization of Cys374 strongly inhibits the binding of vertebrate profilin [744, 745] and *Acanthamoeba* profilin can be cross-linked to Glu364 on the barbed-end side of subdomain IA [840]. These data are consistent with the position of the binding site indicated by the crystal structure and with the weak barbed-end capping effects of profilin [482].

Cofilin family

Actin-depolymerizing factor, destrin, actophorin and depactin are all closely related members of a

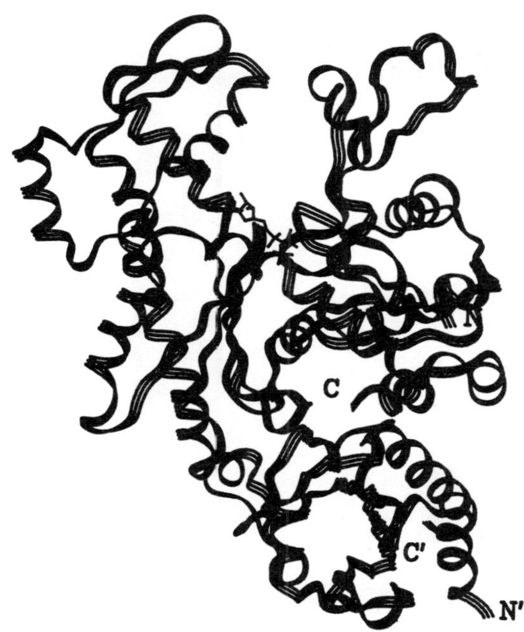

FIGURE 19 Interaction of mammalian profilin with β-non-muscle actin. *The profilin:actin interface as found in the profiliactin crystal. Data from Professors U. Lindberg and C. Schutt. The N and C-termini of both molecules are indicated.*

family of small (about 21 kDa) proteins that can form a binary complex with G-actin, and depending on pH, can also bind to F-actin and bring about subsequent pH-dependent depolymerization [679, 767]. These proteins carry a phosphoinositide-binding site [857] and share some sequence homology at the *C*-terminal end with members of the gelsolin family and with the *C*-terminal part of lower eukaryote profilin [145]. The binding site for cofilins appears to be close to that of the profilins since profilin competes with actophorin for binding to G-actin and no ternary complexes are formed [741]. Members of this family can be cross-linked to acidic residues in the *N*-terminal sequence 1–12 [767, 858] and to 357–375 at the *C*-terminus [826, 827], consistent with the localization of the binding site close to that of profilin (Figure 21).

Actobindin

Actobindin is a small monomer-binding protein that appears to have two actin-binding sites carrying a consensus putative actin-binding sequence 'LKHAET' [838]. Similar sequences are found in fimbrin, plastin, villin, thymosin, myosin, α-actinin and tropomyosin. It may be interesting to note that the homologue of this sequence in tropomyosin forms one or more of the heptad coiled-coil motifs. Actobindin can be cross-linked both to Glu100 and to the *N*-terminal acidic residues 1, 2 or 3 [838] on the barbed end of subdomain IA. Interestingly, mutagenesis of one of the amino acids in the sequence LKHAET motif of either actobindin or thymosin converts an inhibition of assembly into promotion [350]. This 'mutation' appears in some ABPs suggesting a simple evolutionary mechanism for switching function.

Vitamin-D-binding protein

This monomer-binding protein has other synonyms (see Table 1). Vitamin-D-binding protein binds to a 16 kDa *C*-terminal proteolytic fragment of actin [1516]. The site within this fragment that the protein recognizes has been identified with antibodies as the sequence 360–372 [693], at a similar region to other small binding proteins (Figure 21).

Deoxyribonuclease I

The binding site for DNase I is known in some detail from the crystal structure of the DNase I complex. Unlike most other actin-binding proteins, the binding site is at the pointed end of the actin molecule in subdomains IB and IIB (Figure 22). Hydrogen bonds and electrostatic interactions involve residues 39, 41, 43, 45, 61 and 63 in subdomain IB and residues 203 and 207 in subdomain IIB [396]. These observations are consistent with the shielding of lysines 18, 50, 61 and 68 [699] and of the trypsin-sensitive site between Lys68 and Tyr69 [1178] and at Arg62 [1509]. DNase can also be cross-linked to residues 50, 61, 68, 53 and 69 in subdomain IB [823]. DNase-binding is inhibited by derivatization of Tyr69 [1295] and of Lys61 [1403]; this is also explicable in terms of the crystal model. The amino acids identified with the binding site are shown in Figure 21.

Although most actins bind and inhibit DNase I, actins from *Tetrahymena pyriformis* [237, 686], *Entamoeba histolytica* [926, 657] and *Uromyces appendiculatis* [1922] do not despite, in the case of *Entamoeba*, conservation of the residues involved in binding. The 60 kDa actin-related protein also binds to DNase [1155] suggesting conservation of this site in this subfamily of highly diverged actins and further evidence that this sequence has an important biological role.

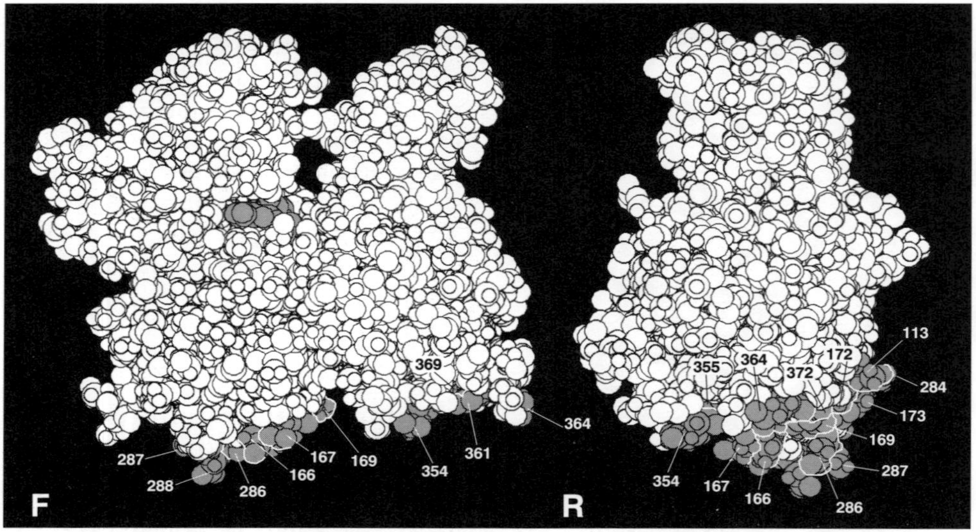

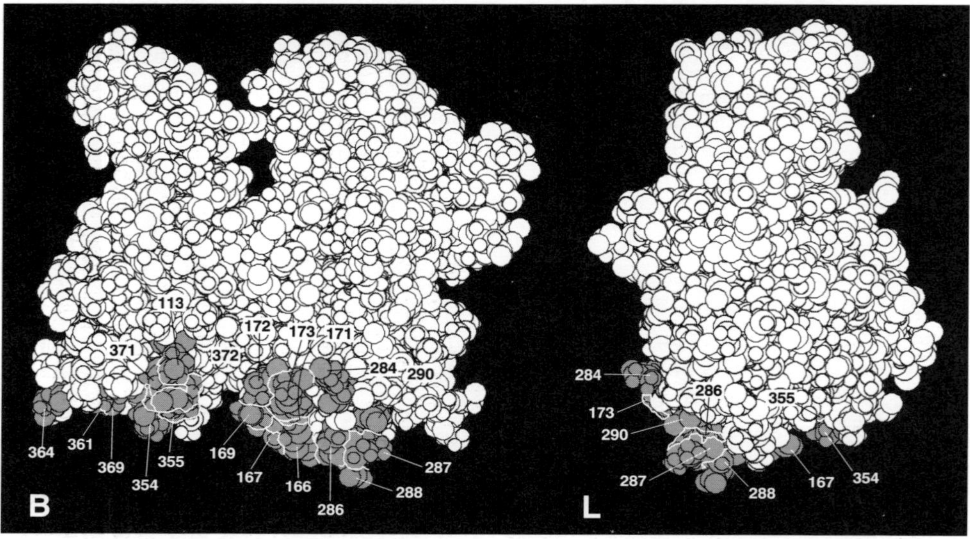

FIGURE 20 Surface binding sites for profilin. *Van der Waals surface images of skeletal muscle actin (lacking 373–375) with amino acids presumed to be involved in binding of profilin. Front (F), right (R), back (B) and left (L) views are shown. See Table 7 and text for discussion.*

DRUG AND TOXIN BINDING

Cytochalasins

The cytochalasins and chaetoglobosins are a group of toxic metabolites produced by certain fungi of the subclasses Ascomycotina and Deuteromycotina. Members of this class of toxin bind differentially to both the glucose transporter of eukaryotic cells and to actin [2051]. Binding to actin is complex in that some cytochalasins (e.g. cytochalasin D) bind to both the monomer to catalyse hydrolysis of bound ATP [1883] and to the barbed ends of actin filaments at a stoichiometry of one molecule per filament to inhibit monomer exchange. Cytochalasin D does not bind significantly to the glucose transporter [2051]. Conversely, cytochalasin B binds with roughly similar affinity to both glucose transporter and the barbed ends of actin filaments [1966, 2051, 2065], but does not significantly affect the monomer ATPase. The biological effects of cytochalasins *in vitro* and on cells is therefore not always easy to interpret [2075]. However, studies on the lack of effect of cytochalasins on cells expressing an actin mutation implicate actin as the major target of cytochalasin. The amino acids substituted in this mutant β-actin are Val139 and Ala295 [2045]. From the X-ray structure [396] it seems unlikely that cytochalasin could reach between these two amino acids and that therefore the barbed end site at either 295 at a site of actin–actin interaction or at 139 close to the 166/169 pocket is a potential interaction site. Derivatization

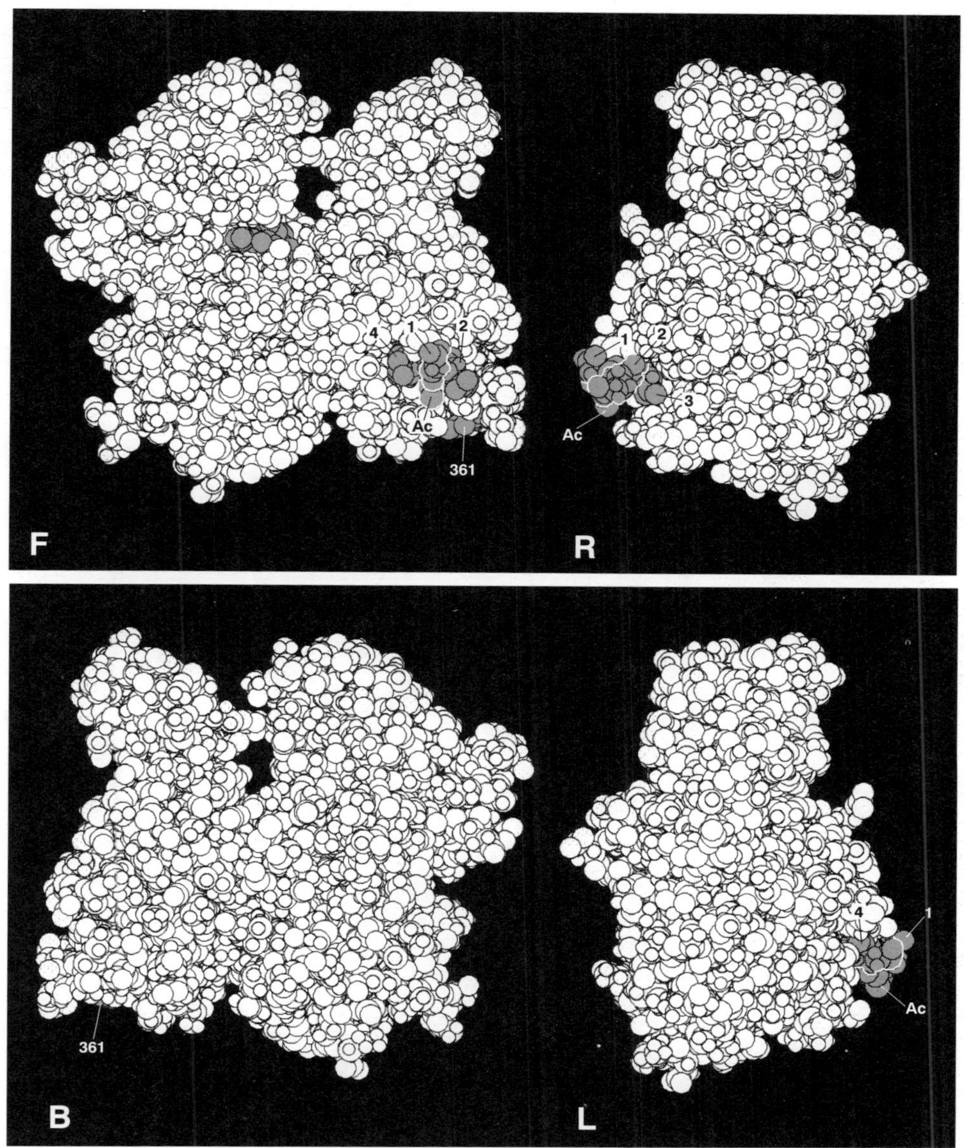

FIGURE 21 Possible surface binding sites for Gc and other small monomer binders. *Van der Waals surface images of skeletal muscle actin (lacking 373–375) with amino acids presumed to be involved in binding at the C-terminus. Front (F), right (R), back (B) and left (L) views are shown. See Table 7 and text for discussion.*

with PCMB, a thiol reagent (see Table 5) inhibits cytochalasin binding, probably via Cys10 [2019]. The binding site has been localized as a single site at the barbed end of the molecule [2065]. These data together suggest that cytochalasins bind on the barbed-end side of the cleft between domains I and II, at a similar site to that at which segment 1 of gelsolin binds. However, cytochalasin can bind to villin-capped filaments [1977]. Since villin is a member of the gelsolin family, this observation suggests either that cytochalasins displace the villin, that their relatively small size does not inhibit binding or that villin-capped filaments use the other (segments 2–3) actin-binding site that cytochalasins do not. Cytochalasin B can be photo-cross-linked to actin, but the site of linkage has not been determined [2063]. The structures of some cytochalasins are given in Figure 24.

Phallotoxins

Phallotoxins are cyclic peptides produced by the basidiomycete *Amanita phalloides*, the death cap mushroom. The toxin binds to polymerized actin at a stoichiometry of one molecule of phalloidin to two of actin [2036] to reduce by two orders of magnitude the critical concentration and to stabilize filaments against high KI concentrations, cytochalasin- and DNaseI-induced depolymerization [1987, 2057]. The toxin does not bind to monomer. Because of the stabilization of filaments and the stoichiometry,

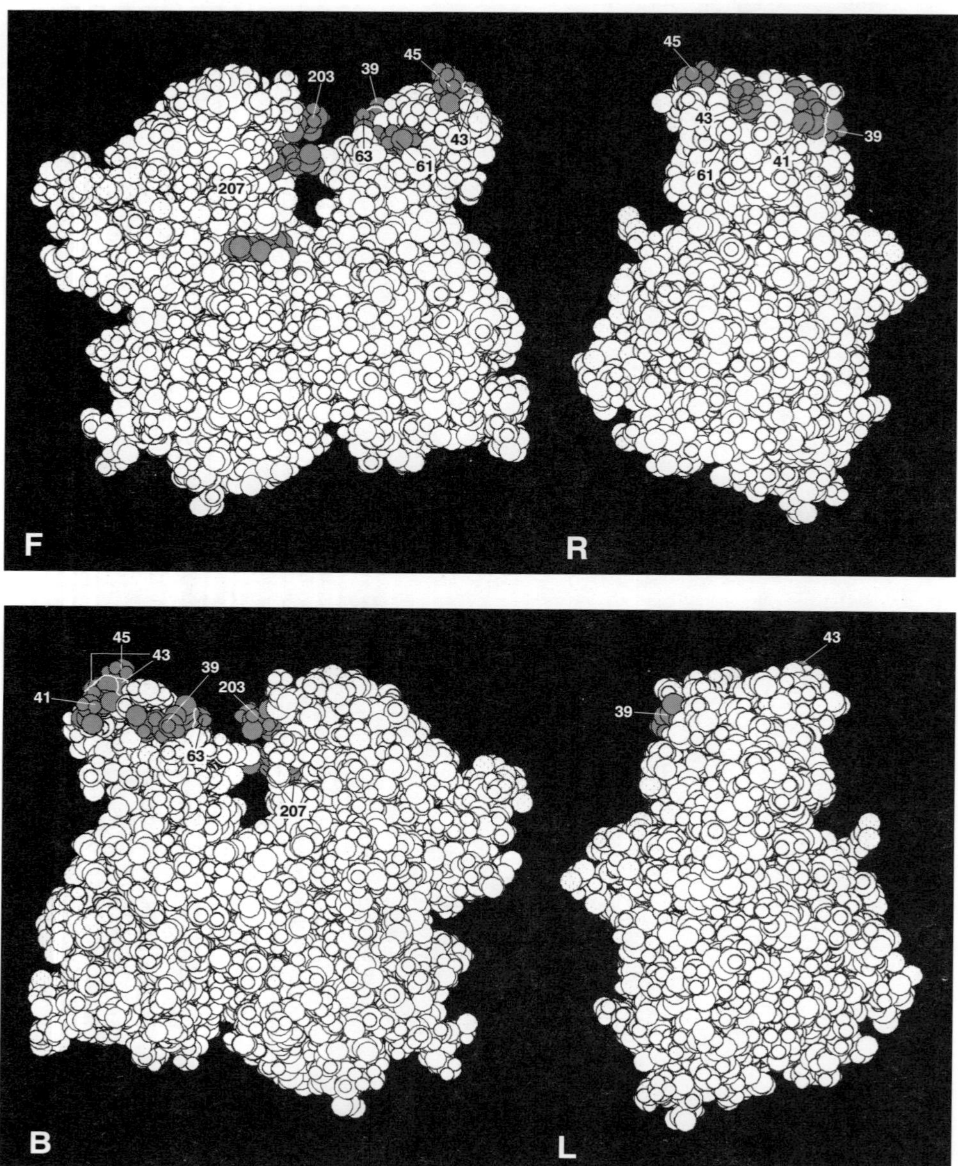

FIGURE 22 Surface binding sites for DNase I. *Van der Waals surface images of skeletal muscle actin (lacking 373–375) with amino acids presumed to be involved in binding of DNase I. Front (F), right (R), back (B) and left (L) views are shown. See Table 7 and text for discussion.*

the toxin may bridge two or more actin monomers in the filament. Cross-linking studies have shown that an affinity-labelled phalloidin can be linked to methionines 119 and 355, and Glu117. Phalloidin coupled to a disulphide linker can interact with Cys10 and Cys374 [1303, 1410, 2072] clustered at the rear face of subdomain IA. It has also been shown recently that the double mutation in yeast actin R177A/D179A prevents the visualization of actin filaments in yeast using a fluorescent phalloidin derivative, suggesting that these residues may also be close to the binding site [637].

A putative phalloidin-binding site has been identified by fitting phalloidin to a refined model of F-actin [1276]. The phalloidin sits in a pocket formed at the junction of three monomers, (Figure 23) consistent with its stoichiometry and effects on critical concentration. The model is also consistent with the biochemical data and places the important tryptophan of phalloidin close to aromatics Trp79 and tyrosines 198 and 279. *Tetrahymena* [237] actins do not bind phalloidin, and have a unique tryptophan to tyrosine substitution at 79 and a rare tyrosine to leucine switch at position 198 [977] that may influence the interactions with the tryptophan of phalloidin (see alignment). The structure of phalloidin is given in Figure 24.

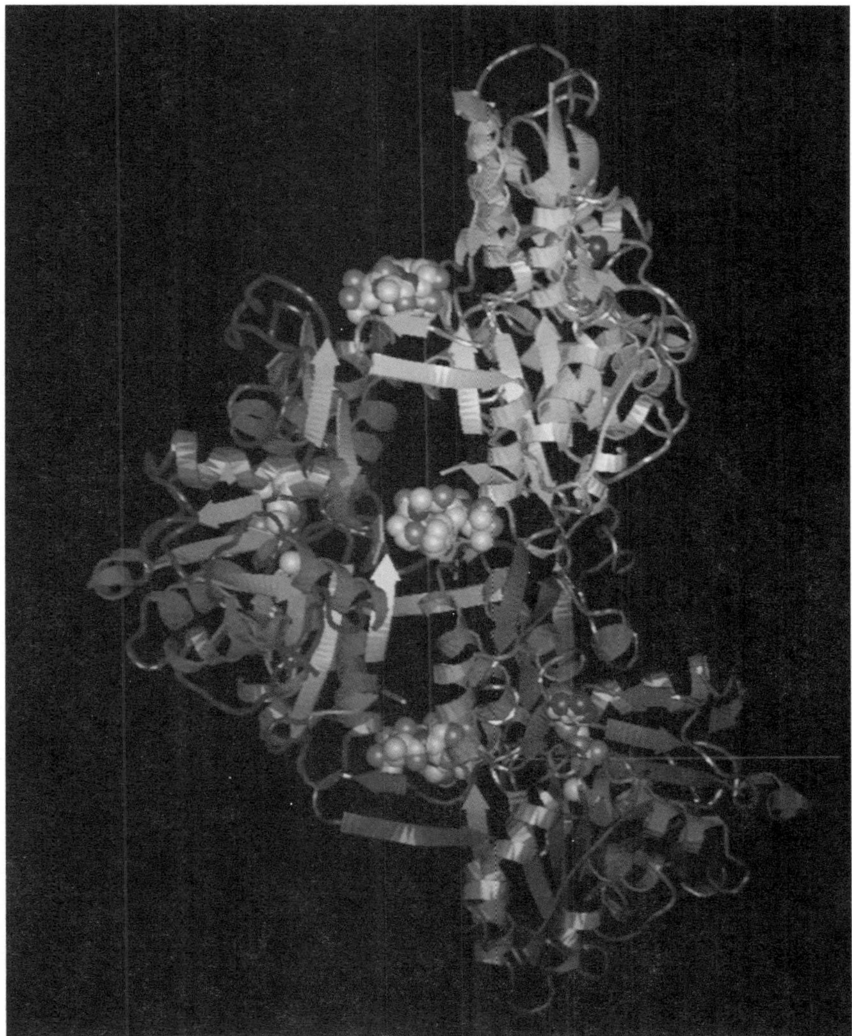

FIGURE 23 Putative binding site for phalloidin. *Phalloidin is imaged as Van der Waals surface in its putative binding position relative to three monomers, represented as secondary structure ribbons, in F-actin [1276]. ADP is also represented as its Van der Waals surface (Courtesy of Dr K. Holmes). See Figure 8 for orientation of monomers and Table 7 and text for discussion.*

Macrolide toxins

Scytophysins

Scytophysins are a group of natural macrolide cytotoxins isolated from cyanobacteria of the family Scytonemataceae [2392]. Tolytoxin, one member of this group, has been shown to disrupt the actin cytoskeleton in cells and to bind directly to monomeric actin *in vitro* [2047] and to sequester non-polymerizable monomers *in situ.* The potential actin-binding activities of other members of this group have not yet been reported. The net effect of these agents is to reduce the concentration of F-actin in a concentration-dependent manner and to cause arborization of cells quite distinct from the effects of cytochalasins. The scytophysins can apparently bypass the P-glycoprotein-mediated multi-drug resistance of cells [2392]. The structure of tolytoxin is given in Figure 24.

Swinholide A

Swinholide A is a natural dilactone macrolide toxin obtained from the marine sponge *Theonella swinhoei* which disrupts the actin cytoskeleton in cells [2381]. Swinholide A stabilizes actin dimers in G-actin solutions and binds with high affinity to experimentally synthesized covalent actin dimers at a stoichiometry of one toxin molecule per dimer. This property, in contrast to the monomer-binding activity of other macrolides, appears to be due to its unusual structure which exhibits a two-fold symmetry [2393] (see Figure 24). In addition, and

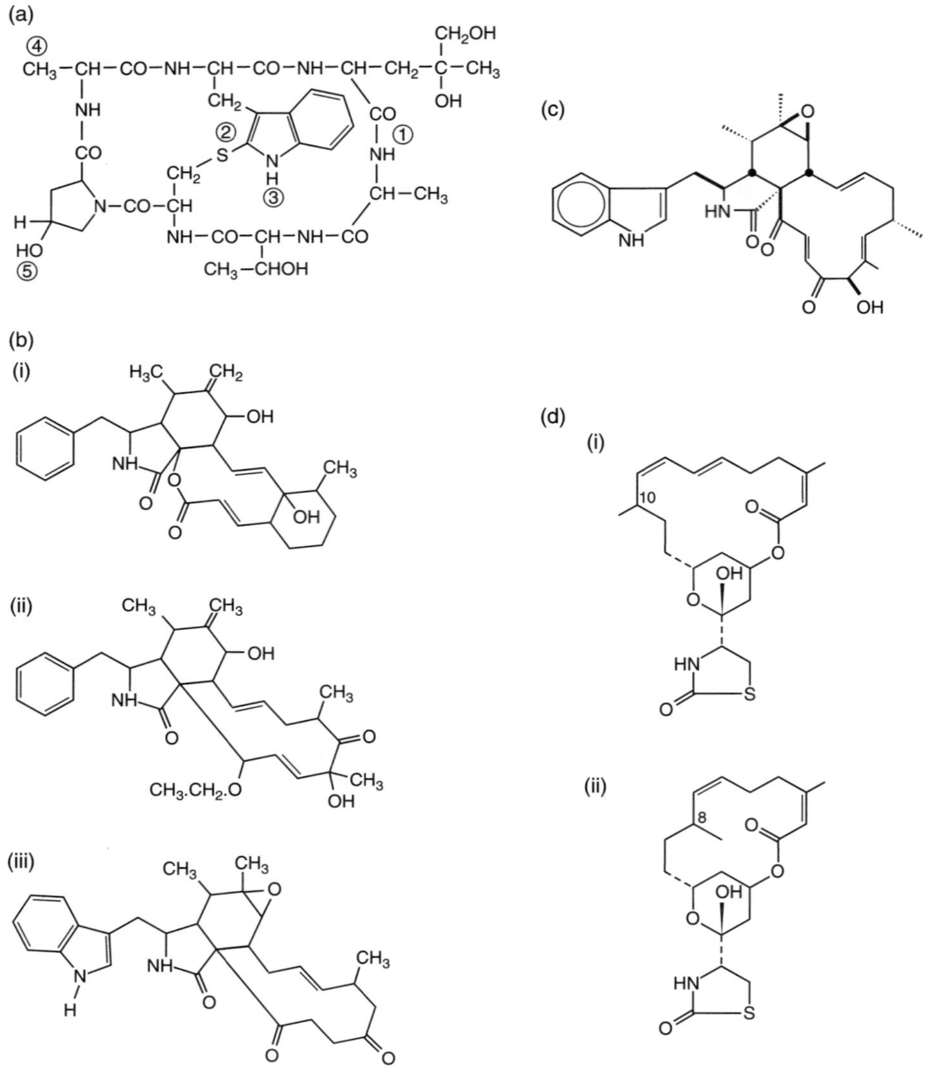

FIGURE 24 The chemical structures of actin-binding toxins. (Continued overleaf) *(a) Structure of a consensus phalloidin. The structural requirements for toxicity are indicated by circled numbers over the structure: 1, breakage of ring abolishes toxicity; 2, breakage of disulphide abolishes toxicity; 3, methy or ethyl only; 4, methyl group required and 5OH necessary and must be in* cis *position. (b) Structures of cytochalasins that (i) bind both glucose transporter and actin, (ii) only actin and (iii) only transporter. Images of cytochalasins B, D and G, respectively. (c) Chaetoglobosin A. (d) Latrunculins A (i) and B (ii). (e) Tolytoxin. (f) Swinholide A. (g) Jasplakinolide.*

again in contrast to other macrolides, swinholide A rapidly severs actin filaments, a property presumably also related to its two-fold structure.

Latrunculin

Latrunculins are macrolide toxins isolated from the Red Sea sponge *Latrunculia magnifica*. Latrunculins A and B (see Figure 24) cause arborization of cells at concentrations close to nanomolar by disrupting the organization of the actin cytoskeleton [1975]. Latrunculin A forms a 1:1 molar complex with G-actin which cannot polymerize and so results in a concentration-dependent decrease in F-actin concentration like the other monovalent macrolides.

Goniodomin A

Goniodomin A is one of the polyether macrolides isolated from the dinoflagellate *Goniodoma pseudogoniaulax*, and was identified as a stimulator of actomyosin ATPase [1995]. This effect appears to be complex, since whilst goniodomin reduces the fluorescence of *N*-1-pyrenyl (Cys374) actin, implying inhibition of assembly, goniodomin–actin complexes can be sedimented by low-speed sedimentation, and gelled actin filaments are visible

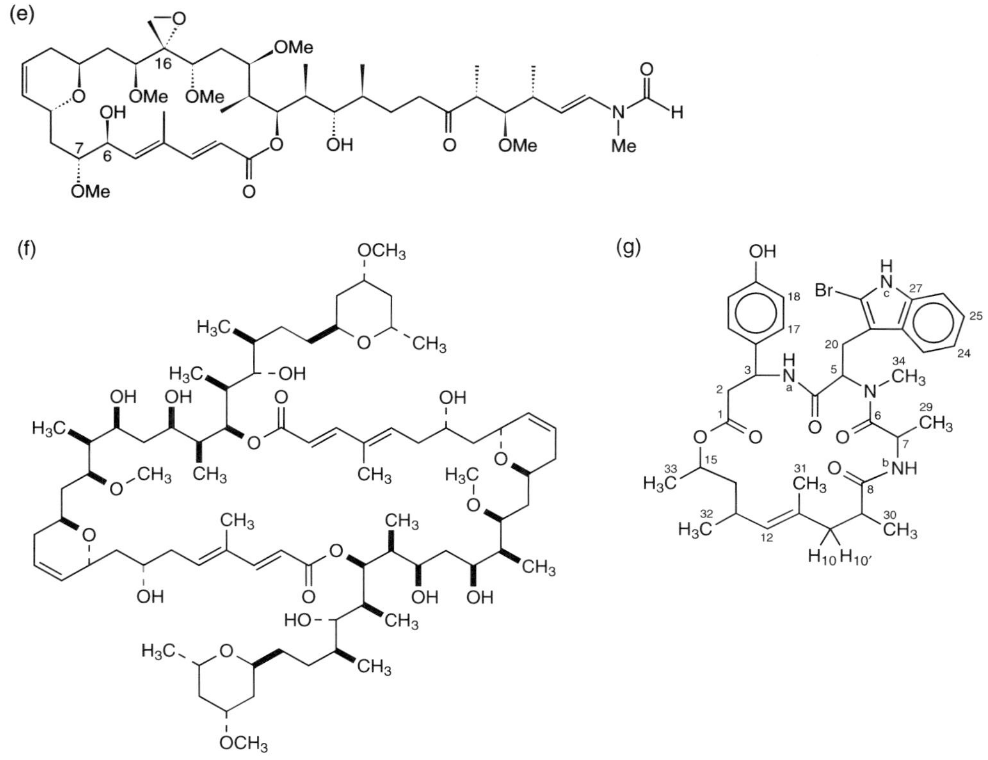

Figure 24 (continued)

by electron microscopy. It appears that goniodomin binding and cross-linking of actin filaments causes a conformational change abolishing pyren fluorescence and enhancing myosin ATPase. The structure of goniodomin A is shown in Figure 24.

Jasplakinolide

Jasplakinolide is a naturally occurring cyclodepsipeptide obtained from the marine sponge *Jaspis johnstoni*. It is a 15-carbon macrocyclic ring containing three amino acids: L-alanine, *N*-methyl-2-bromotryptopham and β-tyrosine [2384] (see Figure 24). Jasplakinolide decreases the critical concentration of rabbit muscle actin assembly, thereby stabilizing F-actin [2380]. Jasplakinolide also competes with phalloidin for binding, and has a slightly higher affinity, but more importantly it appears to be cell permeable, in contrast to phalloidin and the virotoxins, making it a potentially more valuable experimental tool and a potentially useful therapeutic agent.

ACKNOWLEDGEMENTS

The authors thank Uno Lindberg and Clarence Schutt for images of the profilin:actin crystal and Paul McLaughlin and Alan Weeds for images of the gelsolin segment 1:actin crystal. We also thank Nicole Mounier, who prepared the Van der Waals diagrams, Alfonso Valencia for providing images of actin, hsp70 and hexokinase and Ivan Rayment for images of S-1:actin. Clive Lloyd provided the immunocytochemical images of plant cells and Sue Handel the animal cell image. We also thank Bob Read and Sue Sparrow for preparation of some of the diagrams.

ACTIN
3rd EDITION

BIBLIOGRAPHY

REFERENCES UP TO AND INCLUDING 2nd EDITION†

BOOKS

Amos, L.A. and W.B. Amos. 1991. *Molecules of the Cytoskeleton.* Macmillan Molecular Biology, Basingstoke.
Bagshaw, C.R. 1992. *Muscle Contraction*, 2nd edn. Chapman & Hall, London.
Bershadsky, A.D. and J.M. Vasiliev. 1988. *Cytoskeleton.* Plenum Press, New York.
Bray, D. 1992. *Cell Movements.* Garland, New York.
Cappuccinelli, P. 1980. *Motility of Living Cells.* Chapman & Hall, London.
dos Remedios, C.G. and J.A. Barden (eds). 1993. *Actin: Structure and Function in Muscle and Non-muscle Cells.* Academic Press, Sydney.
El Haj E. (ed.). 1992. *Molecular Biology of Muscle.* COB, Cambridge.
Jones, G., C. Wigley, and R. Warn (eds). 1993. *Cell Behaviour: Adhesion and Motility.* COB, Cambridge.
Pollack, G.H. 1990. *Muscles and Molecules: Uncovering the Principles of Biological Motion.* Ebner, Seattle.
Preston, T.M., C.A. King, and J.S. Hyams, 1990. *The Cytoskeleton and Cell Motility.* Blackie, Glasgow.
Schliwa, M. 1986. *The Cytoskeleton: An Introductory Survey.* Springer-Verlag, Wien.
Sheterline, P. 1983. *Mechanisms of Cell Motility: molecular aspects of contractility.* Academic Press, London.
Springer Series in Biophysics, Vol. 3. *Cytoskeletal and Extracellular Proteins* 1989. Springer-Verlag, Berlin.
Stebbings, H. and J.S. Hyams, 1979. *Cell Motility.* Longman, London.
Wilson, L. (ed.). 1982. *Methods in Cell Biology*, Vol. 25. *The Cytoskeleton Part A.* Academic Press, New York.
Wilson, L. (ed.). 1982. *Methods in Cell Biology*, Vol. 25. *The Cytoskeleton Part B.* Academic Press, New York.

REVIEWS ON ACTIN

1.* **Aderem, A.** 1992. Signal transduction and the actin cytoskeleton: the roles of MARCKS and profilin. *Trends Biochem. Sci.* **17**:438–443.
2. **Aktories, K.** 1990. Clostridial ADP-ribosyltransferases–modification of low molecular weight GTP-binding proteins and of actin by clostridial toxins. *Med. Microbiol. Immunol. Berl.* **179**:123–136.
3. **Aktories, K.** 1990. ADP-ribosylation of actin. *J. Muscle Res. Cell Motil.* **11**:95–97.
4. **Aktories, K. and A. Wegner.** 1989. ADP-ribosylation of actin by clostridial toxins. *J. Cell Biol.* **109**:1385–1387.
5. **Aktories, K. and A. Wegner.** 1992. Mechanisms of the cytopathic action of actin-ADP-ribosylating toxins. *Mol. Microbiol.* **6**:2905–2908.
6.* **Aktories, K., M. Wille, and I. Just.** 1992. Clostridial actin-ADP-ribosylating toxins. *Curr. Top. Microbiol. Immunol.* **175**:97–113.
7. **Arpin, M., M. Algrain, and D. Louvard.** 1994. Membrane-actin microfilament connections: an increasing diversity of players related to band-4.1. *Curr. Opin. Cell Biol.* **6**:136–141.
8. **Bearer, E. L.** 1993. Role of actin polymerization in cell locomotion: molecules and models. *Am. J. Respir. Cell Mol. Biol.* **8**:582–591.

★ Key references.
† See p 110 for new references for this edition.

9.* **Blanchard, A., V. Ohanian, and D. Critchley.** 1989. The structure and function of α-actinin. *J. Muscle Res. Cell Motil.* **10**:280–289.
10.* **Bork, P., C. Sander, and A. Valencia.** 1992. An ATPase domain common to prokaryotic cell cycle proteins, sugar kinases, actin, and hsp70 heat shock proteins. *Proc. Natl Acad. Sci. U. S. A.* **89**:7290–7294.
11. **Bray, D.** 1977. Actin and myosin in neurones: a first review. *Biochimie* **59**:1–6.
12. **Bremer, A. and U. Aebi.** 1992. The structure of the F-actin filament and the actin molecule. *Curr. Opin. Cell Biol.* **4**:20–26.
13.* **Buckingham, M., S. Alonso, P. Barton, A. Cohen, P. Daubas, I. Garner, B. Robert, and A. Weydert.** 1986. Actin and myosin multigene families: their expression during the formation and maturation of striated muscle. *Am. J. Med. Genet.* **25**:623–634.
14. **Buckingham, M., S. Alonso, G. Bugaisky, P. Barton, A. Cohen, P. Daubas, A. Minty, B. Robert, and A. Weydert.** 1985. The actin and myosin multigene families. *Adv. Exp. Med. Biol.* **182**:333–344.
15. **Buckingham, M. E.** 1985. Actin and myosin multigene families: their expression during the formation of skeletal muscle. *Essays. Biochem.* **20**:77–109.
16. **Carlier, M. F.** 1988. Role of nucleotide hydrolysis in the polymerization of actin and tubulin. *Cell Biophys.* **12**:105–117.
17. **Carlier, M. F.** 1989. Role of nucleotide hydrolysis in the dynamics of actin filaments and microtubules. *Int. Rev. Cytol.* **115**:139–170.
18.* **Carlier, M. F.** 1990. Actin polymerization and ATP hydrolysis. *Adv. Biophys.* **26**:51–73.
19. **Carlier, M. F.** 1991. Actin: protein structure and filament dynamics. *J. Biol. Chem.* **266**:1–4.
20.* **Carlier, M. F.** 1992. Nucleotide hydrolysis regulates the dynamics of actin filaments and microtubules. *Phil. Trans. R. Soc. Lond. Biol.* **336**:93–97.
21. **Carlier, M. F., D. Pantaloni, and E. D. Korn.** 1987. The mechanisms of ATP hydrolysis accompanying the polymerization of Mg-actin and Ca-actin. *J. Biol. Chem.* **262**:3052–3059.
22. **Clarke, M. and J. A. Spudich.** 1977. Nonmuscle contractile proteins: the role of actin and myosin in cell motility and shape determination. *Annu. Rev. Biochem.* **46**:797–822.
23. **Cleveland, D. W.** 1982. Treadmilling of tubulin and actin. *Cell* **28**:689–691.
24. **Cleveland, D. W., M. A. Lopata, R. J. MacDonald, N. J. Cowan, W. J. Rutter, and M. W. Kirschner.** 1980. Number and evolutionary conservation of α- and β-tubulin and cytoplasmic β- and γ-actin genes using specific cloned cDNA probes. *Cell* **20**:95–105.
25.* **Condeelis, J.** 1993. Life at the leading edge – the formation of cell protrusions. *Annu. Rev. Cell Biol.* **9**:411–444.
26.* **Cooper, J. A.** 1987. Effects of cytochalasin and phalloidin on actin. *J. Cell Biol.* **105**:1473–1478.
27. **Cooper, J. A.** 1991. The role of actin polymerization in cell motility. *Annu. Rev. Physiol.* **53**:585–605.
28.* **dos Remedios, C. G., M. Miki, and J. A. Barden.** 1987. Fluorescence resonance energy transfer measurements of distances in actin and myosin. A critical evaluation. *J. Muscle Res. Cell Motil.* **8**:97–117.
29. **Drubin, D. G.** 1990. Actin and actin-binding proteins in yeast. *Cell Motil. Cytoskeleton.* **15**:7–11.
30. **Dubreuil, R. R.** 1991. Structure and evolution of the actin crosslinking proteins. *Bioessays* **13**:219–226.
31. **Egelman, E. H.** 1985. The structure of F-actin. *J. Muscle Res. Cell Motil.* **6**:129–151.
32. **Elzinga, M. and R. C. Lu.** 1976. Comparative amino acid sequence studies on actins. In: Perry, S.V. and Margreth, A. (eds) *Contractile Systems in Non-muscle Cells*, pp. 29–37. Amsterdam: North-Holland.
33. **Engel, J., H. Fasold, F. W. Hulla, F. Waechter, and A. Wegner.** 1977. The polymerization reaction of muscle actin. *Mol. Cell Biochem.* **18**:3–13.
34.* **Estes, J. E., L. A. Selden, H. J. Kinosian, and L. C. Gershman.** 1992. Tightly-bound divalent cation of actin. *J. Muscle Res. Cell Motil.* **13**:272–284.
35. **Fagraeus, A. and R. Norberg.** 1978. Anti-actin antibodies. *Curr. Top. Microbiol. Immunol.* **82**:1–13.
36.* **Faulstich, H., S. Zobeley, G. Rinnerthaler, and J. V. Small.** 1988. Fluorescent phallotoxins as probes for filamentous actin. *J. Muscle Res. Cell Motil.* **9**:370–383.
37.* **Fechheimer, M. and S. H. Zigmond.** 1993. Focusing on unpolymerized actin. *J. Cell Biol.* **123**:1–5.
38. **Feuer, G., F. Molnar, E. Pettko, and F. B. Straub.** 1993. Studies on the composition and polymerisation of actin. *Acta Physiol.* **1**:150.
39. **Finck, H.** 1968. On the discovery of actin. *Science* **160**:332.
40.* **Firtel, R. A.** 1981. Multigene families encoding actin and tubulin. *Cell* **24**:6–7.
41. **Fornwald, J. A., G. Kuncio, I. Peng, and C. P. Ordahl.** 1982. The complete nucleotide sequence of the chick α-actin gene and its evolutionary relationship to the actin gene family. *Nucl. Acids Res.* **10**:3861–3876.
42.* **Frieden, C.** 1985. Actin and tubulin polymerization: the use of kinetic methods to determine mechanism. *Annu. Rev. Biophys. Biophys. Chem.* **14**:189–210.
43. **Fujimura, K. and A. Kuramoto.** 1985. Actins. *Rinsho. Byori.* Spec No **62**:126–139.
44. **Fyrberg, E. A.** 1984. Structural and functional analyses of *Drosophila melanogaster* actin genes. *Oxf. Surv. Eukaryot. Genes* **1**:61–86.
45. **Fyrberg, E. A., B. J. Bond, N. D. Hershey, K. S. Mixter, and N. Davidson.** 1981. The actin genes of *Drosophila*: protein coding regions are highly conserved but intron positions are not. *Cell* **24**:107–116.
46. **Gaertner, A., K. Ruhnau, E. Schroer, N. Selve, M. Wanger, and A. Wegner.** 1989. Probing nucleation, cutting and capping of actin filaments. *J. Muscle Res. Cell Motil.* **10**:1–9.
47. **Gerisch, G., R. Albrecht, E. De Hostos, E. Wallraff, C. Heizer, M. Kreitmeier, and A. Muller Taubenberger.** 1993. Actin-associated proteins in motility and chemotaxis of *Dictyostelium* cells. *Symp. Soc. Exp. Biol.* **47**:297–315.
48. **Gerisch, G., A. A. Noegel, and M. Schleicher.** 1991. Genetic alteration of proteins in actin-based motility systems. *Annu. Rev. Physiol.* **53**:607–628.
49. **Gershman, L. C., L. A. Selden, H. J. Kinosian, and J. E. Estes.** 1988. Divalent cation exchange on actin. *Biophys. J.* **53**:573.
50. **Goldstein, L., R. Rubin, and C. Ko.** 1977. The presence of actin in nuclei: a critical appraisal. *Cell* **12**:601–608.
51. **Gordon, D. J., J. L. Boyer, and E. D. Korn.** 1977. Comparative biochemistry of non-muscle actins. *J. Biol. Chem.* **252**:8300–8309.
52. **Gurdon, J. B., T. J. Mohun, S. Brennan, and S. Cascio.** 1985. Actin genes in *Xenopus* and their developmental control. *J. Embryol. Exp. Morphol.* **89** Suppl:125–136.

53. **Hamada, H., M. G. Petrino, and T. Kakunaga.** 1982. Molecular structure and evolutionary origin of human cardiac muscle actin gene. *Proc. Natl Acad. Sci. U. S. A.* **79**:5901–5905.
54. **Hambly, B. D., J. A. Barden, M. Miki, and C. G. dos Remedios.** 1986. Structural and functional domains on actin. *Bioessays* **4**:124–128.
55. **Hartwig, J. H.** 1986. Actin filament architecture and movements in macrophage cytoplasm. *Ciba. Found. Symp.* **118**:42–53.
56. **Hartwig, J. H. and D. J. Kwiatkowski.** 1991. Actin-binding proteins. *Curr. Opin. Cell Biol.* **3**:87–97.
57. **Hartwig, J. H., R. Niederman, and S. E. Lind.** 1985. Cortical actin structures and their relationship to mammalian cell movements. *Subcell. Biochem.* **11**:1–49.
58. **Hartwig, J. H. and T. P. Stossel.** 1982. Macrophages: their use in elucidation of the cytoskeletal roles of actin. *Methods Cell Biol.* 25 Pt B:201–225.
59. **Hatano, S.** 1984. Actin and actin-associated proteins. *Cell Struct. Funct.* **9** (Suppl.):s67-s72.
60. **Hayashi, T.** 1967. Reactivities of actin as a contractile protein. *J. Gen. Physiol.* **50**:Suppl.119–Suppl.133.
61. **Heath, J. and B. Holifield.** 1991. Cell locomotion. Actin alone in lamellipodia. *Nature* **352**:107–108.
62.* **Heath, J. P. and B. F. Holifield.** 1993. On the mechanisms of cortical actin flow and its role in cytoskeletal organisation of fibroblasts. *Symp. Soc. Exp. Biol.* **47**:35–56.
63.* **Hennessey, E. S., D. R. Drummond, and J. C. Sparrow.** 1993. Molecular genetics of actin function. *Biochem. J.* **291**:657–671.
64. **Hennessey, E. S., A. Harrison, D. R. Drummond, and J. C. Sparrow.** 1992. Mutant actin: a dead end? *J. Muscle Res. Cell Motil.* **13**:127–131.
65.* **Herman, I. M.** 1993. Actin isoforms. *Curr. Opin. Cell Biol.* **5**:48–55.
66.* **Hightower, R. C. and R. B. Meagher.** 1986. The molecular evolution of actin. *Genetics* **114**:315–332.
67. **Hitchcock DeGregori, S. E.** 1980. Actin assembly. *Nature* **288**:437–438.
68. **Hitt, A. L. and E. J. Luna.** 1994. Membrane interactions with the actin cytoskeleton. *Curr. Opin. Cell Biol.* **6**:120–130.
69.* **Holmes, K. C., D. Popp, W. Gebhard, and W. Kabsch.** 1990. Atomic model of the actin filament. *Nature* **347**:44–49.
70.* **Holmes, K. C., C. Sander, and A. Valencia.** 1993. A new ATP-binding fold in actin, hexokinase and Hsc70. *Trends Cell Biol.* **3**:53–59.
71. **Houk, T. W. J. and K. Ue.** 1974. The measurement of actin concentration in solution: a comparison of methods. *Anal. Biochem.* **62**:66–74.
72. **Janmey, P.** 1993. Cell biology. A slice of the actin. *Nature* **364**:675–676.
73.* **Janmey, P.** 1994. Phosphoinositides and calcium as regulators of cellular actin assembly and disassembly. *Annu. Rev. Physiol.* **56**:169–191.
74.* **Kabsch, W. and J. Vandekerckhove.** 1992. Structure and function of actin. *Annu. Rev. Biophys. Biomol. Struct.* **21**:49–76.
75. **Kane, R. E.** 1986. Components of the actin-based cytoskeleton. *Methods Cell Biol.* **27**:229–242.
76.* **Kedes, L., S. Y. Ng, C. S. Lin, P. Gunning, R. Eddy, T. Shows, and J. Leavitt.** 1985. The human β-actin multigene family. *Trans. Assoc. Am. Physicians.* **98**:42–46.
77.* **Kindle, K. L. and R. A. Firtel.** 1978. Identification and analysis of *Dictyostelium* actin genes, a family of moderately repeated genes. *Cell* **15**:763–778.
78.* **Korn, E. D.** 1982. Actin polymerization and its regulation by proteins from nonmuscle cells. *Physiol. Rev.* **62**:672–737.
79. **Korn, E. D., M. F. Carlier, and D. Pantaloni.** 1987. Actin polymerization and ATP hydrolysis. *Science* **238**:638–644.
80.* **Kovilur, S., J. W. Jacobson, R. L. Beach, W. R. Jeffery, and C. R. Tomlinson.** 1993. Evolution of the chordate muscle actin gene. *J. Mol. Evol.* **36**:361–368.
81. **Laki, K.** 1974. Actin as an energy transducer. *J. Theor. Biol.* **44**:117–130.
82.* **Lorenz, M., D. Popp, and K. C. Holmes.** 1993. Refinement of the F-actin model against X-ray fiber diffraction data by the use of a directed mutation algorithm. *J. Mol. Biol.* **234**:826–836.
83.* **Low, I., W. Jahn, T. Wieland, S. Sekita, K. Yoshihira, and S. Natori.** 1979. Interaction between rabbit muscle actin and several chaetoglobosins or cytochalasins. *Anal. Biochem.* **95**:14–18.
84.* **Lubyphelps, K.** 1994. Physical properties of cytoplasm. *Curr. Opin. Cell Biol.* **6**:3–9.
85. **Mannherz, H. G.** 1992. Crystallization of actin in complex with actin-binding proteins. *J. Biol. Chem.* **267**:11661–11664.
86. **Maruta, H. and G. Isenberg.** 1984. F-actin capping proteins. *Prog. Clin. Biol. Res.* **164**:355–364.
87. **Mauss, S., C. Chaponnier, I. Just, K. Aktories, and G. Gabbiani.** 1990. ADP-ribosylation of actin isoforms by Clostridium botulinum C2 toxin and Clostridium perfringens iota toxin. *Eur. J. Biochem.* **194**:237–241.
88. **McKeown, M., W. C. Taylor, K. L. Kindle, R. A. Firtel, W. Bender, and N. Davidson.** 1978. Multiple, heterogeneous actin genes in *Dictyostelium. Cell* **15**:789–800.
89.* **McLaughlin, P. J., J. T. Gooch, H. G. Mannherz, and A. G. Weeds.** 1993. Structure of gelsolin segment 1-actin complex and the mechanism of filament severing. *Nature* **364**:685–692.
90. **McLean, B. G., S. R. Huang, E. C. McKinney, and R. B. Meagher.** 1990. Plants contain highly divergent actin isovariants. *Cell Motil. Cytoskeleton.* **17**:276–290.
91.* **Meagher, R. B.** 1991. Divergence and differential expression of actin gene families in higher plants. *Int. Rev. Cytol.* **125**:139–163.
92. **Miki, M., S. I. O'Donoghue, and C. G. dos Remedios.** 1992. Structure of actin observed by fluorescence resonance energy transfer spectroscopy. *J. Muscle Res. Cell Motil.* **13**:132–145.
93. **Miki, M. and P. Wahl.** 1984. Fluorescence energy transfers in labelled G-actin and F-actin. *Biochim. Biophys. Acta* **786**:188–196.
94. **Milligan, R. A., M. Whittaker, and D. Safer.** 1990. Molecular structure of F-actin and location of surface binding sites. *Nature* **348**:217–221.
95. **Mitchison, T. J.** 1992. Compare and contrast actin filaments and microtubules. *Mol. Biol. Cell* **3**:1309–1315.
96. **Miwa, T., Y. Manabe, K. Kurokawa, S. Kamada, N. Kanda, G. Bruns, H. Ueyama, and T. Kakunaga.** 1991. Structure, chromosome location, and expression of the human smooth muscle (enteric type) γ-actin gene: evolution of six human actin genes. *Mol. Cell Biol.* **11**:3296–3306.
97. **Mommaerts, W. F.** 1992. Who discovered actin? *Bioessays* **14**:57–59.
98. **Mounier, N., M. Gouy, D. Mouchiroud, and J. C. Prudhomme.** 1992. Insect muscle actins differ distinctly from invertebrate and vertebrate cytoplasmic actins. *J. Mol. Evol.* **34**:406–415.
99.* **Mounier, N. and J. C. Sparrow.** 1993. Muscle actin genes in insects. *Comp. Biochem. Physiol. B.* **105**:231–238.
100.* **Nachmias, V. T.** 1993. Small actin-binding proteins: the β-thymosin family. *Curr. Opin. Cell Biol.* **5**:56–62.
101. **Nakajima Iijima, S., H. Hamada, P. Reddy, and T. Kakunaga.** 1985. Molecular structure of the human cytoplasmic β-actin gene: interspecies homology of sequences in the introns. *Proc. Natl Acad. Sci. U. S. A.* **82**:6133–6137.
102.* **Natori, S.** 1986. Cytochalasins: actin filament modifiers as a group of mycotoxins. *Dev. Toxicol. Environ. Sci.* **12**:291–299.
103. **Neuhaus, J. M., M. Wanger, T. Keiser, and A. Wegner.** 1983. Treadmilling of actin. *J. Muscle Res. Cell Motil.* **4**:507–527.
104. **Ng, S. Y., P. Gunning, R. Eddy, P. Ponte, J. Leavitt, T. Shows, and L. Kedes.** 1985. Evolution of the functional human β-actin gene and its multi-pseudogene family: conservation of noncoding regions and chromosomal dispersion of pseudogenes. *Mol. Cell Biol.* **5**:2720–2732.
105. **Ogievetskaya, M. M.** 1977. Method for the determination of protein evolution rates by amino acid composition. Evolution rate of actins. *Orig. Life.* **8**:145–154.
106. **Oosawa, F.** 1972. Dynamic properties of F-actin and the thin filament. *Nippon. Seirigaku. Zasshi.* **34**:96–97.
107. **Oosawa, F.** 1987. The phenomenological study of the assembly of muscle and non-muscle actin; a history in Japan. *Bioessays* **7**:182–184.
108. **Oosawa, F.** 1993. Physical chemistry of actin: past, present and future. *Biophys. Chem.* **47**:101–111.
109.* **Otto, J. J.** 1994. Actin-bundling proteins. *Curr. Opin. Cell Biol.* **6**:105–109.
110. **Pollard, T. D.** 1986. Assembly and dynamics of the actin filament system in nonmuscle cells. *J. Cell Biochem.* **31**:87–95.
111. **Pollard, T. D.** 1990. Actin. *Curr. Opin. Cell Biol.* **2**:33–40.
112. **Pollard, T. D., U. Aebi, J. A. Cooper, W. E. Fowler, D. P. Kiehart, P. R. Smith, and P. C. Tseng.** 1982. Actin and myosin function in *Acanthamoeba. Phil. Trans. R. Soc. Lond. Biol.* **299**:237–245.
113. **Pollard, T. D., U. Aebi, J. A. Cooper, W. E. Fowler, and P. Tseng.** 1982. Actin structure, polymerization, and gelation. *Cold Spring Harb. Symp. Quant. Biol.* 46 Pt **2**:513–524.
114. **Pollard, T. D. and J. A. Cooper.** 1986. Actin and actin-binding proteins. A critical evaluation of mechanisms and functions. *Annu. Rev. Biochem.* **55**:987–1035.
115. **Pollard, T. D. and R. R. Weihing.** 1974. Actin and myosin and cell movement. *CRC. Crit. Rev. Biochem.* **2**:1–65.
116. **Quax, W. J., H. J. Dodemont, J. A. Lenstra, F. C. Ramaekers, P. Soriano, M. E. van Workum, G. Bernardi, and H. Bloemendal.** 1982. Genes coding for vimentin and actin in mammals and birds. *Adv. Exp. Med. Biol.* **158**:349–357.
117.* **Rampal, A. L., H. B. Pinkofsky, and C. Y. Jung.** 1980. Structure of cytochalasins and cytochalasin B binding sites in human erythrocyte membranes. *Biochemistry* **19**:679–683.
118. **Reedy, M. K.** 1994. Myosin–actin motors: the partnership goes atomic. *Structure* **1**:1–5.
119.* **Reisler, E.** 1993. Actin molecular structure and function. *Curr. Opin. Cell Biol.* **5**:41–47.
120. **Ridley, A. J. and A. Hall.** 1992. Distinct patterns of actin organization regulated by the small GTP-binding proteins Rac and Rho. *Cold Spring Harb. Symp. Quant. Biol.* **57**:661–671.
121.* **Romans, P. and R. A. Firtel.** 1985. Organization of the actin multigene family of *Dictyostelium discoideum* and analysis of variability in the protein coding regions. *J. Mol. Biol.* **186**:321–335.
122. **Rozycki, M. D., J. C. Myslik, C. E. Schutt, and U. Lindberg.** 1994. Structural aspects of actin-binding proteins. *Curr. Opin. Cell Biol.* **6**:87–95.
123. **Rubenstein, P. A.** 1990. The functional importance of multiple actin isoforms. *Bioessays* **12**:309–315.
124. **Sanger, J. W. and J. M. Sanger.** 1992. Cell motility. Beads, bacteria and actin. *Nature* **357**:442.
125. **Sawtell, N. M., A. L. Hartman, and J. L. Lessard.** 1990. Conserved tissue-restricted expression of HUC 1–1 actin phenotype among eumetazoan organisms. *J. Exp. Zool.* **256**:54–62.
126. **Schutt, C. E. and U. Lindberg.** 1992. Actin as the generator of tension during muscle contraction. *Proc. Natl Acad. Sci. U. S. A.* **89**:319–323.
127.* **Seagull, R. W.** 1989. Plant actins. *CRC Crit. Rev. Plant Sci.* **8**:131–167.
128.* **Sekita, S., K. Yoshihira, S. Natori, F. Harada, K. Iida, and I. Yahara.** 1985. Structure–activity relationship of thirty-nine cytochalasans observed in the effects on cellular structures and cellular events and on actin polymerization *in vitro. J. Pharmacobiodyn.* **8**:906–916.

129. **Shah, D. M., R. C. Hightower, and R. B. Meagher.** 1983. Genes encoding actin in higher plants: intron positions are highly conserved but the coding sequences are not. *J. Mol. Appl. Genet.* **2**:111–126.
130. **Sheterline, P.** 1993. Mechanisms of actin filament turnover in animal cells. *Symp. Soc. Exp. Biol.* **47**:339–352.
131. **Sheterline, P., S. E. Handel, C. Molloy, and K. A. Hendry.** 1991. The nature and regulation of actin filament turnover in cells. *Acta Histochem. Suppl.* **41**:303–309.
132. **Sheterline, P., S. H. Handel, and K. A. Hendry.** 1991. Reorganization and turnover of actin filament architectures in cells. *Biochem. Soc. Trans.* **19**:1120–1124.
133. **Singh, V. B. and M. S. Kanungo.** 1983. A comparative study of actins of different vertebrates. *Indian J. Biochem. Biophys.* **20**:46–49.
134. **Small, J. V., A. Rohlfs, and M. Herzog.** 1993. Actin and cell movement. *Symp. Soc. Exp. Biol.* **47**:57–71.
135.* **Sparrow, J. C., D. R. Drummond, E. S. Hennessey, J. D. Clayton, and F. B. Lindegaard.** 1992. *Drosophila* actin mutants and the study of myofibrillar assembly and function. *Symp. Soc. Exp. Biol.* **46**:111–129.
136. **Spudich, J. A., J. D. Pardee, P. A. Simpson, K. Yamamoto, E. R. Kuczmarski, and L. Stryer.** 1982. Actin and myosin: control of filament assembly. *Phil. Trans. R. Soc. Lond. Biol.* **299**:247–261.
137. **Stossel, T. P.** 1984. Contribution of actin to the structure of the cytoplasmic matrix. *J. Cell Biol.* **99**:15s-21s.
138. **Stossel, T. P.** 1986. The actin system and the rheology of peripheral cytoplasm. *Biorheology.* **23**:621–631.
139.* **Stossel, T. P.** 1989. From signal to pseudopod. How cells control cytoplasmic actin assembly. *J. Biol. Chem.* **264**:18261–18264.
140. **Stossel, T. P., C. Chaponnier, R. M. Ezzell, J. H. Hartwig, P. A. Janmey, D. J. Kwiatkowski, S. E. Lind, D. B. Smith, F. S. Southwick, H. L. Yin, and et al.** 1985. Nonmuscle actin-binding proteins. *Annu. Rev. Cell Biol.* **1**:353–402.
141. **Strzelecka-Golsaszewska, H., G. Boguta, S. Zmorzynski, and J. Moraczewska.** 1989. Biochemical and theoretical approach to localization of metal-ion-binding sites in the actin primary structure. *Eur. J. Biochem.* **182**:299–305.
142. **Tilney, L. G.** 1975. The role of actin in non-muscle cell motility. *Soc. Gen. Physiol. Ser.* **30**:339–388.
143. **Titus, M. A.** 1993. Myosins. *Curr. Opin. Cell Biol.* **5**:77–81.
144. **Travis, J.** 1992. How cells get their actin together. *Science* **256**:177–178.
145. **Vandekerckhove, J.** 1989. Structural principles of actin-binding proteins. *Curr. Opin. Cell Biol.* **1**:15–22.
146.* **Vandekerckhove, J.** 1990. Actin-binding proteins. *Curr. Opin. Cell Biol.* **2**:41–50.
147. **Vandekerckhove, J., W. W. Franke, and K. Weber.** 1981. Diversity of expression of non-muscle actin in amphibia. *J. Mol. Biol.* **152**:413–426.
148. **Vandekerckhove, J. and K. Vancompernolle.** 1992. Structural relationships of actin-binding proteins. *Curr. Opin. Cell Biol.* **4**:36–42.
149. **Vandekerckhove, J. and K. Weber.** 1978. At least six different actins are expressed in a higher mammal: an analysis based on the amino acid sequence of the amino-terminal tryptic peptide. *J. Mol. Biol.* **126**:783–802.
150. **Vandekerckhove, J. and K. Weber.** 1978. Mammalian cytoplasmic actins are the products of at least two genes and differ in primary structure in at least 25 identified positions from skeletal muscle actins. *Proc. Natl Acad. Sci. U. S. A.* **75**:1106–1110.
151. **Vandekerckhove, J. and K. Weber.** 1978. Comparison of the amino acid sequences of three tissue-specific cytoplasmic actins with rabbit skeletal muscle actin. *Arch. Int. Physiol. Biochim.* **86**:891–892.
152. **Vandekerckhove, J. and K. Weber.** 1978. Actin amino-acid sequences. Comparison of actins from calf thymus, bovine brain, and SV40-transformed mouse 3T3 cells with rabbit skeletal muscle actin. *Eur. J. Biochem.* **90**:451–462.
153. **Vandekerckhove, J. and K. Weber.** 1979. Amino-acid sequence analysis of the amino-terminal tryptic peptides of different actins from the same mammal. *Arch. Int. Physiol. Biochim.* **87**:210–212.
154. **Vandekerckhove, J. and K. Weber.** 1984. Chordate muscle actins differ distinctly from invertebrate muscle actins. The evolution of the different vertebrate muscle actins. *J. Mol. Biol.* **179**:391–413.
155. **Walling, E. A., G. A. Krafft, and B. R. Ware.** 1988. Actin assembly activity of cytochalasins and cytochalasin analogs assayed using fluorescence photobleaching recovery. *Arch. Biochem. Biophys.* **264**:321–332.
156. **Wanger, M., T. Keiser, J. M. Neuhaus, and A. Wegner.** 1985. The actin treadmill. *Can. J. Biochem. Cell Biol.* **63**:414–421.
157. **Way, M. and A. Weeds.** 1990. Actin-binding proteins. Cytoskeletal ups and downs *Nature* **344**:292–294.
158.* **Weber, K. and W. Kabsch.** 1994. Intron positions in actin genes seem unrelated to the secondary structure of the protein. *EMBO J.* **13**:1280–1286.
159. **Weeds, A. and S. Maciver.** 1993. F-actin capping proteins. *Curr. Opin. Cell Biol.* **5**:63–69.
160.* **Weeds, A. G., J. Gooch, M. Hawkins, B. Pope, and M. Way.** 1992. Role of actin-binding proteins in cytoskeletal dynamics. *Biochem. Soc. Trans.* **19**:1016–1020.
161. **Wegner, A.** 1985. Subtleties of actin assembly. *Nature* **313**:97–98.
162. **Wegner, A. and J. Engel.** 1975. Kinetics of the cooperative association of actin to actin filaments. *Biophys. Chem.* **3**:215–225.
163. **Weisenberg, R. C.** 1981. The role of nucleotide triphosphate in actin and tubulin assembly and function. *Cell Motil.* **1**:485–497.
164.* **Welch, M. D., D. A. Holtzman, and D. G. Drubin.** 1994. The yeast actin cytoskeleton. *Curr. Opin. Cell Biol.* **6**:110–119.
165. **Wertman, K., D. G. Drubin, and D. Botstein.** 1992. Systematic mutational analysis at the yeast ACT1 gene. *Genetics* **132**:337–350.
166. **Wertman, K. F. and D. G. Drubin.** 1992. Actin constitution: guaranteeing the right to assemble. *Science* **258**:759–760.
167.* **Wieland, T.** 1976. Interaction of phallotoxins with actin. *Adv. Enzyme. Regul.* **15**:285–300.
168. **Wieland, T.** 1977. Modification of actins by phallotoxins. *Naturwissenschaften.* **64**:303–309.
169. **Yin, H. L.** 1989. Calcium and polyphosphoinositide regulation of actin network structure by gelsolin. *Adv. Exp. Med. Biol.* **255**:315–323.
170. **Zafar, R. S. and A. Sodja.** 1983. Homology between actin coding and its adjacent sequences in widely divergent species. *Biochem. Biophys. Res. Commun.* **111**:67–73.

ACTIN POLYMERIZATION

171. **Aharoni, D., A. Dantes, and A. Amsterdam.** 1993. Cross-talk between adenylate cyclase activation and tyrosine phosphorylation leads to modulation of the actin cytoskeleton and to acute progesterone secretion in ovarian granulosa cells. *Endocrinology* **133**:1426–1436.
172. **Arisaka, F., M. Kawamura, and K. Murayama.** 1973. Changes in the distribution of particle length of F-actin transformed from Mg-polymer. *J. Biochem. Tokyo* **73**:1211–1215.
173. **Arisaka, F., H. Noda, and K. Maruyama.** 1975. Kinetic analysis of the polymerization process of actin. *Biochim. Biophys. Acta* **400**:263–274.
174. **Attri, A. K., M. S. Lewis, and E. D. Korn.** 1991. The formation of actin oligomers studied by analytical ultracentrifugation. *J. Biol. Chem.* **266**:6815–6824.
175. **Bailin, G. and M. Barany.** 1967. Studies on actin-actin and actin-myosin interaction. *Biochim. Biophys. Acta* **140**:208–221.
176.* **Blikstad, I., F. Markey, L. Carlsson, T. Persson, and U. Lindberg.** 1978. Selective assay of monomeric and filamentous actin in cell extracts, using inhibition of deoxyribonuclease I. *Cell* **15**:935–943.
177.* **Bonder, E. M., D. J. Fishkind, and M. S. Mooseker.** 1983. Direct measurement of critical concentrations and assembly rate constants at the two ends of an actin filament. *Cell* **34**:491–501.
178. **Borejdo, J., A. Muhlrad, S. J. Leibovich, and A. Oplatka.** 1981. Polymerization of G-actin by hydrodynamic shear stresses. *Biochim. Biophys. Acta* **667**:118–131.
179. **Bray, D.** 1985. Computer modelling of the behaviour of actin gels. *Nature* **313**:738.
180. **Brenner, S. L. and E. D. Korn.** 1983. On the mechanism of actin monomer-polymer subunit exchange at steady state. *J. Biol. Chem.* **258**:5013–5020.
181. **Brown, S. S.** 1982. Directional growth of actin off polylysine-coated beads. *Methods Cell Biol.* **24**:291–300.
182. **Burlacu, S., P. A. Janmey, and J. Borejdo.** 1992. Distribution of actin filament lengths measured by fluorescence microscopy. *Am. J. Physiol.* **262**:C569-C577.
183. **Cano, M. L., D. A. Lauffenburger, and S. H. Zigmond.** 1991. Kinetic analysis of F-actin depolymerization in polymorphonuclear leukocyte lysates indicates that chemoattractant stimulation increases actin filament number without altering the filament length distribution. *J. Cell Biol.* **115**:677–687.
184.* **Carlier, M. F.** 1990. Actin polymerization and ATP hydrolysis. *Adv. Biophys.* **26**:51–73.
185. **Carlier, M. F.** 1992. Nucleotide hydrolysis regulates the dynamics of actin filaments and microtubules. *Phil. Trans. R. Soc. Lond. Biol.* **336**:93–97.
186. **Carlier, M. F., D. Didry, I. Erk, J. Lepault, and D. Pantaloni.** 1994. Myosin subfragment-1-induced polymerization of G-actin. Formation of partially decorated filaments at high actin-S1 ratios. *J. Biol. Chem.* **269**:3829–3837.
187. **Carlier, M. F., D. Pantaloni, and E. D. Korn.** 1984. Steady state length distribution of F-actin under controlled fragmentation and mechanism of length redistribution following fragmentation. *J. Biol. Chem.* **259**:9987–9991.
188. **Cerven, E.** 1987. Facilitation of actin polymerization by interfilamentous factors. *J. Muscle Res. Cell Motil.* **8**:23–29.
189. **Cleveland, D. W.** 1982. Treadmilling of tubulin and actin. *Cell* **28**:689–691.
190. **Colombo, R., A. Milzani, P. Contini, and D. I. Donne.** 1991. Effects of lithium ions on actin polymerization in the presence of magnesium ions. *Biochem. J.* **274**:421–425.
191. **Colombo, R., A. Milzani, and I. D. Donne.** 1991. Lithium increases actin polymerization rates by enhancing the nucleation step. *J. Mol. Biol.* **217**:401–404.
192. **Coluccio, L. M. and L. G. Tilney.** 1983. Under physiological conditions actin disassembles slowly from the non-preferred end of an actin filament. *J. Cell Biol.* **97**:1629–1634.
193. **Condeelis, J. and A. L. Hall.** 1991. Measurement of actin polymerization and cross-linking in agonist-stimulated cells. *Methods Enzymol.* **196**:486–496.
194. **Cooke, R.** 1975. The role of the bound nucleotide in the polymerization of actin. *Biochemistry* **14**:3250–3256.
195. **Cooper, J. A., E. L. J. Buhle, S. B. Walker, T. Y. Tsong, and T. D. Pollard.** 1983. Kinetic evidence for a monomer activation step in actin polymerization. *Biochemistry* **22**:2193–2202.
196.* **Cooper, J. A., S. B. Walker, and T. D. Pollard.** 1983. Pyrene actin: documentation of the validity of a sensitive assay for actin polymerization. *J. Muscle Res. Cell Motil.* **4**:253–262.

197. **Cortese, J. D., B. Schwab, C. Frieden, and E. L. Elson.** 1989. Actin polymerization induces a shape change in actin-containing vesicles. *Proc. Natl Acad. Sci. U. S. A.* **86**:5773–5777.

198. **DasGupta, G., J. White, M. Phillips, J. C. Bulinski, and E. Reisler.** 1990. Immunochemical probing of the N-terminal segment on actin: the polymerization reaction. *Biochemistry* **29**:3319–3324.

199.* **Detmers, P., A. Weber, M. Elzinga, and R. E. Stephens.** 1981. 7-Chloro-4-nitrobenzeno-2-oxa-1,3-diazole actin as a probe for actin polymerization. *J. Biol. Chem.* **256**:99–105.

200.* **Drenckhahn, D. and T. D. Pollard.** 1986. Elongation of actin filaments is a diffusion-limited reaction at the barbed end and is accelerated by inert macromolecules. *J. Biol. Chem.* **261**:12754–12758.

201. **Drewes, G. and H. Faulstich.** 1993. Cooperative effects on filament stability in actin modified at the C-terminus by substitution or truncation. *Eur. J. Biochem.* **212**:247–253.

202. **Fazekas, S., M. Nagy, V. Szekessy Hermann, and E. Wolfram.** 1970. Viscometric investigation of actin G and actin F. *Acta Biochim. Biophys. Acad. Sci. Hung.* **5**:121–129.

203. **Fox, J. E. and D. R. Phillips.** 1983. Polymerization and organization of actin filaments within platelets. *Semin. Hematol.* **20**:243–260.

204. **Frieden, C.** 1983. Polymerization of actin: mechanism of the Mg^{2+}-induced process at pH 8 and 20 degrees C. *Proc. Natl Acad. Sci. U. S. A.* **80**:6513–6517.

205. **Frieden, C.** 1985. Actin and tubulin polymerization: the use of kinetic methods to determine mechanism. *Annu. Rev. Biophys. Biophys. Chem.* **14**:189–210.

206. **Frieden, C. and D. W. Goddette.** 1983. Polymerization of actin and actin-like systems: evaluation of the time course of polymerization in relation to the mechanism. *Biochemistry* **22**:5836–5843.

207. **Fujime, S. and S. Ishiwata.** 1971. Dynamic study of F-actin by quasielastic scattering of laser light. *J. Mol. Biol.* **62**:251–265.

208. **Fung, B. M. and E. Eyob.** 1983. The effect of ATP concentration on the rate of actin polymerization. *Arch. Biochem. Biophys.* **220**:370–378.

209.* **Gaertner, A., K. Ruhnau, E. Schroer, N. Selve, M. Wanger, and A. Wegner.** 1989. Probing nucleation, cutting and capping of actin filaments. *J. Muscle Res. Cell Motil.* **10**:1–9.

210. **Gaertner, A. and A. Wegner.** 1991. Mechanism of the insertion of actin monomers between the barbed ends of actin filaments and barbed end-bound insertin. *J. Muscle Res. Cell Motil.* **12**:27–36.

211. **Gershman, L. C., L. A. Selden, H. J. Kinosian, and J. E. Estes.** 1989. Preparation and polymerization properties of monomeric ADP-actin. *Biochim. Biophys. Acta* **995**:109–115.

212. **Giuliano, K. A. and D. L. Taylor.** 1994. Fluorescent actin analogs with a high affinity for profilin *in vitro* exhibit an enhanced gradient of assembly in living cells. *J. Cell Biol.* **124**:971–983.

213. **Glacy, S. D.** 1983. Pattern and time course of rhodamine-actin incorporation in cardiac myocytes. *J. Cell Biol.* **96**:1164–1167.

214. **Glacy, S. D.** 1983. Subcellular distribution of rhodamine-actin microinjected into living fibroblastic cells. *J. Cell Biol.* **97**:1207–1213.

215. **Goddette, D. W., E. C. Uberbacher, G. J. Bunick, and C. Frieden.** 1986. Formation of actin dimers as studied by small angle neutron scattering. *J. Biol. Chem.* **261**:2605–2609.

216. **Gordon, D. J., Y. Z. Yang, and E. D. Korn.** 1976. Polymerization of *Acanthamoeba* actin. Kinetics, thermodynamics, and co-polymerization with muscle actin. *J. Biol. Chem.* **251**:7474–7479.

217. **Grazi, E.** 1985. Polymerization of N-(1-pyrenyl) iodoacetamide-labelled actin: the fluorescence signal is not directly proportional to the incorporation of the monomer into the polymer. *Biochem. Biophys. Res. Commun.* **128**:1058–1063.

218. **Grazi, E.** 1986. Treadmilling or fragmentation? A proposed repair mechanism for actin filaments. *Ital. J. Biochem.* **35**:284–286.

219. **Grazi, E., A. Ferri, and S. Cino.** 1983. The polymerization of actin. A study of the nucleation reaction. *Biochem. J.* **213**:727–732.

220. **Grazi, E. and V. Lanzara.** 1987. The experimental basis of some recent hypotheses on the mechanism of the polymerization of actin: a reappraisal. *Arch. Biochem. Biophys.* **257**:115–122.

221. **Grazi, E. and E. Magri.** 1987. Kinetic heterogeneity of F-actin polymers. Further evidence that the elongation reaction may occur through condensation of the actin filaments with small aggregates. *Biochem. J.* **248**:721–725.

222. **Grazi, E. and G. Trombetta.** 1986. Evaluation of the actin filament length from the time course of the depolymerization process. *Biochem. Biophys. Res. Commun.* **139**:109–114.

223. **Greer, C. and R. Schekman.** 1982. Calcium control of Saccharomyces cerevisiae actin assembly. *Mol. Cell Biol.* **2**:1279–1286.

224. **Griffith, L. M. and T. D. Pollard.** 1982. Cross-linking of actin filament networks by self-association and actin-binding macromolecules. *J. Biol. Chem.* **257**:9135–9142.

225. **Hama, H., M. Kasai, K. Maruyama, and H. Noda.** 1969. Natural F-actin. IV. Length distribution and its change studied by electron microscopy. *Biochim. Biophys. Acta* **194**:470–477.

226. **Hama, H., K. Maruyama, and H. Noda.** 1967. Natural F-actin. II. Natural F-actin and its transformation to Straub F-actin. *Biochim. Biophys. Acta* **133**:251–262.

226a. **Handel, S. E., K. A. Hendry, and P. Sheterline.** 1990. Microinjection of covalently cross-linked actin oligomers causes disruption of existing actin filament architecture in PtK2 cells. *J. Cell Sci.* **97**:325–333.

227. **Hartwig, J. H., R. Niederman, and S. E. Lind.** 1985. Cortical actin structures and their relationship to mammalian cell movements. *Subcell. Biochem.* **11**:1–49.

228. **Hayashi, T. and W. Ip.** 1976. Polymerization polarity of actin. *J. Mechanochem. Cell Motil.* **3**:163–169.

229. **Heacock, C. S. and J. R. Bamburg.** 1983. The quantitation of G- and F-actin in cultured cells. *Anal. Biochem.* **135**:22–36.

230. **Heacock, C. S. and J. R. Bamburg.** 1983. Levels of filamentous and globular actin in Chinese hamster ovary cells throughout the cell cycle. *Exp. Cell Res.* **147**:240–246.

231. **Hegyi, G. and S. Y. Venyaminov.** 1980. Absence of gross changes in the secondary structure of actin at G–F transition. *FEBS Lett.* **109**:134–136.

232. **Hill, T. L.** 1980. Bioenergetic aspects and polymer length distribution in steady-state head-to-tail polymerization of actin or microtubules. *Proc. Natl Acad. Sci. U. S. A.* **77**:4803–4807.

233. **Hill, T. L.** 1981. Steady-state head-to-tail polymerization of actin or microtubules. II. Two-state and three-state kinetic cycles. *Biophys. J.* **33**:353–371.

234.* **Hill, T. L. and M. W. Kirschner.** 1982. Subunit treadmilling of microtubules or actin in the presence of cellular barriers: possible conversion of chemical free energy into mechanical work. *Proc. Natl Acad. Sci. U. S. A.* **79**:490–494.

235. **Hill, T. L. and M. W. Kirschner.** 1982. Bioenergetics and kinetics of microtubule and actin filament assembly-disassembly. *Int. Rev. Cytol.* **78**:1–125.

236. **Hill, T. L. and M. W. Kirschner.** 1983. Regulation of microtubule and actin filament assembly-disassembly by associated small and large molecules. *Int. Rev. Cytol.* **84**:185–234.

237. **Hirono, M., R. Tanaka, and Y. Watanabe.** 1990. *Tetrahymena* actin: copolymerization with skeletal muscle actin and interactions with muscle actin-binding proteins. *J. Biochem. Tokyo* **107**:32–36.

238. **Hitchcock De Gregori, S. E., S. Mandala, and G. A. Sachs.** 1982. Changes in actin lysine reactivities during polymerization detected using a competitive labeling method. *J. Biol. Chem.* **257**:12573–12580.

239. **Howard, T. H. and C. O. Oresajo.** 1985. A method for quantifying F-actin in chemotactic peptide activated neutrophils: study of the effect of tBOC peptide. *Cell Motil.* **5**:545–557.

240. **Ikkai, T. and T. Ooi.** 1966. The effects of pressure on F-G transformation of actin. *Biochemistry* **5**:1551–1560.

241. **Ikkai, T., T. Ooi, and H. Noguchi.** 1966. Actin: volume change on transformation of G-form to F-form. *Science* **152**:1756–1757.

242. **Janmey, P. A.** 1991. A torsion pendulum for measurement of the viscoelasticity of biopolymers and its application to actin networks. *J. Biochem. Biophys. Methods* **22**:41–53.

243. **Janmey, P. A., S. Hvidt, J. Peetermans, J. Lamb, J. D. Ferry, and T. P. Stossel.** 1988. Viscoelasticity of F-actin and F-actin/gelsolin complexes. *Biochemistry* **27**:8218–8227.

244. **Janmey, P. A. and T. P. Stossel.** 1986. Kinetics of actin monomer exchange at the slow growing ends of actin filaments and their relation to the elongation of filaments shortened by gelsolin. *J. Muscle Res. Cell Motil.* **7**:446–454.

245. **Jen, C. J., L. V. McIntire, and J. Bryan.** 1982. The viscoelastic properties of actin solutions. *Arch. Biochem. Biophys.* **216**:126–132.

246. **Kasai, M.** 1969. Thermodynamical aspect of G-F transformations of actin. *Biochim. Biophys. Acta* **180**:399–409.

247. **Kasai, M., E. Nakano, and F. Oosawa.** 1965. Polymerisation of actin free from nucleotides and divalent cations. *Biochim. Biophys. Acta* **94**:494–503.

248. **Kawamura, M. and K. Maruyama.** 1970. Electron microscopic particle length of F-actin polymerized *in vitro*. *J. Biochem. Tokyo* **67**:437–457.

249. **Kawamura, M. and K. Maruyama.** 1972. Length distribution of F-actin transformed from Mg-polymer. *Biochim. Biophys. Acta* **267**:422–434.

250. **Kawamura, M. and K. Maruyama.** 1972. A further study of electron microscopic particle length of F-actin polymerized *in vitro*. *J. Biochem. Tokyo* **72**:179–188.

251. **Keiser, T., A. Schiller, and A. Wegner.** 1986. Nonlinear increase of elongation rate of actin filaments with actin monomer concentration. *Biochemistry* **25**:4899–4906.

251a. **Kellogg, D. R., T. J. Mitchison, and B. M. Alberts.** 1988. Behaviour of microtubules and actin filaments in living *Drosophila* embryos. *Development.* **103**:675–686.

252. **Khaitlina, S. Y.** 1986. Polymerization of β-like actin from scallop adductor muscle. *FEBS Lett.* **198**:221–224.

253. **Khaitlina, S. Y., J. Moraczewska, and H. Strzelecka Golaszewska.** 1993. The actin/actin interactions involving the N-terminus of the DNase-I-binding loop are crucial for stabilization of the actin filament. *Eur. J. Biochem.* **218**:911–920.

254. **Kikuchi, M., K. Maruyama, and H. Noda.** 1974. Sedimentation behavior of F-actin of various types. *J. Biochem. Tokyo* **75**:1103–1111.

255. **Kinosian, H. J., L. A. Selden, J. E. Estes, and L. C. Gershman.** 1991. Thermodynamics of actin polymerization; influence of the tightly bound divalent cation and nucleotide. *Biochim. Biophys. Acta* **1077**:151–158.

256. **Kinosian, H. J., L. A. Selden, J. E. Estes, and L. C. Gershman.** 1993. Actin filament annealing in the presence of ATP and phalloidin. *Biochemistry* **32**:12353–12357.

257. **Kirschner, M. W.** 1980. Implications of treadmilling for the stability and polarity of actin and tubulin polymers *in vivo*. *J. Cell Biol.* **86**:330–334.

258. **Kocks, C., E. Gouin, M. Tabouret, P. Berche, H. Ohayon, and P. Cossart.** 1992. L. monocytogenes-induced actin assembly requires the actA gene product, a surface protein. *Cell* **68**:521–531.

259. **Koffer, A., W. B. Gratzer, G. D. Clarke, and A. Hales.** 1983. Phase equilibria of cytoplasmic actin of cultured epithelial (BHK) cells. *J. Cell Sci.* **61**:191–218.

260. **Kondo, H. and S. Ishiwata.** 1976. Unidirectional growth of F-actin. *J. Biochem. Tokyo* **79**:159–171.

261.* **Korn, E. D.** 1982. Actin polymerization and its regulation by proteins from nonmuscle cells. *Physiol. Rev.* **62**:672–737.

262.* **Korn, E. D., M. F. Carlier, and D. Pantaloni.** 1987. Actin polymerization and ATP hydrolysis. *Science* **238**:638–644.

263. **Kouchi, K., H. Takahashi, and Y. Shimada.** 1993. Incorporation of microinjected biotin-labelled actin into nascent myofibrils of cardiac myocytes: an immunoelectron microscopic study. *J. Muscle Res. Cell Motil.* **14**:292–301.

264.* **Kouyama, T. and K. Mihashi.** 1981. Fluorimetry study of N-(1-pyrenyl)-iodoacetamide-labelled F-actin. Local structural change of actin protomer both on polymerization and on binding of heavy meromyosin. *Eur. J. Biochem.* **114**:33–38.

265.* **Kreis, T. E., B. Geiger, and J. Schlessinger.** 1982. Mobility of microinjected rhodamine actin within living chicken gizzard cells determined by fluorescence photobleaching recovery. *Cell* **29**:835–845.

266. **Kreis, T. E., K. H. Winterhalter, and W. Birchmeier.** 1979. *In vivo* distribution and turnover of fluorescently labeled actin microinjected into human fibroblasts. *Proc. Natl Acad. Sci. U. S. A.* **76**:3814–3818.

267. **Kuckel, C. L., B. W. Lubit, P. K. Lambooy, and P. N. Farnsworth.** 1993. Methylisocyanate and actin polymerization: the *in vitro* effects of carbamylation. *Biochim. Biophys. Acta* **1162**:143–148.

268. **Kukulies, J., W. Stockem, and F. Achenbach.** 1984. Distribution and dynamics of fluorochromed actin in living stages of Physarum polycephalum. *Eur. J. Cell Biol.* **35**:235–245.

269. **Lal, A. A., S. L. Brenner, and E. D. Korn.** 1984. Preparation and polymerisation of skeletal muscle ADP-actin. *J. Biol. Chem.* **259**:13061–13065.

270.* **Lal, A. A., E. D. Korn, and S. L. Brenner.** 1984. Rate constants for actin polymerisation in ATP determined using cross-linked actin trimers as nuclei. *J. Biol. Chem.* **259**:8794–8800.

271. **Lanni, F., D. L. Taylor, and B. R. Ware.** 1981. Fluorescence photobleaching recovery in solutions of labeled actin. *Biophys. J.* **35**:351–364.

272. **Leffers, H., M. S. Nielsen, A. H. Andersen, B. Honore, P. Madsen, J. Vandekerckhove, and J. E. Celis.** 1993. Identification of two human Rho GDP dissociation inhibitor proteins whose overexpression leads to disruption of the actin cytoskeleton. *Exp. Cell Res.* **209**:165–174.

273. **Lheureux, K., T. Forne, and P. Chaussepied.** 1993. Interaction and polymerization of the G-actin-myosin head complex: effect of DNase I. *Biochemistry* **32**:10005–10014.

274. **Lu, R. C. and L. Szilagyi.** 1981. Change of reactivity of lysine residues upon actin polymerization. *Biochemistry* **20**:5914–5919.

275. **Marriott, G., H. Miyata, and K. J. Kinosita.** 1992. Photomodulation of the nucleating activity of a photocleavable crosslinked actin dimer. *Biochem. Int.* **26**:943–951.

276. **Maruyama, K.** 1981. Sonic vibration induces the nucleation of actin in the absence of magnesium ions and cytochalasins inhibit the elongation of the nuclei. *J. Biol. Chem.* **256**:1060–1062.

277. **Maruyama, K. and K. Tsukagoshi.** 1984. Effects of KCl, $MgCl_2$, and $CaCl_2$ concentrations on the monomer-polymer equilibrium of actin in the presence and absence of cytochalasin D. *J. Biochem. Tokyo* **96**:605–611.

278. **Maruyama, K., N. Yamada, and I. Mabuchi.** 1984. Capping one end of an actin filament affects elongation at the other end. *J. Biochem. Tokyo* **96**:613–620.

279. **Masai, J., S. Ishiwata, and S. Fujime.** 1986. Dynamic light-scattering study on polymerization process of muscle actin. *Biophys. Chem.* **25**:253–269.

280. **Masuhara, M., H. Yokoyama, and N. Tatsumi.** 1983. Theoretical studies on capillary microviscometry of skeletal muscle actin. *Biosci. Rep.* **3**:195–206.

281. **Matsudaira, P., J. Bordas, and M. H. Koch.** 1987. Synchrotron X-ray diffraction studies of actin structure during polymerization. *Proc. Natl Acad. Sci. U. S. A.* **84**:3151–3155.

282. **Merkler, I., C. Stournaras, and H. Faulstich.** 1987. Changes in OD at 235 nm do not correspond to the polymerization step of actin. *Biochem. Biophys. Res. Commun.* **145**:46–51.

283. **Miki Noumura, T. and H. Kondo.** 1970. Polymerization of actin from sea urchin eggs. *Exp. Cell Res.* **61**:31–41.

284. **Miki, M. and T. Iio.** 1984. Fluorescence energy transfer measurements between the nucleotide binding site and Cys-373 in actin and their application to the kinetics of actin polymerization. *Biochim. Biophys. Acta* **790**:201–207.

285. **Mockrin, S. C. and E. D. Korn.** 1983. Kinetics of polymerization and ATP hydrolysis by covalently crosslinked actin dimer. *J. Biol. Chem.* **258**:3215–3221.

286. **Monk, P. N. and P. Banks.** 1991. Evidence for the involvement of multiple signalling pathways in C5a-induced actin polymerization and nucleation in human monocyte-like cells. *J. Mol. Endocrinol.* **6**:241–247.

287. **Mooseker, M. S., T. D. Pollard, and K. A. Wharton.** 1982. Nucleated polymerization of actin from the membrane-associated ends of microvillar filaments in the intestinal brush border. *J. Cell Biol.* **95**:223–233.

288. **Mossakowska, M., J. Moraczewska, S. Khaitlina, and H. Strzelecka Golaszewska.** 1993. Proteolytic removal of three C-terminal residues of actin alters the monomer-monomer interactions. *Biochem. J.* **289**:897–902.

289. **Mozo Villarias, A. and B. R. Ware.** 1984. Distinctions between mechanisms of cytochalasin D activity for Mg^{2+}- and K^+-induced actin assembly. *J. Biol. Chem.* **259**:5549–5554.

290. **Murphy, D. B., R. O. Gray, W. A. Grasser, and T. D. Pollard.** 1988. Direct demonstration of actin filament annealing *in vitro*. *J. Cell Biol.* **106**:1947–1954.

291. **Murthy, V. K., L. E. Crevasse, and J. C. Shipp.** 1968. Actin polymerization and its relationship to adenosine triphosphatase activity. *Circ. Res.* **22**:65–73.

292. **Nachmias, V. T.** 1976. Adenosine 5′–phosphate and assembly of actin filaments *in vitro*. *J. Mol. Biol.* **107**:623–629.

293.* **Neuhaus, J. M., M. Wanger, T. Keiser, and A. Wegner.** 1983. Treadmilling of actin. *J. Muscle Res. Cell Motil.* **4**:507–527.

294. **Newman, J., J. E. Estes, L. A. Selden, and L. C. Gershman.** 1985. Presence of oligomers at sub-critical actin concentrations. *Biochemistry* **24**:1538–1544.

295. **Nishida, E. and H. Sakai.** 1983. Kinetic analysis of actin polymerization. *J. Biochem. Tokyo* **93**:1011–1020.

296. **Nunnally, M. H., L. D. Powell, and S. W. Craig.** 1981. Reconstitution and regulation of actin gel–sol transformation with purified filamin and villin. *J. Biol. Chem.* **256**:2083–2086.

297. **Ohm, T. and A. Wegner.** 1991. Random copolymerization of ATP-actin and ADP-actin. *Biochemistry* **30**:11193–11197.

298.* **Okabe, S. and N. Hirokawa.** 1989. Incorporation and turnover of biotin-labeled actin microinjected into fibroblastic cells: an immunoelectron microscopic study. *J. Cell Biol.* **109**:1581–1595.

299. **Okabe, S. and N. Hirokawa.** 1991. Actin dynamics in growth cones. *J. Neurosci.* **11**:1918–1929.

300. **Oosawa, F.** 1972. Dynamic properties of F-actin and the thin filament. *Nippon. Seirigaku. Zasshi.* **34**:96–97.

301. **Oosawa, F.** 1977. Actin–actin bond strength and the conformational change of F-actin. *Biorheology* **14**:11–19.

302. **Oosawa, M. and K. Maruyama.** 1986. An intermediate state of G-actin between native and denatured: polymerization rate decreases but extent of polymerization remains unchanged. *J. Biochem. Tokyo* **100**:1001–1008.

303. **Oriol Audit, C.** 1978. Polyamine-induced actin polymerization. *Eur. J. Biochem.* **87**:371–376.

304. **Pan, X. X. and B. R. Ware.** 1988. Actin assembly by lithium ions. *Biophys. J.* **53**:11–16.

305. **Pantaloni, D. and M. F. Carlier.** 1993. How profilin promotes actin filament assembly in the presence of thymosin β 4. *Cell* **75**:1007–1014.

306. **Pantaloni, D., M. F. Carlier, and E. D. Korn.** 1985. The interaction between ATP-actin and ADP-actin. A tentative model for actin polymerization. *J. Biol. Chem.* **260**:6572–6578.

307. **Pantaloni, D., T. L. Hill, M. F. Carlier, and E. D. Korn.** 1985. A model for actin polymerization and the kinetic effects of ATP hydrolysis. *Proc. Natl Acad. Sci. U. S. A.* **82**:7207–7211.

308. **Pardee, J. D., P. A. Simpson, L. Stryer, and J. A. Spudich.** 1982. Actin filaments undergo limited subunit exchange in physiological salt conditions. *J. Cell Biol.* **94**:316–324.

309. **Pardee, J. D. and J. A. Spudich.** 1982. Mechanism of K^+-induced actin assembly. *J. Cell Biol.* **93**:648–654.

310. **Plank, L. and B. R. Ware.** 1986. Assembly of *Acanthamoeba* actin in the presence of *Acanthamoeba* profilin measured by fluorescence photobleaching recovery. *Biochem. Biophys. Res. Commun.* **140**:308–312.

311.* **Pollard, T. D.** 1983. Measurement of rate constants for actin filament elongation in solution. *Anal. Biochem.* **134**:406–412.

312. **Pollard, T. D.** 1984. Polymerization of ADP-actin. *J. Cell Biol.* **99**:769–777.

313.* **Pollard, T. D.** 1986. Rate constants for the reactions of ATP- and ADP-actin with the ends of actin filaments. *J. Cell Biol.* **103**:2747–2754.

314. **Pollard, T. D.** 1986. Assembly and dynamics of the actin filament system in nonmuscle cells. *J. Cell Biochem.* **31**:87–95.

315.* **Pollard, T. D. and M. S. Mooseker.** 1981. Direct measurement of actin polymerisation rate constants by electron microscopy of actin filaments nucleated by isolated microvillus cores. *J. Cell Biol.* **88**:654–659.

316. **Prochniewicz, E. and T. Yanagida.** 1981. Comparison of intermonomer interactions within polymers of chicken gizzard and rabbit skeletal muscle actins. *J. Biochem. Tokyo* **89**:1215–1221.

317. **Quirion, F. and C. Gicquaud.** 1993. Changes in molar volume and heat capacity of actin upon polymerization. *Biochem. J.* **295**:671–672.

318.* **Ridley, A. J. and A. Hall.** 1992. Distinct patterns of actin organization regulated by the small GTP-binding proteins Rac and Rho. *Cold Spring Harb. Symp. Quant. Biol.* **57**:661–671.

319. **Ridley, A. J. and A. Hall.** 1992. The small GTP-binding protein rho regulates the assembly of focal adhesions and actin stress fibers in response to growth factors. *Cell* **70**:389–399.

320. **Rouayrenc, J. F. and F. Travers.** 1981. The first step in the polymerisation of actin. *Eur. J. Biochem.* **116**:73–77.

321.* **Sanders, M. C., A. L. Goldstein, and Y. L. Wang.** 1992. Thymosin β 4 (Fx peptide) is a potent regulator of actin polymerization in living cells. *Proc. Natl Acad. Sci. U. S. A.* **89**:4678–4682.

322.* **Sanders, M. C. and Y. L. Wang.** 1991. Assembly of actin-containing cortex occurs at distal regions of growing neurites in PC12 cells. *J. Cell Sci.* **100**:771–780.

323. **Sanger, J. M. and J. W. Sanger.** 1980. Banding and polarity of actin filaments in interphase and cleaving cells. *J. Cell Biol.* **86**:568–575.

324. **Schmit, A. C. and A. M. Lambert.** 1990. Microinjected fluorescent phalloidin *in vivo* reveals the F-actin dynamics and assembly in higher plant mitotic cells. *Plant. Cell* **2**:129–138.

325.* **Selve, N. and A. Wegner.** 1986. Rate of treadmilling of actin filaments *in vitro*. *J. Mol. Biol.* **187**:627–631.

326. **Shanbhag, V. P., L. Backman, and G. Pinaev.** 1981. Effect of Cibachron Blue F3GA on the polymerization of actin. *Mol. Cell Biochem.* **37**:71–74.

327. **Sheterline, P., J. E. Rickard, and R. C. Richards.** 1984. Fc receptor-directed phagocytic stimuli induce transient actin assembly at an early stage of phagocytosis in neutrophil leukocytes. *Eur. J. Cell Biol.* **34**:80–87.
328. **Simpson, P. A. and J. A. Spudich.** 1980. ATP-driven steady-state exchange of monomeric and filamentous actin from *Dictyostelium discoideum. Proc. Natl Acad. Sci. U. S. A.* **77**:4610–4613.
329.* **Small, J. V., G. Isenberg, and J. E. Celis.** 1978. Polarity of actin at the leading edge of cultured cells. *Nature* **272**:638–639.
330. **Spudich, J. A., E. R. Kuczmarski, J. D. Pardee, P. A. Simpson, K. Yamamoto, and L. Stryer.** 1982. Control of assembly of *Dictyostelium* myosin and actin filaments. *Cold Spring Harb. Symp. Quant. Biol.* 46 Pt **2**:553–561.
331. **Spudich, J. A., J. D. Pardee, P. A. Simpson, K. Yamamoto, E. R. Kuczmarski, and L. Stryer.** 1982. Actin and myosin: control of filament assembly. *Phil. Trans. R. Soc. Lond. Biol.* **299**:247–261.
332.* **Stossel, T. P.** 1989. From signal to pseudopod. How cells control cytoplasmic actin assembly. *J. Biol. Chem.* **264**:18261–18264.
333. **Stozharov, A.** 1985. Relationship between metabolic states of liver mitochondria and the G/F-actin ratio. *Biomed. Biochim. Acta* **44**:1287–1294.
334. **Strzelecka Golaszewska, H., S. Zmorzynski, and M. Mossakowska.** 1985. Bovine aorta actin. Development of an improved purification procedure and comparison of polymerization properties with actins from other types of muscle. *Biochim. Biophys. Acta* **828**:13–21.
335.* **Svitkina, T. M., A. A. J. Neyfakh, and A. D. Bershadsky.** 1986. Actin cytoskeleton of spread fibroblasts appears to assemble at the cell edges. *J. Cell Sci.* **82**:235–248.
336. **Swezey, R. R. and G. N. Somero.** 1985. Pressure effects on actin self-assembly: interspecific differences in the equilibrium and kinetics of the G to F transformation. *Biochemistry* **24**:852–860.
337.* **Symons, M. H. and T. J. Mitchison.** 1991. Control of actin polymerization in live and permeabilized fibroblasts. *J. Cell Biol.* **114**:503–513.
338. **Tait, J. F. and C. Frieden.** 1982. Polymerization and gelation of actin studied by fluorescence photobleaching recovery. *Biochemistry* **21**:3666–3674.
339. **Tait, J. F. and C. Frieden.** 1982. Polymerization-induced changes in the fluorescence of actin labeled with iodoacetamidotetramethylrhodamine. *Arch. Biochem. Biophys.* **216**:133–141.
340. **Tait, J. F. and C. Frieden.** 1982. Chemical modification of actin. Acceleration of polymerization and reduction of network formation by reaction with N-ethylmaleimide, (iodoacetamido)tetramethylrhodamine, or 7-chloro-4-nitro-2,1,3-benzoxadiazole. *Biochemistry* **21**:6046–6053.
341. **Taylor, D. L., J. Reidler, J. A. Spudich, and L. Stryer.** 1981. Detection of actin assembly by fluorescence energy transfer. *J. Cell Biol.* **89**:362–367.
342. **Taylor, D. L. and Y. L. Wang.** 1978. Molecular cytochemistry: incorporation of fluorescently labeled actin into living cells. *Proc. Natl Acad. Sci. U. S. A.* **75**:857–861.
342a. **Taylor, D. L., Y. L. Wang, and J. M. Heiple.** 1980. Contractile basis of ameboid movement. VII. The distribution of fluorescently labeled actin in living amebas. *J. Cell Biol.* **86**:590–598.
343. **Tellam, R.** 1985. Mechanism of $CaCl_2$-induced actin polymerization. *Biochemistry* **24**:4455–4460.
344. **Tellam, R. and C. Frieden.** 1982. Cytochalasin D and platelet gelsolin accelerate actin polymer formation. A model for regulation of the extent of actin polymer formation *in vivo. Biochemistry* **21**:3207–3214.
345. **Tellam, R. L., M. J. Sculley, L. W. Nichol, and P. R. Wills.** 1983. The influence of poly(ethylene glycol) 6000 on the properties of skeletal-muscle actin. *Biochem. J.* **213**:651–659.
346.* **Theriot, J. A. and T. J. Mitchison.** 1991. Actin microfilament dynamics in locomoting cells. *Nature* **352**:126–131.
347. **Theriot, J. A., T. J. Mitchison, L. G. Tilney, and D. A. Portnoy.** 1992. The rate of actin-based motility of intracellular Listeria monocytogenes equals the rate of actin polymerization. *Nature* **357**:257–260.
348.* **Tobacman, L. S. and E. D. Korn.** 1983. The kinetics of actin nucleation and polymerization. *J. Biol. Chem.* **258**:3207–3214.
349. **Tsuchiya, T. and Y. Nagai.** 1983. Theoretical description of release, uptake, and pulse chase of labeled subunits of actin or a microtubule that undergoes head-to-tail polymerization. *Biophys. J.* **41**:275–281.
350. **Vancompernolle, K., M. Goethals, C. Huet, D. Louvard, and J. Vandekerckhove.** 1992. G- to F-actin modulation by a single amino acid substitution in the actin binding site of actobindin and thymosin β4. *EMBO J.* **11**:4739–4746.
351. **Wachsstock, D. H., W. H. Schwarz, and T. D. Pollard.** 1994. Cross-linker dynamics determine the mechanical properties of actin gels. *Biophys. J.* **66**:801–809.
352. **Wallace, P. J., R. P. Wersto, C. H. Packman, and M. A. Lichtman.** 1984. Chemotactic peptide-induced changes in neutrophil actin conformation. *J. Cell Biol.* **99**:1060–1065.
353. **Wang, F., R. V. Sampogna, and B. R. Ware.** 1989. pH dependence of actin self-assembly. *Biophys. J.* **55**:293–298.
354. **Wang, Y. L.** 1985. Exchange of actin subunits at the leading edge of living fibroblasts: possible role of treadmilling. *J. Cell Biol.* **101**:597–602.
355. **Wang, Y. L. and D. L. Taylor.** 1981. Exchange of 1,N6-etheno-ATP with actin-bound nucleotides as a tool for studying the steady-state exchange of subunits in F-actin solutions. *Proc. Natl Acad. Sci. U. S. A.* **78**:5503–5507.
356. **Wanger, M., T. Keiser, J. M. Neuhaus, and A. Wegner.** 1985. The actin treadmill. *Can. J. Biochem. Cell Biol.* **63**:414–421.
357. **Weber, A., J. Northrop, M. F. Bishop, F. A. Ferrone, and M. S. Mooseker.** 1987. Nucleation of actin polymerization by villin and elongation at subcritical monomer concentration. *Biochemistry* **26**:2528–2536.
358.* **Weber, A., J. Northrop, M. F. Bishop, F. A. Ferrone, and M. S. Mooseker.** 1987. Kinetics of actin elongation and depolymerization at the pointed end. *Biochemistry* **26**:2537–2544.
359. **Weber, A., C. R. Pennise, and M. Pring.** 1994. DNAseI increases the rate constant of depolymerisation at the pointed (−) end of actin filament. *Biochemistry* **33**:4780–4786.
360. **Wegner, A.** 1976. Head to tail polymerization of actin. *J. Mol. Biol.* **108**:139–150.
361. **Wegner, A.** 1977. The mechanism of ATP hydrolysis by polymer actin. *Biophys. Chem.* **7**:51–58.
362. **Wegner, A.** 1982. Spontaneous fragmentation of actin filaments in physiological conditions. *Nature* **296**:266–267.
363.* **Wegner, A.** 1982. Treadmilling of actin at physiological salt concentrations. An analysis of the critical concentrations of actin filaments. *J. Mol. Biol.* **161**:607–615.
364. **Wegner, A.** 1985. Subtleties of actin assembly. *Nature* **313**:97–98.
365.* **Wegner, A. and G. Isenberg.** 1983. 12-fold difference between the critical monomer concentrations of the two ends of actin filaments in physiological salt conditions. *Proc. Natl Acad. Sci. U. S. A.* **80**:4922–4925.
366. **Wegner, A. and J. M. Neuhaus.** 1981. Requirement of divalent cations for fast exchange of actin monomers and actin filament subunits. *J. Mol. Biol.* **153**:681–693.
367. **Wegner, A. and P. Savko.** 1982. Fragmentation of actin filaments. *Biochemistry* **21**:1909–1913.
368. **Wendel, H. and P. Dancker.** 1986. Kinetics of actin depolymerization: influence of ions, temperature, age of F-actin, cytochalasin B and phalloidin. *Biochim. Biophys. Acta* **873**:387–396.
369. **Wertman, K. F. and D. G. Drubin.** 1992. Actin constitution: guaranteeing the right to assemble. *Science* **258**:759–760.
370.* **Woodrum, D. T., S. A. Rich, and T. D. Pollard.** 1975. Evidence for biased bidirectional polymerization of actin filaments using heavy meromyosin prepared by an improved method. *J. Cell Biol.* **67**:231–237.
371. **Yin, H. L., K. S. Zaner, and T. P. Stossel.** 1980. Ca^{2+} control of actin gelation. Interaction of gelsolin with actin filaments and regulation of actin gelation. *J. Biol. Chem.* **255**:9494–9500.
372. **Zimmerle, C. T. and C. Frieden.** 1986. Effect of temperature on the mechanism of actin polymerization. *Biochemistry* **25**:6432–6438.

CATION BINDING TO ACTIN

373. **Barden, J. A., P. M. Curmi, and C. G. dos Remedios.** 1981. Crystalline actin tubes. III. The interaction of scandium and yttrium with skeletal muscle actin. *Biochim. Biophys. Acta* **671**:25–32.
374. **Barden, J. A. and C. G. dos Remedios.** 1980. Crystalline actin tubes. I. Is the conformation of the lanthanide-induced actin tube monomer more like F-actin than G-actin? *Biochim. Biophys. Acta* **624**:163–173.
375. **Burtnick, L. D.** 1982. Tb^{3+} as a luminescent probe of actin structure: effects of polymerization, KI, and the binding of deoxyribonuclease I. *Arch. Biochem. Biophys.* **216**:81–87.
376.* **Carlier, M. F., D. Pantaloni, and E. D. Korn.** 1986. The effects of Mg^{2+} at the high-affinity and low-affinity sites on the polymerization of actin and associated ATP hydrolysis. *J. Biol. Chem.* **261**:10785–10792.
377.* **Carlier, M. F., D. Pantaloni, and E. D. Korn.** 1986. Fluorescence measurements of the binding of cations to high-affinity and low-affinity sites on ATP-G-actin. *J. Biol. Chem.* **261**:10778–10784.
378. **Chrambach, A., M. Barany, and F. Finkelman.** 1961. The bound calcium of actin. *Arch. Biochem. Biophys.* **98**:28–45.
379. **Curmi, P. M., J. A. Barden, and C. G. dos Remedios.** 1982. Conformational studies of G-actin containing bound lanthanide. *Eur. J. Biochem.* **122**:239–244.
380. **DalleDonne, I., A. Milzani, U. Fascio, A. Ratti, and R. Colombo.** 1994. Lithium preserves F-actin from the disarrangement induced by either DNAseI of cytochalasin D. *Biochem. Cell Biol.* **71**:440–446.
381. **Diebler, H., M. Eigen, G. Ilgenfritz, G. Maas, and R. Winkler.** 1969. Kinetics and mechanism of reaction of main-group metal ions with biological carriers. *Pure Appl. Chem.* **20**:93–115.
382. **dos Remedios, C. G. and J. A. Barden.** 1977. Effects of Gd(III) on G-actin: inhibition of polymerization of G-actin and activation of myosin ATPase activity by Gd-G-actin. *Biochem. Biophys. Res. Commun.* **77**:1339–1346.
383. **Drabikowski, W., S. Lehrer, B. Nagy, and J. Gergely.** 1977. Loss of Cu^{2+}-binding to actin upon removal of the C-terminal phenylalanine by carboxypeptidase A. *Arch. Biochem. Biophys.* **181**:359–361.
384. **Drabikowski, W. and H. Strzelecka-Golaszewska.** 1963. The exchange of actin-bound calcium with various bivalent cations. *Biochim. Biophys. Acta* **71**:486–487.
385.* **Estes, J. E., L. A. Selden, and L. C. Gershman.** 1987. Tight binding of divalent cations to monomeric actin. Binding kinetics support a simplified model. *J. Biol. Chem.* **262**:4952–4957.
386.* **Estes, J. E., L. A. Selden, H. J. Kinosian, and L. C. Gershman.** 1992. Tightly-bound divalent cation of actin. *J. Muscle Res. Cell Motil.* **13**:272–284.
387. **Frieden, C., D. Lieberman, and H. R. Gilbert.** 1980. A fluorescent probe for conformational changes in skeletal muscle G-actin. *J. Biol. Chem.* **255**:8991–8993.
388. **Gershman, L. C., J. E. Estes, and L. A. Selden.** 1988. Polymerisation characteristics of divalent cation-free actin. *Ann. N. Y. Acad. Sci.* **529**:264–267.

389. **Gershman, L. C., J. Newman, L. A. Selden, and J. E. Estes.** 1984. Bound-cation exchange affects the lag phase in actin polymerization. *Biochemistry* **23**:2199–2203.
390. **Gershman, L. C., L. A. Selden, and J. E. Estes.** 1979. On the interaction of muscle actin with gadolinium. *Biochem. Biophys. Res. Commun.* **91**:1280–1287.
391.* **Gershman, L. C., L. A. Selden, and J. E. Estes.** 1986. High affinity binding of divalent cation to actin monomer is much stronger than previously reported. *Biochem. Biophys. Res. Commun.* **135**:607–614.
392.* **Gershman, L. C., L. A. Selden, and J. E. Estes.** 1991. Cation binding to the high-affinity site regulates nucleotide binding to actin. *Biophys. J.* **59**:53.
393.* **Gershman, L. C., L. A. Selden, and J. E. Estes.** 1991. High affinity divalent cation exchange on actin. Association rate measurements support the simple competitive model. *J. Biol. Chem.* **266**:76–82.
394. **Gershman, L. C., L. A. Selden, H. J. Kinosian, and J. E. Estes.** 1988. Divalent cation exchange on actin. *Biophys. J.* **53**:573.
395. **Hozumi, T.** 1988. Effect of divalent cation on the structure of skeletal muscle G-actin molecule. *Biochem. Int.* **16**:59–67.
396. **Kabsch, W., H. G. Mannherz, D. Suck, E. F. Pai, and K. C. Holmes.** 1990. Atomic structure of the actin:DNase I complex. *Nature* **347**:37–44.
397. **Kasai, M.** 1969. The divalent cation bound to actin and thin filament. *Biochim. Biophys. Acta* **172**:171–173.
398. **Kasai, M. and F. Oosawa.** 1968. The exchangeability of actin-bound calcium with various divalent cations. *Biochim. Biophys. Acta* **154**:520–528.
399. **Kasai, M. and F. Oosawa.** 1969. Behavior of divalent cations and nucleotides bound to F-actin. *Biochim. Biophys. Acta* **172**:300–310.
400. **Kasprzak, A. A.** 1993. Myosin subfragment 1 inhibits dissociation of nucleotide and calcium from G-actin. *J. Biol. Chem.* **268**:13261–13266.
401.* **Kinosian, H. J., L. A. Selden, J. E. Estes, and L. C. Gershman.** 1993. Nucleotide binding to actin. Cation dependence of nucleotide dissociation and exchange rates. *J. Biol. Chem.* **268**:8683–8691.
402. **Kitazawa, T., H. Shuman, and A. P. Somlyo.** 1982. Calcium and magnesium binding to thin and thick filaments in skinned muscle fibres: electron-probe analysis. *J. Muscle Res. Cell Motil.* **3**:437–454.
403. **Konno, K. and M. F. Morales.** 1985. Exposure of actin thiols by the removal of tightly held calcium ions. *Proc. Natl Acad. Sci. U. S. A.* **82**:7904–7908.
404. **Kuehl, W. M. and J. Gergely.** 1969. The kinetics of exchange of adenosine triphosphate and calcium with G-actin. *J. Biol. Chem.* **244**:4720–4729.
405. **Kushmerick, M. J., P. F. Dillon, R. A. Meyers, T. R. Brown, J. M. Krisanda, and H. L. Sweeney.** 1986. 31P-NMR spectroscopy, chemical analysis and free Mg^{++} of rabbit bladder and uterine smooth muscle. *J. Biol. Chem.* **261**:14420–14429.
406.* **Larsson, H. and U. Lindberg.** 1988. The effect of divalent cations on the interaction between calf spleen profilin and different actins. *Biochim. Biophys. Acta* **953**:95–105.
407. **Lehrer, S. S., B. Nagy, and J. Gergely.** 1972. The binding of Cu^{2+} to actin without loss of polymerizability: the involvement of the rapidly reacting -SH group. *Arch. Biochem. Biophys.* **150**:164–174.
408. **Martonosi, A., C. M. Molino, and J. Gergely.** 1964. The binding of divalent cations to actin. *J. Biol. Chem.* **239**:1057–1064.
409. **Maruyama, K. and J. Gergely.** 1961. Removal of the bound calcium of G-actin by EDTA. *Biochem. Biophys. Res. Commun.* **6**:245–249.
410. **Maruyama, K. and A. Martonosi.** 1961. Protective action of nucleoside triphosphates against the inactivation of G-actin by EDTA. *Biochem. Biophys. Res. Commun.* **5**:85–87.
411. **Mejean, C., H. K. Hue, F. Pons, C. Roustan, and Y. Benyamin.** 1988. Cation binding sites on actin: a structural relationship between antigenic epitopes and cation exchange. *Biochem. Biophys. Res. Commun.* **152**:368–375.
412. **Miki, M., J. A. Barden, and C. G. dos Remedios.** 1986. The distance separating Cys-10 from the high-affinity metal binding site in actin. *Biochem. Int.* **12**:807–813.
413.* **Nowak, E., H. Strzelecka Golaszewska, and R. S. Goody.** 1988. Kinetics of nucleotide and metal ion interaction with G-actin. *Biochemistry* **27**:1785–1792.
414. **Roustan, C., Y. Benyamin, M. Boyer, R. Bertrand, E. Audemard, and J. Jauregui Adell.** 1985. Conformational changes induced by Mg^{2+} on actin monomers. An immunologic attempt to localize the affected region. *FEBS Lett.* **181**:119–123.
415. **Selden, L. A., J. E. Estes, and L. C. Gershman.** 1983. The tightly bound divalent cation regulates actin polymerization. *Biochem. Biophys. Res. Commun.* **116**:478–485.
416.* **Selden, L. A., J. E. Estes, and L. C. Gershman.** 1989. High affinity divalent cation binding to actin. Effect of low affinity salt binding. *J. Biol. Chem.* **264**:9271–9277.
417. **Selden, L. A., L. C. Gershman, and J. E. Estes.** 1986. A kinetic comparison between Mg-actin and Ca-actin. *J. Muscle Res. Cell Motil.* **7**:215–224.
418. **Selden, L. A., L. C. Gershman, H. J. Kinosian, and J. E. Estes.** 1987. Conversion of ATP-actin to ADP-actin reverses the affinity of monomeric actin for Ca^{2+} vs Mg^{2+}. *FEBS Lett.* **217**:89–93.
419. **Selden, L. A., L. C. Gershman, H. J. Kinosian, and J. E. Estes.** 1990. Mg^{++} bound at the high-affinity site on actin is a cofactor for actin ATPase activity. *Biophys. J.* **57**:325.
420.* **Strzekecka Golaszewska, H.** 1973. Relative affinities of divalent cations to the site of the tight calcium binding in G-actin. *Biochim. Biophys. Acta* **310**:60–69.
421. **Strzelecka Golaszewska, H. and W. Drabikowski.** 1967. Correlation between the binding of calcium and ATP by G-actin. *Acta Biochim. Pol.* **14**:195–208.
422. **Strzelecka Golaszewska, H. and W. Drabikowski.** 1968. Studies on the exchange of G-actin-bound calcium with bivalent cations. *Biochim. Biophys. Acta* **162**:581–595.
423. **Strzelecka Golaszewska, H., E. Prochniewicz, and W. Drabikowski.** 1978. Interaction of actin with divalent cations. 2. Characterization of protein-metal complexes. *Eur. J. Biochem.* **88**:229–237.
424. **Strzelecka Golaszewska, H., E. Prochniewicz, and W. Drabikowski.** 1978. Interaction of actin with divalent cations. 1. The effect of various cations on the physical state of actin. *Eur. J. Biochem.* **88**:219–227.
425. **Strzelecka-Golaszewska, H.** 1973. Relative affinities of divalent cations to the site of the tight calcium binding in G-actin. *Biochim. Biophys. Acta* **370**:60–69.
426. **Strzelecka-Golaszewska, H. and W. Drabikowski.** 1968. Studies on the exchange of G-actin-bound calcium with bivalent cations. *Biochim. Biophys. Acta* **162**:518–595.
427. **Strzelecka-Golaszewska, H. and W. Drabikowski.** 1978. Interaction of actin with divalent cations **2**: characterisation of protein-metal complexes. *Eur. J. Biochem.* **88**:229–237.
428. **Strzelecka-Golsaszewska, H., G. Boguta, S. Zmorzynski, and J. Moraczewska.** 1989. Biochemical and theoretical approach to localization of metal-ion-binding sites in the actin primary structure. *Eur. J. Biochem.* **182**:299–305.
429. **Tellam, R. L. and J. A. Turner.** 1984. The binding of Ca^{2+} to actin monomer is monitored by the fluorescence of actin-bound auramine O. *Biochem. J.* **224**:269–276.
430. **Tonomura, Y. and J. Yoshimura.** 1961. Removal of bound nucleotide and calcium of G-actin by treatment with EDTA. *J. Biochem. Tokyo* **50**:79–80.
431. **Tsien, R. Y., T. Pozzan, and T. J. Rink.** 1982. Calcium homeostasis in intact lymphocytes: cytoplasmic free calcium monitored with a new, intracellularly-trapped fluorescent indicator. *J. Cell Biol.* **94**:325–334.
432. **Ue, K., A. Muhlrad, C. G. Edmonds, D. Bivin, A. Clark, W. V. Piechowski, and M. F. Morales.** 1992. The sequence location of the actin metal. *Eur. J. Biochem.* **203**:493–498.
433*. Valentin-Ranc, C. and M. F. Carlier. 1989. Evidence for the direct interaction between tightly bound divalent metal ion and ATP on actin. Binding of the lambda isomers of $\beta\gamma$-bidentate Cr-ATP to actin. *J. Biol. Chem.* **264**:20871–20880.
434. **Valentin-Ranc, C., C. Combeau, M. F. Carlier, and D. Pantaloni.** 1991. Myosin subfragment-1 interacts with two G-actin molecules in the absence of ATP. *J. Biol. Chem.* **266**:17872–17879.
435. **Zimmerle, C. T. and C. Frieden.** 1988. pH-induced changes in G-actin conformation and metal affinity. *Biochemistry* **27**:7759–7765.
436. **Zimmerle, C. T., K. Patane, and C. Frieden.** 1987. Divalent cation binding to the high- and low-affinity sites on G-actin. *Biochemistry* **26**:6545–6552.

NUCLEOTIDE BINDING AND ACTIN ATPASE

437. **Anderson, N. L.** 1979. The β and cytoplasmic actins are differentially thermostabilized by MgADP; γ actin binds MgADP more strongly. *Biochem. Biophys. Res. Commun.* **89**:486–490.
438. **Barany, M., A. F. Tucci, and T. E. Conover.** 1966. The removal of the bound ADP of F-actin. *J. Mol. Biol.* **19**:483–502.
439. **Barden, J. A.** 1987. Conformation of the ATP binding peptide in actin revealed by proton NMR spectroscopy. *Biochemistry* **26**:6023–6030.
440. **Barden, J. A. and B. E. Kemp.** 1987. NMR of a synthetic peptide spanning the triphosphate binding site of adenosine 5′-triphosphate in actin. *Biochemistry* **26**:1471–1478.
441. **Bender, N., H. Fasold, and M. Rack.** 1974. Interaction of rabbit muscle actin and chemically modified actin with ATP, ADP, and protein reactive analogues: role of the nucleotide. *FEBS Lett.* **44**:209–212.
442. **Blackholm, H. and H. Faulstich.** 1981. Protection of one of the two reactive thiol groups in F-actin by ATP and phalloidin. *Biochem. Biophys. Res. Commun.* **103**:125–130.
443. **Brenner, S. L. and E. D. Korn.** 1981. Stimulation of actin ATPase activity by cytochalasins provides evidence for a new species of monomeric actin. *J. Biol. Chem.* **256**:8663–8670.
444.* **Brenner, S. L. and E. D. Korn.** 1984. Evidence that F-actin can hydrolyze ATP independent of monomer-polymer end interactions. *J. Biol. Chem.* **259**:1441–1446.
445. **Burt, C. T., T. Glonek, and M. Barany.** 1977. Analysis of living tissue by phosphorus-31 magnetic resonance. *Science* **195**:145–149.
446.* **Carlier, M. F.** 1987. Measurement of Pi dissociation from actin filaments following ATP hydrolysis using a linked enzyme assay. *Biochem. Biophys. Res. Commun.* **143**:1069–1075.
447.* **Carlier, M. F.** 1988. Role of nucleotide hydrolysis in the polymerization of actin and tubulin. *Cell Biophys.* **12**:105–117.
448.* **Carlier, M. F.** 1989. Role of nucleotide hydrolysis in the dynamics of actin filaments and microtubules. *Int. Rev. Cytol.* **115**:139–170.
449.* **Carlier, M. F.** 1990. Actin polymerization and ATP hydrolysis. *Adv. Biophys.* **26**:51–73.
450.* **Carlier, M. F.** 1992. Nucleotide hydrolysis regulates the dynamics of actin filaments and microtubules. *Phil. Trans. R. Soc. Lond. Biol.* **336**:93–97.

451.* **Carlier, M. F., C. Jean, K. J. Rieger, M. Lenfant, and D. Pantaloni.** 1993. Modulation of the interaction between G-actin and thymosin β4 by the ATP/ADP ratio: possible implication in the regulation of actin dynamics. *Proc. Natl Acad. Sci. U. S. A.* **90**:5034–5038.

452.* **Carlier, M. F. and D. Pantaloni.** 1986. Direct evidence for ADP-Pi-F-actin as the major intermediate in ATP-actin polymerization. Rate of dissociation of Pi from actin filaments. *Biochemistry* **25**:7789–7792.

453.* **Carlier, M. F. and D. Pantaloni.** 1988. Binding of phosphate to F-ADP-actin and role of F-ADP-Pi-actin in ATP-actin polymerization. *J. Biol. Chem.* **263**:817–825.

454. **Carlier, M. F., D. Pantaloni, J. A. Evans, P. K. Lambooy, E. D. Korn, and M. R. Webb.** 1988. The hydrolysis of ATP that accompanies actin polymerization is essentially irreversible. *FEBS Lett.* **235**:211–214.

455.* **Carlier, M. F., D. Pantaloni, and E. D. Korn.** 1984. Evidence for an ATP cap at the ends of actin filaments and its regulation of the F-actin steady state. *J. Biol. Chem.* **259**:9983–9986.

456. **Carlier, M. F., D. Pantaloni, and E. D. Korn.** 1985. Polymerization of ADP-actin and ATP-actin under sonication and characteristics of the ATP-actin equilibrium polymer. *J. Biol. Chem.* **260**:6565–6571.

457. **Carlier, M. F., D. Pantaloni, and E. D. Korn.** 1987. The mechanisms of ATP hydrolysis accompanying the polymerization of Mg-actin and Ca-actin. *J. Biol. Chem.* **262**:3052–3059.

458. **Cheesman, D. F. and A. Priston.** 1972. Exchange of actin-bound nucleotide in brief electrical stimulation of muscle. *Biochem. Biophys. Res. Commun.* **48**:552–558.

459. **Cohen, L. B.** 1966. Viscosity of G-ADP and G-ATP actin. *Arch. Biochem. Biophys.* **117**:289–295.

460.* **Combeau, C. and M. F. Carlier.** 1988. Probing the mechanism of ATP hydrolysis on F-actin using vanadate and the structural analogs of phosphate BeF_3 and AlF_4. *J. Biol. Chem.* **263**:17429–17436.

461.* **Combeau, C. and M. F. Carlier.** 1989. Characterization of the aluminum and beryllium fluoride species bound to F-actin and microtubules at the site of the γ-phosphate of the nucleotide. *J. Biol. Chem.* **264**:19017–19021.

462. **Cooke, R.** 1975. The bound nucleotide of actin. *J. Supramol. Struct.* **3**:146–153.

463.* **Cooke, R. and L. Murdoch.** 1973. Interaction of actin with analogs of adenosine triphosphate. *Biochemistry* **12**:3927–3932.

464. **Coue, M. and E. D. Korn.** 1986. Interaction of plasma gelsolin with ADP-actin. *J. Biol. Chem.* **261**:3628–3631.

465.* **Coue, M. and E. D. Korn.** 1986. ATP hydrolysis by the gelsolin-actin complex and at the pointed ends of gelsolin-capped filaments. *J. Biol. Chem.* **261**:1588–1593.

466. **Dancker, P. and S. Fischer.** 1985. The influence of adenosine triphosphate, adenosine diphosphate and cytochalasin B on nucleotide exchange of F-actin. Evidence that treadmilling is not involved. *Biochim. Biophys. Acta* **838**:6–11.

467. **Dancker, P. and S. Fischer.** 1989. Stabilization of actin filaments by ATP and inorganic phosphate. *Z. Naturforsch. C.* **44**:698–704.

468. **Dancker, P. and L. Hess.** 1990. Phalloidin reduces the release of inorganic phosphate during actin polymerization. *Biochim. Biophys. Acta* **1035**:197–200.

469. **Dancker, P. and A. Kliche.** 1981. Cytochalasin B-induced ATPase activity of actin: dependence on monomer concentration. *Z. Naturforsch. C.* **36**:1050–1055.

470. **Daniel, J. L., L. Robkin, I. R. Molish, and H. Holmsen.** 1979. Determination of the ADP concentration available to participate in energy metabolism in an actin-rich cell, the platelet. *J. Biol. Chem.* **254**:7870–7873.

471. **Engel, J., H. Fasold, F. W. Hulla, F. Waechter, and A. Wegner.** 1977. The polymerization reaction of muscle actin. *Mol. Cell Biochem.* **18**:3–13.

472. **Estes, J. E. and C. Moos.** 1969. Effect of bound-nucleotide substitution on the properties of F-actin. *Arch. Biochem. Biophys.* **132**:388–396.

473. **Faulstich, H., I. Merkler, H. Blackholm, and C. Stournaras.** 1984. Nucleotide in monomeric actin regulates the reactivity of the thiol groups. *Biochemistry* **23**:1608–1612.

474. **Faust, U., H. Fasold, and F. Ortanderl.** 1974. Synthesis of a protein-reactive ATP analog and its application for the affinity labeling of rabbit-muscle actin. *Eur. J. Biochem.* **43**:273–279.

475.* **Frieden, C. and K. Patane.** 1988. Mechanism for nucleotide exchange in monomeric actin. *Biochemistry* **27**:3812–3820.

476. **Gaertner, A., G. W. Mayr, and A. Wegner.** 1991. Binding of sugar phosphates, inositol phosphates and phosphorylated amino acids to actin. *Eur. J. Biochem.* **198**:67–71.

477.* **Geipel, U., I. Just, B. Schering, D. Haas, and K. Aktories.** 1989. ADP-ribosylation of actin causes increase in the rate of ATP exchange and inhibition of ATP hydrolysis. *Eur. J. Biochem.* **179**:229–232.

478. **Geisbuhler, T., R. A. Altschuld, R. W. Trewyn, A. Z. Ansel, K. Lamka, and G. P. Brierley.** 1984. Adenine nucleotide metabolism and compartmentalisation in isolated adult rat heart cells. *Circ. Res.* **54**:536–546.

479.* **Gershman, L. C., L. A. Selden, and J. E. Estes.** 1991. Cation binding to the high-affinity site regulates nucleotide binding to actin. *Biophys. J.* **59**:53.

480. **Gillies, R. J., T. Ogino, R. G. Shulman, and D. C. Ward.** 1982. 31P nuclear magnetic resonance evidence for the regulation of intracellular pH by Ehrlich ascites tumour cells. *J. Cell Biol.* **95**:24–28.

481.* **Goldschmidt Clermont, P. J., M. I. Furman, D. Wachsstock, D. Safer, V. T. Nachmias, and T. D. Pollard.** 1992. The control of actin nucleotide exchange by thymosin β 4 and profilin. A potential regulatory mechanism for actin polymerization in cells. *Mol. Biol. Cell* **3**:1015–1024.

482.* **Goldschmidt Clermont, P. J., L. M. Machesky, S. K. Doberstein, and T. D. Pollard.** 1991. Mechanism of the interaction of human platelet profilin with actin. *J. Cell Biol.* **113**:1081–1089.

483. **Grazi, E., G. Trombetta, and E. Magri.** 1984. A mechanism for the selective preservation of homogeneous. F(ATP) actin. *Biochem. Int.* **9**:669–674.

484. **Grubhofer, N. and H. H. Weber.** 1961. Actin-Nucleotide interactions: the function and binding of nucleotides to G- and F-actin. *Z. Naturforsch. C.* **16**:435–444.

485. **Hayden, S. M., P. S. Miller, A. Brauweiler, and J. R. Bamburg.** 1993. Analysis of the interactions of actin depolymerizing factor with G- and F-actin. *Biochemistry* **32**:9994–10004.

486. **Hegyi, G., L. Szilagyi, and J. Belagyi.** 1988. Influence of the bound nucleotide on the molecular dynamics of actin. *Eur. J. Biochem.* **175**:271–274.

487. **Hegyi, G., L. Szilagyi, and M. Elzinga.** 1986. Photoaffinity labeling of the nucleotide binding site of actin. *Biochemistry* **25**:5793–5798.

488. **Hendry, K. A. K., A. McGregor, D. Critchley, and P. Sheterline.** 1991. F-actin structure is modulated by inorganic phosphate. *J. Muscle Res. Cell Motil.* **12**:483.

489. **Higashi, S. and F. Oosawa.** 1965. Conformational changes associated with polymerization and nucleotide binding in actin molecules. *J. Mol. Biol.* **12**:843–865.

490. **Hill, T. L.** 1986. Theoretical study of a model for the ATP cap at the end of an actin filament. *Biophys. J.* **49**:981–986.

491. **Hitchcock, S. E.** 1980. Actin deoxyribonuclease I interaction. Depolymerization and nucleotide exchange. *J. Biol. Chem.* **255**:5668–5673.

492.* **Holmes, K. C., C. Sander, and A. Valencia.** 1993. A new ATP-binding fold in actin, hexokinase and Hsc70. *Trends Cell Biol.* **3**:53–59.

493. **Hozumi, T.** 1988. Structural aspects of skeletal muscle G-actin molecule as studied by proteolytic digestion: effect of nucleotide. *Biochem. Int.* **17**:171–178.

494. **Hozumi, T.** 1988. Structural aspects of skeletal muscle F-actin as studied by tryptic digestion: evidence for a second nucleotide interacting site. *J. Biochem. Tokyo* **104**:285–288.

495. **Hozumi, T.** 1990. A hydrophobicity on skeletal muscle actin molecule modulated by nucleotide binding at a second site. *Biochem. Int.* **20**:45–51.

496. **Huang, C. K., J. M. J. Hill, B. J. Bormann, W. M. Mackin, and E. L. Becker.** 1983. The photoaffinity probe 8-$N^3[\alpha$-$^{32}P]$ATP labels the ATP-binding sites of rabbit neutrophil and skeletal muscle actin. *FEBS Lett.* **159**:145–149.

497. **Ingwall, J. S. and R. A. Weiss.** 1993. 31P NMR Spectroscopy. The non-invasive tool for the study of the biochemistry of the cardiovascular system. *Trends Cardiovasc. Med.* **3**:29–35.

498. **Jacobson, G. R. and J. P. Rosenbusch.** 1976. ATP binding to a protease-resistant core of actin. *Proc. Natl Acad. Sci. U. S. A.* **73**:2742–2746.

499. **Janmey, P. A., S. Hvidt, G. F. Oster, J. Lamb, T. P. Stossel, and J. H. Hartwig.** 1990. Effect of ATP on actin filament stiffness. *Nature* **347**:95–99.

500. **Kabsch, W., H. G. Mannherz, D. Suck, E. F. Pai, and K. C. Holmes.** 1990. Atomic structure of the actin:DNase I complex. *Nature* **347**:37–44.

501. **Kasai, M. and F. Oosawa.** 1969. Behavior of divalent cations and nucleotides bound to F-actin. *Biochim. Biophys. Acta* **172**:300–310.

502. **Kasprzak, A. A.** 1993. Myosin subfragment 1 inhibits dissociation of nucleotide and calcium from G-actin. *J. Biol. Chem.* **268**:13261–13266.

503. **Kerbey, A. L., M. McPherson, P. Sheterline, and J. G. Schofield.** 1974. Cofactor requirements for the stimulation of growth hormone release *in vitro. Symp. Soc. Exp. Biol.* **28**:375–397.

504. **Kiessling, P., B. Polzar, and H. G. Mannherz.** 1993. Evidence that the presumptive second nucleotide interacting site on actin is of low specificity and affinity. *Biol. Chem. Hoppe Seylers* **374**:183–192.

505.* **Kinosian, H. J., L. A. Selden, J. E. Estes, and L. C. Gershman.** 1993. Nucleotide binding to actin. Cation dependence of nucleotide dissociation and exchange rates. *J. Biol. Chem.* **268**:8683–8691.

506. **Kitagawa, S., W. Drabikowski, and J. Gergely.** 1968. Exchange and release of the bound nucleotide of F-actin. *Arch. Biochem. Biophys.* **125**:706–714.

507. **Kopp, S. J., J. T. Barron, and J. P. Tow.** 1990. Phosphate metabolites, intracellular pH and free $[Mg^{++}]$ in single, intact carotid artery segments studied by 31P-NMR. *Biochim. Biophys. Acta* **1055**:27–35.

508.* **Korn, E. D., M. F. Carlier, and D. Pantaloni.** 1987. Actin polymerization and ATP hydrolysis. *Science* **238**:638–644.

509. **Kuehl, W. M. and J. Gergely.** 1969. The kinetics of exchange of adenosine triphosphate and calcium with G-actin. *J. Biol. Chem.* **244**:4720–4729.

510. **Kuwayama, H., M. Miki, and C. G. dos Remedios.** 1989. The effect of the replacement of ADP with a photoaffinity ATP analogue, 2-azido-ADP, in F-actin on its function. *FEBS Lett.* **250**:328–330.

511.* **Laham, L. E., J. A. Lamb, P. G. Allen, and P. A. Janmey.** 1993. Selective binding of gelsolin to actin monomers containing ADP. *J. Biol. Chem.* **268**:14202–14207.

512. **Laki, K. and R. E. Alving.** 1973. Sulfhydryl groups and the ATPase activity of actin. *J. Mechanochem. Cell Motil.* **2**:45–49.

513. **Laki, K. and A. M. Clark.** 1951. On the nucleotide content of actin preparations. *J. Biol. Chem.* **191**:599–606.

514. **Mannherz, H. G., H. Brehme, and U. Lamp.** 1975. Depolymerisation of F-actin to G-actin and its repolymerisation in the presence of analogs of adenosine triphosphate. *Eur. J. Biochem.* **60**:109–116.

515. **Mannherz, H. G., R. S. Goody, M. Konrad, and E. Nowak.** 1980. The interaction of bovine pancreatic deoxyribonuclease I and skeletal muscle actin. *Eur. J. Biochem.* **104**:367–379.
516. **Martonosi, A. and M. A. Gouvea.** 1961. Studies on actin IV: the interaction of nucleoside triphosphates with actin. *J. Biol. Chem.* **236**:1345–1352.
517. **Maruyama, K. and A. Martonosi.** 1961. Protective action of nucleoside triphosphates against the inactivation of G-actin by EDTA. *Biochem. Biophys. Res. Commun.* **5**:85–87.
518. **Miki, M.** 1990. Interaction of F-actin-AMPPNP with myosin subfragment-1 and troponin–tropomyosin: influence of an extra phosphate at the nucleotide binding site in F-actin on its function. *J. Biochem. Tokyo* **108**:457–461.
519. **Miki, M., J. A. Barden, and C. G. dos Remedios.** 1986. Fluorescence resonance energy transfer between the nucleotide binding site and Cys-10 in G-actin and F-actin. *Biochim. Biophys. Acta* **872**:76–82.
520. **Miki, M., B. D. Hambly, and C. G. dos Remedios.** 1986. Fluorescence energy transfer between nucleotide binding sites in an F-actin filament. *Biochim. Biophys. Acta* **871**:137–141.
521. **Miki, M., H. Onuma, and K. Mihashi.** 1974. Interaction of actin water ϵ-ATP. *FEBS Lett.* **46**:17–19.
522. **Mockrin, S. C. and E. D. Korn.** 1983. Kinetics of polymerization and ATP hydrolysis by covalently crosslinked actin dimer. *J. Biol. Chem.* **258**:3215–3221.
523.* **Moos, C., E. Eisenberg, and J. E. Estes.** 1967. Bound-nucleotide exchange in actin and actomyosin. *Biochim. Biophys. Acta* **147**:536–545.
524. **Moos, C., J. E. Estes, and E. Eisenberg.** 1966. Exchange of F-actin-bound nucleotide in the presence and absence of myosin. *Biochem. Biophys. Res. Commun.* **23**:347–351.
525.* **Muhlrad, A., P. Cheung, B. C. Phan, C. Miller, and E. Reisler.** 1994. Dynamic properties of actin: structural changes induced by beryllium fluoride. *J. Biol. Chem.* **269**:11852–11858.
526. **Murthy, V. K., L. E. Crevasse, and J. C. Shipp.** 1968. Actin polymerization and its relationship to adenosine triphosphatase activity. *Circ. Res.* **22**:65–73.
527. **Naber, N. and R. Cooke.** 1994. Mobility and orientation of spin probes attached to nucleotides incorporated into actin. *Biochemistry* **33**:3855–3861.
528. **Nakaoka, Y.** 1972. The adenosine triphosphate splitting of actin modified with salyrgan. *Biochim. Biophys. Acta* **283**:364–372.
529. **Nakaoka, Y. and M. Kasai.** 1969. Behaviour of sonicated actin polymers: adenosine triphosphate splitting and polymerization. *J. Mol. Biol.* **44**:319–332.
530. **Neidl, C. and J. Engel.** 1979. Exchange of ADP, ATP and **1**: N6-ethenoadenosine 5′-triphosphate at G-actin. Equilibrium and kinetics. *Eur. J. Biochem.* **101**:163–169.
531.* **Newman, J., K. S. Zaner, K. L. Schick, L. C. Gershman, L. A. Selden, H. J. Kinosian, J. L. Travis, and J. E. Estes.** 1993. Nucleotide exchange and rheometric studies with F-actin prepared from ATP- or ADP-monomeric actin. *Biophys. J.* **64**:1559–1566.
532.* **Nishida, E.** 1985. Opposite effects of cofilin and profilin from porcine brain on rate of exchange of actin-bound adenosine 5′-triphosphate. *Biochemistry* **24**:1160–1164.
533. **Nonomura, Y., E. Katayama, and S. Ebashi.** 1975. Effect of phosphates on the structure of the actin filament. *J. Biochem. Tokyo* **78**:1101–1104.
534. **Nowak, E. and R. S. Goody.** 1988. Kinetics of adenosine 5′-triphosphate and adenosine 5′-diphosphate interaction with G-actin. *Biochemistry* **27**:8613–8617.
535. **Nowak, E., H. Strzelecka Golaszewska, and R. S. Goody.** 1988. Kinetics of nucleotide and metal ion interaction with G-actin. *Biochemistry* **27**:1785–1792.
536. **Pantaloni, D., M. F. Carlier, M. Coue, A. A. Lal, S. L. Brenner, and E. D. Korn.** 1984. The critical concentration of actin in the presence of ATP increases with the number concentration of filaments and approaches the critical concentration of actin.ADP. *J. Biol. Chem.* **259**:6274–6283.
537.* **Pollard, T. D. and A. G. Weeds.** 1984. The rate constant for ATP hydrolysis by polymerized actin. *FEBS Lett.* **170**:94–98.
538.* **Rickard, J. E. and P. Sheterline.** 1986. Cytoplasmic concentrations of inorganic phosphate affect the critical concentration for assembly of actin in the presence of cytochalasin D or ADP. *J. Mol. Biol.* **191**:273–280.
539.* **Rickard, J. E. and P. Sheterline.** 1988. Effect of ATP removal and inorganic phosphate on length redistribution of sheared actin filament populations. Evidence for a mechanism of end-to-end annealing. *J. Mol. Biol.* **201**:675–681.
540. **Root, D. D. and E. Reisler.** 1992. The accessibility of etheno-nucleotides to collisional quenchers and the nucleotide cleft in G- and F-actin. *Protein. Sci.* **1**:1014–1022.
541. **Schliwa, M., R. M. Ezzell, and U. Euteneuer.** 1984. erythro-9-[3-(2-Hydroxynonyl)]adenine is an effective inhibitor of cell motility and actin assembly. *Proc. Natl Acad. Sci. U. S. A.* **81**:6044–6048.
542. **Selden, L. A., L. C. Gershman, H. J. Kinosian, and J. E. Estes.** 1990. Mg^{++} bound at the high-affinity site on actin is a cofactor for actin ATPase activity. *Biophys. J.* **57**:325.
543. **Siegmund, B., A. Koop, T. Klietz, P. Schwartz, and H. M. Piper.** 1990. Sarcolemmal integrity and metabolic competence of cardiomyocytes under anoxia-reoxygenation. *Am. J. Physiol.* **258**:285–291.
544. **Soboll, S. and R. Bunger.** 1981. Compartmentation of adenine nucleotides in the isolated working guinea pig heart stimulated by noradrenaline. *Hoppe Seylers Z. Physiol. Chem.* **362**:125–132.
545. **Spychala, J. and G. Van den Bergh.** 1987. Adenine nucleotide metabolism in isolated chicken hepatocytes. *Biochem. J.* **242**:551–558.
546. **Stournaras, C., I. Merkler, and H. Faulstich.** 1988. Thiol group reactivity and polymerization of actin in the presence of ATP analogs. *Biochem. Biophys. Res. Commun.* **155**:962–970.
547. **Straub, F. B.** 1989. Note on the work of F. Bruno Straub concerning 'Adenosine Triphosphate. The Functional Group of Actin'. *Biochim. Biophys. Acta* **1000**:179.
548. **Straub, F. B. and G. Feuer.** 1950. Adenosinetriphosphate functional group of actin. *Biochim. Biophys. Acta* **4**:455–470.
549. **Straub, F. B. and G. Feuer.** 1989. Adenosinetriphosphate. The functional group of actin. 1950. *Biochim. Biophys. Acta* **1000**:180–195.
550. **Strohman, R. C. and A. J. Samorodin.** 1962. The requirements for adenosine triphosphate binding to globular actin. *J. Biol. Chem.* **237**:363–369.
551.* **Strzelecka Golaszewska, H.** 1973. Effect of tightly bound divalent cation on the equilibria between G-actin-bound and free ATP. *Eur. J. Biochem.* **37**:434–440.
552. **Strzelecka Golaszewska, H. and W. Drabikowski.** 1967. Correlation between the binding of calcium and ATP by G-actin. *Acta Biochim. Pol.* **14**:195–208.
553. **Szent Gyorgyi, A. G. and G. Prior.** 1966. Exchange of adenosine diphosphate bound to actin in superprecipitated actomyosin and contracted myofibrils. *J. Mol. Biol.* **15**:515–538.
554. **Tellam, R. L.** 1986. Gelsolin inhibits nucleotide exchange from actin. *Biochemistry* **25**:5799–5804.
555. **Thames, K. E., H. C. Cheung, and S. C. Harvey.** 1974. Binding of 1,N6-ethanoadenosine triphosphate to actin. *Biochem. Biophys. Res. Commun.* **60**:1252–1261.
556. **Tokiwa, T., T. Shimada, and Y. Tonomura.** 1967. Role of ADP of F-actin in superprecipitation and enzymatic activity of actomyosin. *J. Biochem. Tokyo.* **61**:108–122.
557. **Tonomura, Y., T. Tokiwa, and T. Shimada.** 1966. Role of ADP bound to F-actin in superprecipitation and enzymatic activity of actomyosin. *J. Biochem. Tokyo* **59**:322–324.
558. **Tonomura, Y. and J. Yoshimura.** 1961. Removal of bound nucleotide and calcium of G-actin by treatment with EDTA. *J. Biochem. Tokyo* **50**:79–80.
559. **Totsuka, T. and S. Hatano.** 1970. ATPase activity of plasmodium actin polymer formed in the presence of Mg^{2+}. *Biochim. Biophys. Acta* **223**:189–197.
560. **Tsuboi, K. K.** 1968. Actin and bound-nucleotide stoichiometry. *Biochim. Biophys. Acta* **160**:420–434.
561. **Valentin-Ranc, C. and M. F. Carlier.** 1989. Evidence for the direct interaction between tightly bound divalent metal ion and ATP on actin. Binding of the lambda isomers of $\beta\gamma$-bidentate CrATP to actin. *J. Biol. Chem.* **264**:20871–20880.
562. **Verhoeven, A. J., C. A. Cook, and H. Holmsen.** 1988. Use of actin-bound adenosine 5′-diphosphate as a method to determine the specific ^{32}P-radioactivity of the γ-phosphoryl group of adenosine 5′-triphosphate in a highly compartmentalized cell, the platelet. *Anal. Biochem.* **174**:672–678.
563. **Waechter, F.** 1975. The influence of Ca^{2+} on the dissociation of 1,N6-ethenoadenosine 5′-triphosphate from actin. *Hoppe Seylers Z. Physiol. Chem.* **356**:1821–1822.
564. **Waechter, F. and J. Engel.** 1975. The kinetics of the exchange of G-actin-bound **1**: N6-ethenoadenosine 5′-triphosphate with ATP as followed by fluorescence. *Eur. J. Biochem.* **57**:453–459.
565.* **Waechter, F. and J. Engel.** 1977. Association kinetics and binding constants of nucleoside triphosphates with G-actin. *Eur. J. Biochem.* **74**:227–232.
566.* **Walsh, T. P. and R. L. Tellam.** 1986. Erythro-9-[3-(2-hydroxynonyl)]adenine accelerates actin polymerization and nucleotide exchange. *Biochem. Biophys. Res. Commun.* **137**:1181–1186.
567. **Wanger, M. and A. Wegner.** 1983. Similar affinities of ADP and ATP for G-actin at physiological salt concentrations. *FEBS Lett.* **162**:112–116.
568. **Wanger, M. and A. Wegner.** 1987. Binding of phosphate ions to actin. *Biochim. Biophys. Acta* **914**:105–113.
569. **Ward, L. C.** 1979. The turnover of F-actin-bound ADP *in vivo*. *Experientia.* **35**:1145–1146.
570. **West, J. J.** 1970. Binding of adenosine diphosphate to G-actin. *Biochemistry* **9**:1239–1246.
571. **West, J. J.** 1971. Binding of nucleotide to cation-free actin. *Biochemistry* **10**:3547–3553.
572. **West, J. J., B. Nagy, and J. Gergely.** 1967. Free adenosine diphosphate as an intermediary in the phosphorylation by creatine phosphate of adenosine diphosphate bound to actin. *J. Biol. Chem.* **242**:1140–1145.

BINDING OF PROTEINS AND OTHER LIGANDS TO ACTIN

573. **Adams, M. E., L. S. Minamide, G. Duester, and J. R. Bamburg.** 1990. Nucleotide sequence and expression of a cDNA encoding chick brain actin depolymerizing factor. *Biochemistry* **29**:7414–7420.
574.* **Adams, S., G. DasGupta, J. M. Chalovich, and E. Reisler.** 1990. Immunochemical evidence for the binding of caldesmon to the NH_2-terminal segment of actin. *J. Biol. Chem.* **265**:19652–19657.
575.* **Adams, S. and E. Reisler.** 1993. Role of sequence 18–29 on actin in actomyosin interactions. *Biochemistry* **32**:5051–5056.
576. **Adamski, J., B. Husen, H. H. Thole, U. Groeschel Stewart, and P. W. Jungblut.** 1993. Linkage of 17 β-oestradiol dehydrogenase to actin by epsilon-(γ-glutamyl)-lysine in porcine endometrial cells. *Biochem. J.* **296**:797–802.

577.* **Aderem, A.** 1992. Signal transduction and the actin cytoskeleton: the roles of MARCKS and profilin. *Trends Biochem. Sci.* **17**:438–443.

578. **Amatruda, J. F., D. J. Gattermeir, T. S. Karpova, and J. A. Cooper.** 1992. Effects of null mutations and overexpression of capping protein on morphogenesis, actin distribution and polarized secretion in yeast. *J. Cell Biol.* **119**:1151–1162.

579. **Amos, L. A., H. E. Huxley, K. C. Holmes, R. S. Goody, and K. A. Taylor.** 1982. Structural evidence that myosin heads may interact with two sites on F-actin. *Nature* **299**:467–469.

580. **Andreev, O. A., A. L. Andreeva, V. S. Markin, and J. Borejdo.** 1993. Two different rigor complexes of myosin subfragment 1 and actin. *Biochemistry* **32**:12046–12053.

581.* **Andreev, O. A. and J. Borejdo.** 1991. The myosin head can bind two actin monomers. *Biochem. Biophys. Res. Commun.* **177**:350–356.

582. **Andreeva, A. L., O. A. Andreev, and J. Borejdo.** 1993. Structure of the 265-kilodalton complex formed upon EDC cross-linking of subfragment 1 to F-actin. *Biochemistry* **32**:13956–13960.

583. **Arata, T.** 1986. Structure of the actin–myosin complex produced by crosslinking in the presence of ATP. *J. Mol. Biol.* **191**:107–116.

584. **Arata, T.** 1991. Interaction of nonpolymerizable actins with myosin. *J. Biochem. Tokyo* **109**:335–340.

585.* **Aspenstrom, P. and R. Karlsson.** 1991. Interference with myosin subfragment-1 binding by site-directed mutagenesis of actin. *Eur. J. Biochem.* **200**:35–41.

586.* **Aspenstrom, P., U. Lindberg, and R. Karlsson.** 1992. Site-specific amino-terminal mutants of yeast-expressed β-actin. Characterization of the interaction with myosin and tropomyosin. *FEBS Lett.* **303**:59–63.

587. **Bader, M. F., J. M. Trifaro, O. K. Langley, D. Thierse, and D. Aunis.** 1986. Secretory cell actin-binding proteins: identification of a gelsolin-like protein in chromaffin cells. *J. Cell Biol.* **102**:636–646.

588. **Bahler, M. and P. Greengard.** 1987. Synapsin I bundles F-actin in a phosphorylation-dependent manner. *Nature* **326**:704–707.

589. **Barden, J. A., M. Miki, B. D. Hambly, and C. G. dos Remedios.** 1987. Localization of the phalloidin and nucleotide-binding sites on actin. *Eur. J. Biochem.* **162**:583–588.

590. **Barden, J. A. and L. Phillips.** 1990. 19F NMR study of the myosin and tropomyosin binding sites on actin. *Biochemistry* **29**:1348–1354.

591. **Bartegi, A., A. Fattoum, and R. Kassab.** 1990. Cross-linking of smooth muscle caldesmon to the NH_2-terminal region of skeletal F-actin. *J. Biol. Chem.* **265**:2231–2237.

592. **Bearer, E. L.** 1992. An actin-associated protein present in the microtubule organizing center and the growth cones of PC-12 cells. *J. Neurosci.* **12**:750–761.

593. **Bennardini, F., A. Wrzosek, and M. Chiesi.** 1992. B-crystallin in cardiac tissue. Association with actin and desmin filaments. *Circ. Res.* **71**:288–294.

594.* **Bennett, V., K. Gardner, and J. P. Steiner.** 1988. Brain adducin: a protein kinase C substrate that may mediate site-directed assembly at the spectrin-actin junction. *J. Biol. Chem.* **263**:5860–5869.

595. **Bertrand, R., P. Chaussepied, R. Kassab, M. Boyer, C. Roustan, and Y. Benyamin.** 1988. Cross-linking of the skeletal myosin subfragment 1 heavy chain to the N-terminal actin segment of residues 40–113. *Biochemistry* **27**:5728–5736.

596. **Bettache, N., R. Bertrand, and R. Kassab.** 1992. Specific cross-linking of the SH1 thiol of skeletal myosin subfragment 1 to F-actin and G-actin. *Biochemistry* **31**:389–395.

597.* **Blanchard, A., V. Ohanian, and D. Critchley.** 1989. The structure and function of α-actinin. *J. Muscle Res. Cell Motil.* **10**:280–289.

598. **Bonafe, N., P. Chaussepied, J. P. Capony, J. Derancourt, and R. Kassab.** 1993. Photochemical cross-linking of the skeletal myosin head heavy chain to actin subdomain-1 at Arg95 and Arg28. *Eur. J. Biochem.* **213**:1243–1254.

599. **Bonafe, N., M. Mathieu, R. Kassab, and P. Chaussepied.** 1994. Tropomyosin inhibits the glutaraldehyde-induced cross-link between the central 48-kDa fragment of myosin head and segment 48–67 in actin subdomain 2. *Biochemistry* **33**:2594–2603.

600. **Boyer, M., J. Feinberg, H. K. Hue, J. P. Capony, Y. Benyamin, and C. Roustan.** 1987. Antigenic probes locate a serum-gelsolin-interaction site on the C-terminal part of actin. *Biochem. J.* **248**:359–364.

601. **Bryan, J., R. Edwards, P. Matsudaira, J. Otto, and J. Wulfkuhle.** 1993. Fascin, an echinoid actin-bundling protein, is a homolog of the *Drosophila singed gene product. Proc. Natl Acad. Sci. U. S. A.* **90**:9115–9119.

602. **Buss, M. R., M. S. Lewis, and E. D. Korn.** 1991. The interaction of monomeric actin with two binding sites on *Acanthamoeba actobindin. J. Biol. Chem.* **266**:3820–3826.

603. **Burtnick, L. D. and K. W. Chan.** 1983. Fluorescence of actin-bound hydrophobic molecules. *Can. J. Biochem. Cell Biol.* **61**:981–988.

604. **Cartoux, L., T. Chen, G. DasGupta, P. B. Chase, M. J. Kushmerick, and E. Reisler.** 1992. Antibody and peptide probes of interactions between the SH1-SH2 region of myosin subfragment 1 and actin's N-terminus. *Biochemistry* **31**:10929–10935.

605.* **Casella, J. F. and M. A. Torres.** 1994. Interaction of Cap Z with actin. The NH_2-terminal domains of the $\alpha1$ and β subunits are not required for actin capping, and $\alpha1\beta$ and $\alpha2\beta$ heterodimers bind differentially to actin. *J. Biol. Chem.* **269**:6992–6998.

606.* **Ceccaldi, P. E., F. Benfenati, E. Chieregatti, P. Greengard, and F. Valtorta.** 1993. Rapid binding of synapsin I to F- and G-actin. A study using fluorescence resonance energy transfer. *FEBS Lett.* **329**:301–305.

607.* **Chen, M. J., C. L. Shih, and K. Wang.** 1993. Nebulin as an actin zipper. A two-module nebulin fragment promotes actin nucleation and stabilizes actin filaments. *J. Biol. Chem.* **268**:20327–20334.

608. **Chen, T., M. J. Haigentz, and E. Reisler.** 1992. Myosin subfragment 1 and structural elements of G-actin: effects of S-1(A2) on sequences 39–52 and 61–69 in subdomain 2 of G-actin. *Biochemistry* **31**:2941–2946.

609. **Chen, Y. D. and J. M. Chalovich.** 1992. A mosaic multiple-binding model for the binding of caldesmon and myosin subfragment-1 to actin. *Biophys. J.* **63**:1063–1070.

610. **Cheney, R. E., R. A. Riley, and M. S. Mooseker.** 1993. Phylogenetic analysis of the myosin superfamily. *Cell Motil. Cytoskeleton.* **24**:215–223.

611.* **Chia, C. P., A. Shariff, S. A. Savage, and E. J. Luna.** 1993. The integral membrane protein, ponticulin, acts as a monomer in nucleating actin assembly. *J. Cell Biol.* **120**:909–922.

612. **Combeau, C., D. Didry, and M. F. Carlier.** 1992. Interaction between G-actin and myosin subfragment-1 probed by covalent cross-linking. *J. Biol. Chem.* **267**:14038–14046.

613. **Condeelis, J., M. Vahey, J. M. Carboni, J. DeMey, and S. Ogihara.** 1984. Properties of the 120,000- and 95,000-dalton actin-binding proteins from *Dictyostelium discoideum* and their possible functions in assembling the cytoplasmic matrix. *J. Cell Biol.* **99**:119s-126s.

614. **Cooke, N. E.** 1986. Rat vitamin D-binding protein. Determination of the full length primary structure from cloned cDNA. *J. Biol. Chem.* **261**:3441–3450.

615. **Correas, I., T. L. Leto, D. W. Speicher, and V. T. Marchesi.** 1986. Identification of the functional site of erythrocyte protein 4.1 involved in spectrin-actin associations. *J. Biol. Chem.* **261**:3310–3315.

616. **Correas, I., R. Padilla, and J. Avila.** 1990. The tubulin-binding sequence of brain microtubule-associated proteins, τ and MAP-2, is also involved in actin binding. *Biochem. J.* **269**:61–64.

617. **Correas, I., D. W. Speicher, and V. T. Marchesi.** 1986. Structure of the spectrin-actin binding site of erythrocyte protein 4.1. *J. Biol. Chem.* **261**:13362–13366.

618. **Criddle, A. H., M. A. Geeves, and T. Jeffries.** 1985. The use of actin labelled with N-(1-pyrenyl)iodoacetamide to study the interaction of actin with myosin subfragments and troponin/tropomyosin. *Biochem. J.* **232**:343–349.

619. **Crosbie, R., S. Adams, J. M. Chalovich, and E. Reisler.** 1991. The interaction of caldesmon with the COOH terminus of actin. *J. Biol. Chem.* **266**:20001–20006.

620. **Crosbie, R. H., J. M. Chalovich, and E. Reisler.** 1992. Interaction of caldesmon and myosin subfragment 1 with the C-terminus of actin. *Biochem. Biophys. Res. Commun.* **184**:239–245.

621. **DalleDonne, I., A. Milzani, P. Contini, G. Bernardini, and R. Colombo.** 1992. Interaction of cardiac α-actinin and actin in the presence of doxorubicin. *Exp. Mol. Pathol.* **56**:229–238.

622. **Daoud, E. W., S. M. Hayden, and J. R. Bamburg.** 1988. Inhibition of deoxyribonuclease I activity by actin covalently cross-linked to chick brain actin depolymerizing factor through exposed sulfhydryls. *Biochem. Biophys. Res. Commun.* **155**:890–894.

623. **DasGupta, G. and E. Reisler.** 1989. Antibody against the amino terminus of α-actin inhibits actomyosin interactions in the presence of ATP. *J. Mol. Biol.* **207**:833–836.

624. **DasGupta, G. and E. Reisler.** 1991. Nucleotide-induced changes in the interaction of myosin subfragment 1 with actin: detection by antibodies against the N-terminal segment of actin. *Biochemistry* **30**:9961–9966.

625. **DasGupta, G., J. White, P. Cheung, and E. Reisler.** 1990. Interactions between G-actin and myosin subfragment **1**: immunochemical probing of the NH_2-terminal segment on actin. *Biochemistry* **29**:8503–8508.

626. **de Arruda, M. V., H. Bazari, M. Wallek, and P. Matsudaira.** 1992. An actin footprint on villin. Single site substitutions in a cluster of basic residues inhibit the actin severing but not capping activity of villin. *J. Biol. Chem.* **267**:13079–13085.

627. **de Hostos, E. L., B. Bradtke, F. Lottspeich, and G. Gerisch.** 1993. Coactosin, a 17 kDa F-actin binding protein from *Dictyostelium discoideum. Cell Motil. Cytoskeleton* **26**:181–191.

628. **de Hostos, E. L., B. Bradtke, F. Lottspeich, R. Guggenheim, and G. Gerisch.** 1991. Coronin, an actin binding protein of *Dictyostelium discoideum* localized to cell surface projections, has sequence similarities to G protein β subunits. *EMBO J.* **10**:4097–4104.

629. **De, B. P., A. L. Burdsall, and A. K. Banerjee.** 1993. Role of cellular actin in human parainfluenza virus type 3 genome transcription. *J. Biol. Chem.* **268**:5703–5710.

630. **den Hartigh, J. C., P. M. van Bergen en Henegouwen, A. J. Verkleij, and J. Boonstra.** 1992. The EGF receptor is an actin-binding protein. *J. Cell Biol.* **119**:349–355.

631. **DeRosier, D. J. and K. T. Edds.** 1980. Evidence for fascin cross-links between the actin filaments in coelomocyte filopodia. *Exp. Cell Res.* **126**:490–494.

632. **DeRosier, D. J., L. G. Tilney, E. M. Bonder, and P. Frankl.** 1982. A change in twist of actin provides the force for the extension of the acrosomal process in *Limulus* sperm: the false-discharge reaction. *J. Cell Biol.* **93**:324–337.

633. **Dharmawardhane, S., M. Demma, F. Yang, and J. Condeelis.** 1991. Compartmentalization and actin binding properties of ABP-**50**: the elongation factor-1α of *Dictyostelium. Cell Motil. Cytoskeleton* **20**:279–288.

634. **Doi, Y., M. Higashida, and S. Kido.** 1987. Plasma-gelsolin-binding sites on the actin sequence. *Eur. J. Biochem.* **164**:89–94.

635. **Doi, Y., Y. Kanatani, and F. Kim.** 1992. The amino-terminal fragment of gelsolin is cross-linked to Cys-374 of actin in the EGTA-resistant actin–gelsolin complex. *FEBS Lett.* **301**:99–102.

636. **Doi, Y. K., M. Banba, and A. Vertut Doi.** 1991. Cysteine-374 of actin resides at the gelsolin contact site in the EGTA-resistant actin–gelsolin complex. *Biochemistry* **30**:5769–5777.
637. **Drubin, D. G., H. D. Jones, and K. F. Wertman.** 1993. Actin structure and function – roles in mitochondrial organisation and morphogenesis in budding yeast and identification of the phalloidin-binding site. *Mol. Biol. Cell* **4**:1277–1294.
638. **Drubin, D. G., J. Mulholland, Z. M. Zhu, and D. Botstein.** 1990. Homology of a yeast actin-binding protein to signal transduction proteins and myosin-I. *Nature* **343**:288–290.
639. **Drummond, D. R., E. S. Hennessey, and J. C. Sparrow.** 1992. The binding of mutant actins to profilin, ATP and DNase I. *Eur. J. Biochem.* **209**:171–179.
640. **Duke, J., R. Takashi, K. Ue, and M. F. Morales.** 1976. Reciprocal reactivities of specific thiols when actin binds to myosin. *Proc. Natl Acad. Sci. U. S. A.* **73**:302–306.
641. **Eddy, R. J., R. A. Sauterer, and J. S. Condeelis.** 1993. Aginactin, an agonist-regulated F-actin capping activity is associated with an Hsc70 in *Dictyostelium*. *J. Biol. Chem.* **268**:23267–23274.
642. **Edmonds, B. T.** 1993. ABP50: an actin-binding elongation factor 1 α from *Dictyostelium discoideum*. *J. Cell Biochem.* **52**:134–139.
643. **Eichinger, L., A. A. Noegel, and M. Schleicher.** 1991. Domain structure in actin-binding proteins: expression and functional characterization of truncated severin. *J. Cell Biol.* **112**:665–676.
644. **el Saleh, S. C., R. Thieret, P. Johnson, and J. D. Potter.** 1984. Modification of Lys-237 on actin by 2,4-pentanedione. Alteration of the interaction of actin with tropomyosin. *J. Biol. Chem.* **259**:11014–11021.
645. **Estes, J. E., L. A. Selden, H. J. Kinosian, and L. C. Gershman.** 1992. Tightly-bound divalent cation of actin. *J. Muscle Res. Cell Motil.* **13**:272–284.
646. **Eto, M., R. Suzuki, F. Morita, H. Kuwayama, N. Nishi, and S. Tokura.** 1990. Roles of the amino acid side chains in the actin-binding S-site of myosin heavy chain. *J. Biochem. Tokyo* **108**:499–504.
647. **Fabbrizio, E., A. Bonet Kerrache, J. J. Leger, and D. Mornet.** 1993. Actin-dystrophin interface. *Biochemistry* **32**:10457–10463.
648. **Fabbrizio, E., J. Leger, M. Anoal, J. J. Leger, and D. Mornet.** 1993. Monoclonal antibodies targeted against the C-terminal domain of dystrophin or utrophin. *FEBS Lett.* **322**:10–14.
649. **Fabbrizio, E., U. Nudel, G. Hugon, A. Robert, F. Pons, and D. Mornet.** 1994. Characterisation and localisation of a 77kDa protein related to the dystrophin gene family. *Biochem. J.* **299**:359–365.
650. **Fechheimer, M. and R. Furukawa.** 1993. A 27,000-D core of the *Dictyostelium* 34,000-D protein retains Ca^{2+}-regulated actin cross-linking but lacks bundling activity. *J. Cell Biol.* **120**:1169–1176.
651. **Feinberg, J., J. P. Capony, Y. Benyamin, and C. Roustan.** 1993. Definition of the EGTA-independent interface involved in the serum gelsolin–actin complex. *Biochem. J.* **293**:813–817.
652. **Field, V. L. and W. J. Bowen.** 1968. The C-terminal residue of actin and its role in reactions of actin and myosin. *Arch. Biochem. Biophys.* **127**:59–64.
653. **Franck, Z., R. Gary, and A. Bretscher.** 1993. Moesin, like ezrin, colocalizes with actin in the cortical cytoskeleton in cultured cells, but its expression is more variable. *J. Cell Sci.* **105**:219–231.
654. **Frappier, T., J. Derancourt, and L. A. Pradel.** 1992. Actin and neurofilament binding domain of brain spectrin β subunit. *Eur. J. Biochem.* **205**:85–91.
655. **Fujii, T.** 1991. Purification and characterization of an actin-, calmodulin- and tropomyosin-binding protein from chicken gizzard smooth muscle. *Chem. Pharm. Bull. Tokyo* **39**:2622–2626.
656. **Fujii, T., M. Watanabe, Y. Ogoma, Y. Kondo, and T. Arai.** 1993. Microtubule-associated proteins, MAP 1A and MAP 1B, interact with F-actin *in vitro*. *J. Biochem. Tokyo* **114**:827–829.
657. **Gadasi, H.** 1982. Isolated *Entamoeba histolytica* actin does not inhibit DNAse-I activity. *Biochem. Biophys. Res. Commun.* **104**:158–164.
658. **Gicquaud, C.** 1993. Actin conformation is drastically altered by direct interaction with membrane lipids: a differential scanning calorimetry study. *Biochemistry* **32**:11873–11877.
659. **Gieselmann, R. W. and K. Mann.** 1992. ASP-56, a new actin sequestering protein from pig platelets with homology to CAP, an adenylate cyclase-associated protein from yeast. *FEBS Lett.* **298**:149–153.
660. **Glenney, J. R. J., P. Glenney, and K. Weber.** 1982. F-actin-binding and cross-linking properties of porcine brain fodrin, a spectrin-related molecule. *J. Biol. Chem.* **257**:9781–9787.
661. **Glenney, J. R. J., B. Tack, and M. A. Powell.** 1987. Calpactins: two distinct Ca^{2+}-regulated phospholipid- and actin-binding proteins isolated from lung and placenta. *J. Cell Biol.* **104**:503–511.
662. **Goldmann, W. H. and G. Isenberg.** 1991. Kinetic determination of talin-actin binding. *Biochem. Biophys. Res. Commun.* **178**:718–723.
663. **Goldmann, W. H. and G. Isenberg.** 1993. Analysis of filamin and α-actinin binding to actin by the stopped flow method. *FEBS Lett.* **336**:408–410.
664. **Goldschmidt Clermont, P. J., R. M. Galbraith, D. L. Emerson, F. Marsot, A. E. Nel, and P. Arnaud.** 1985. Distinct sites on the G-actin molecule bind group-specific component and deoxyribonuclease I. *Biochem. J.* **228**:471–477.
665. **Goldschmidt Clermont, P. J. and P. A. Janmey.** 1991. Profilin, a weak CAP for actin and RAS. *Cell* **66**:419–421.
666. **Goldschmidt Clermont, P. J., L. M. Machesky, J. J. Baldassare, and T. D. Pollard.** 1990. The actin-binding protein profilin binds to PIP2 and inhibits its hydrolysis by phospholipase C. *Science* **247**:1575–1578.
667. **Gopalakrishnan, S. and L. Takemoto.** 1992. Binding of actin to lens α crystallins. *Curr. Eye Res.* **11**:929–933.
668. **Gorlin, J. B., R. Yamin, S. Egan, M. Stewart, T. P. Stossel, D. J. Kwiatkowski, and J. H. Hartwig.** 1990. Human endothelial actin-binding protein (ABP-280, nonmuscle filamin): a molecular leaf spring. *J. Cell Biol.* **111**:1089–1105.
669. **Grabarek, Z. and J. Gergely.** 1987. Location of the Tn-I binding site in the primary structure of actin. *Acta Biochim. Biophys. Acad. Sci. Hung.* **22**:307–316.
670. **Graceffa, P., L. P. Adam, and W. Lehman.** 1993. Disulphide cross-linking of smooth-muscle and non-muscle caldesmon to the C-terminus of actin in reconstituted and native thin filaments. *Biochem. J.* **294**:63–67.
671. **Graceffa, P. and A. Jancso.** 1991. Disulfide cross-linking of caldesmon to actin. *J. Biol. Chem.* **266**:20305–20310.
672. **Grazi, E., P. Cuneo, E. Magri, C. Schwienbacher, and G. Trombetta.** 1993. Diffusion hindrance and geometry of filament crossings account for the complex interactions of F-actin with α-actinin from chicken gizzard. *Biochemistry* **32**:8896–8901.
673. **Griffith, L. M. and T. D. Pollard.** 1978. Evidence for actin filament–microtubule interaction mediated by microtubule-associated proteins. *J. Cell Biol.* **78**:958–965.
674. **Griffith, L. M. and T. D. Pollard.** 1982. The interaction of actin filaments with microtubules and microtubule-associated proteins. *J. Biol. Chem.* **257**:9143–9151.
675. **Grimard, R., P. Tancrede, and C. Gicquaud.** 1993. Interaction of actin with positively charged phospholipids: a monolayer study. *Biochem. Biophys. Res. Commun.* **190**:1017–1022.
676. **Habazettl, J., D. Gondol, R. Wiltscheck, J. Otlewski, M. Schleicher, and T. A. Holak.** 1992. Structure of hisactophilin is similar to interleukin-1 β and fibroblast growth factor. *Nature* **359**:855–858.
677. **Haddad, J. G., Y. Z. Hu, M. A. Kowalski, C. Laramore, K. Ray, P. Robzyk, and N. E. Cooke.** 1992. Identification of the sterol- and actin-binding domains of plasma vitamin D binding protein (Gc-globulin). *Biochemistry* **31**:7174–7181.
678. **Haus, U., P. Trommler, P. R. Fisher, H. Hartmann, F. Lottspeich, A. A. Noegel, and M. Schleicher.** 1993. The heat shock cognate protein from *Dictyostelium* affects actin polymerization through interaction with the actin-binding protein cap32/34. *EMBO J.* **12**:3763–3771.
679. **Hawkins, M., B. Pope, S. K. Maciver, and A. G. Weeds.** 1993. Human actin depolymerizing factor mediates a pH-sensitive destruction of actin filaments. *Biochemistry* **32**:9985–9993.
680. **Hayakawa, K., T. Okagaki, S. Higashi Fujime, and K. Kohama.** 1994. Bundling of actin filaments by myosin light chain kinase from smooth muscle. *Biochem. Biophys. Res. Commun.* **199**:786–791.
681. **Heintz, D., A. Reichert, M. Mihelic, W. Voelter, and H. Faulstich.** 1993. Use of bimanyl actin derivative (TMB-actin) for studying complexation of β-thymosins. Inhibition of actin polymerization by thymosin β 9. *FEBS Lett.* **329**:9–12.
682. **Hemmings, L., P. A. Kuhlman, and D. R. Critchley.** 1992. Analysis of the actin-binding domain of α-actinin by mutagenesis and demonstration that dystrophin contains a functionally homologous domain. *J. Cell Biol.* **116**:1369–1380.
683. **Hens, J. J., F. Benfenati, H. B. Nielander, F. Valtorta, W. H. Gispen, and P. N. De Graan.** 1993. B-50/GAP-43 binds to actin filaments without affecting actin polymerization and filament organization. *J. Neurochem.* **61**:1530–1533.
684. **Hesterkamp, T., A. G. Weeds, and H. G. Mannherz.** 1993. The actin monomers in the ternary gelsolin: 2 actin complex are in an antiparallel orientation. *Eur. J. Biochem.* **218**:507–513.
685. **Hill, L. E., J. P. Mehegan, C. A. Butters, and L. S. Tobacman.** 1992. Analysis of troponin–tropomyosin binding to actin. Troponin does not promote interactions between tropomyosin molecules. *J. Biol. Chem.* **267**:16106–16113.
686. **Hirono, M., K. Sutoh, Y. Watanabe, and T. Ohno.** 1992. A chimeric actin carrying N-terminal portion of *Tetrahymena* actin does not bind to DNase I. *Biochem. Biophys. Res. Commun.* **184**:1511–1516.
687. **Hock, R. S. and J. S. Condeelis.** 1987. Isolation of a 240-kilodalton actin-binding protein from *Dictyostelium discoideum*. *J. Biol. Chem.* **262**:394–400.
688. **Hofmann, A., L. Eichinger, E. Andre, D. Rieger, and M. Schleicher.** 1992. Cap100, a novel phosphatidylinositol 4,5-bisphosphate-regulated protein that caps actin filaments but does not nucleate actin assembly. *Cell Motil. Cytoskeleton.* **23**:133–144.
689. **Hofmann, A., A. A. Noegel, L. Bomblies, F. Lottspeich, and M. Schleicher.** 1993. The 100 kDa F-actin capping protein of *Dictyostelium amoebae* is a villin prototype ('protovillin'). FEBS Lett. **328**:71–76.
690. **Holmes, K. C. and R. S. Goody.** 1984. The nature of the actin cross-bridge interaction. *Adv. Exp. Med. Biol.* **170**:373–384.
691. **Hori, K. and F. Morita.** 1992. Actin–actin contact: inhibition of actin-polymerization by subdomain 4 peptide fragments. *J. Biochem. Tokyo* **112**:401–408.
692. **Hosoya, H., R. Kobayashi, S. Tsukita, and F. Matsumura.** 1992. $Ca(^{2+})$-regulated actin and phospholipid binding protein (68 kD-protein) from bovine liver: identification as a homologue for annexin VI and intracellular localization. *Cell Motil. Cytoskeleton* **22**:200–210.
693. **Houmeida, A., V. Hanin, J. Constans, Y. Benyamin, and C. Roustan.** 1992. Localization of a vitamin-D-binding protein interaction site in the COOH-terminal sequence of actin. *Eur. J. Biochem.* **203**:499–503.

694. **Hu, G. F. and J. F. Riordan.** 1993. Angiogenin enhances actin acceleration of plasminogen activation. *Biochem. Biophys. Res. Commun.* **197**:682–687.
695. **Hu, G. F., D. J. Strydom, J. W. Fett, J. F. Riordan, and B. L. Vallee.** 1993. Actin is a binding protein for angiogenin. *Proc. Natl Acad. Sci. U. S. A.* **90**:1217–1221.
696. **Igarashi, M., T. Tashiro, and Y. Komiya.** 1992. Actin-binding proteins in the growth cone particles (GCP) from fetal rat brain: a 44 kDa actin-binding protein is enriched in the fetal GCP fraction. *Brain Res. Dev. Brain Res.* **67**:197–203.
697. **Iida, K., K. Moriyama, S. Matsumoto, H. Kawasaki, E. Nishida, and I. Yahara.** 1993. Isolation of a yeast essential gene, COF1, that encodes a homologue of mammalian cofilin, a low-M_r actin-binding and depolymerizing protein. *Gene* **124**:115–120.
698. **Ishikawa, R., T. Okagaki, and K. Kohama.** 1992. Regulation by Ca^{2+}-calmodulin of the actin-bundling activity of Physarum 210-kDa protein. *J. Muscle Res. Cell Motil.* **13**:321–328.
699. **Jancso, A., L. Szilagyi, and R. C. Lu.** 1986. Changes of lysine reactivities of actin in complex with DNAase I. *Biochim. Biophys. Acta* **873**:331–334.
700. **Jin, J. P. and K. Wang.** 1991. Nebulin as a giant actin-binding template protein in skeletal muscle sarcomere. Interaction of actin and cloned human nebulin fragments. *FEBS Lett.* **281**:93–96.
701. **Jones, P. G., G. J. Moore, and D. M. Waisman.** 1992. A nonapeptide to the putative F-actin binding site of annexin-II tetramer inhibits its calcium-dependent activation of actin filament bundling. *J. Biol. Chem.* **267**: 13993–13997.
702. **Jongstra-Bilen, J., P. A. Janmey, J. H. Hartwig, S. Galea, and J. Jongstra.** 1992. The lymphocyte-specific protein LSP1 binds to F-actin and to the cytoskeleton through its COOH-terminal basic domain. *J. Cell Biol.* **118**:1443–1453.
703. **Joubert, R., M. Caron, V. Avellana Adalid, D. Mornet, and D. Bladier.** 1992. Human brain lectin: a soluble lectin that binds actin. *J. Neurochem.* **58**:200–203.
704. **Jung, G. and J. A. Hammer.** 1994. The actin binding site in the tail domain of *Dictyostelium* myosin IC (myoC) resides within the glycine- and proline-rich sequence (tail homology region 2). *FEBS Lett.* **342**:197–202.
705. **Kabsch, W., H. G. Mannherz, D. Suck, E. F. Pai, and K. C. Holmes.** 1990. Atomic structure of the actin:DNase I complex. *Nature* **347**:37–44.
706. **Kanoh, S., M. Ito, E. Niwa, Y. Kawano, and D. J. Hartshorne.** 1993. Actin-binding peptide from smooth muscle myosin light chain kinase. *Biochemistry* **32**:8902–8907.
707. **Katakami, Y., N. Katakami, P. A. Janmey, J. H. Hartwig, and T. P. Stossel.** 1992. Isolation of the phosphatidylinositol 4-monophosphate dissociable high-affinity profilin-actin complex. *Biochim. Biophys. Acta* **1122**:123–135.
708. **Katoh, T. and F. Morita.** 1993. Actin-binding peptides obtained from the C-terminal 24-kDa fragment of porcine aorta smooth muscle myosin subfragment-1 heavy chain. *J. Biol. Chem.* **268**:2380–2388.
709. **Kaufmann, S., J. Kas, W. H. Goldmann, E. Sackmann, and G. Isenberg.** 1992. Talin anchors and nucleates actin filaments at lipid membranes. A direct demonstration. *FEBS Lett.* **314**:203–205.
710. **Keane, A. M., I. P. Trayer, B. A. Levine, C. Zeugner, and J. C. Ruegg.** 1990. Peptide mimetics of an actin-binding site on myosin span two functional domains on actin. *Nature* **344**:265–268.
711. **Kim, S. R., Y. Kim, and G. An.** 1993. Molecular cloning and characterization of anther-preferential cDNA encoding a putative actin-depolymerizing factor. *Plant. Mol. Biol.* **21**:39–45.
712. **Kinosian, H. J., L. A. Selden, J. E. Estes, and L. C. Gershman.** 1993. Nucleotide binding to actin. Cation dependence of nucleotide dissociation and exchange rates. *J. Biol. Chem.* **268**:8683–8691.
713. **Kitano, Y., N. Okada, and J. Adachi.** 1986. TPA-induced alteration of actin organization in cultured human keratinocytes. *Exp. Cell Res.* **167**:369–375.
714. **Knull, H. R., W. W. Bronstein, P. DesJardins, and W. G. J. Niehaus.** 1980. Interaction of selected brain glycolytic enzymes with an F-actin-tropomyosin complex. *J. Neurochem.* **34**:222–225.
715. **Kobayashi, R., T. Kubota, and H. Hidaka.** 1994. Purification, characterization, and partial sequence analysis of a new 25-kDa actin-binding protein from bovine aorta: a SM22 homolog. *Biochem. Biophys. Res. Commun.* **198**:1275–1280.
716. **Kogler, H., A. J. Moir, I. P. Trayer, and J. C. Ruegg.** 1991. Peptide competition of actin activation of myosin-subfragment 1 ATPase by an amino terminal actin fragment. *FEBS Lett.* **294**:31–34.
717. **Kolakowski, J., R. Makuch, and R. Dabrowska.** 1992. Lys-373 of actin is involved in binding to caldesmon. *FEBS Lett.* **309**:65–67.
718. **Kouyama, T. and K. Mihashi.** 1981. Fluorimetry study of N-(1-pyrenyl)iodoacetamide-labelled F-actin. Local structural change of actin protomer both on polymerization and on binding of heavy meromyosin. *Eur. J. Biochem.* **114**:33–38.
719. **Koyasu, S., E. Nishida, Y. Miyata, H. Sakai, and I. Yahara.** 1989. HSP100, a 100-kDa heat shock protein, is a Ca^{2+}-calmodulin-regulated actin-binding protein. *J. Biol. Chem.* **264**:15083–15087.
720. **Kuckel, C. L., P. K. Lambooy, and P. N. Farnsworth.** 1993. Modulation of actin polymerization by an exogenous protein, lysozyme. *Biochem. Cell Biol.* **71**:65–72.
721. **Labbe, J. P., M. Boyer, C. Mejean, C. Roustan, and Y. Benyamin.** 1993. Localization of two myosin-subfragment-1 binding contacts in the 96–132 region of actin subdomain-1. *Eur. J. Biochem.* **215**:17–24.
722. **Labbe, J. P., M. Boyer, C. Roustan, and Y. Benyamin.** 1992. Localization of a myosin subfragment-1 interaction site on the C-terminal part of actin. *Biochem. J.* **284**:75–79.
723. **Labbe, J. P., S. Lelievre, M. Boyer, and Y. Benyamin.** 1994. Interaction of skeletal muscle myosin subfragment-1 with actin (338–348) peptide. *Biochem. J.* **299**:875–879.
724. **Labbe, J. P., C. Mejean, Y. Benyamin, and C. Roustan.** 1990. Characterization of an actin-myosin head interface in the 40–113 region of actin using specific antibodies as probes. *Biochem. J.* **271**:407–413.
725. **Lakatos, S. and A. P. Minton.** 1991. Interactions between globular proteins and F-actin in isotonic saline solution. *J. Biol. Chem.* **266**:18707–18713.
726. **Lanzara, V., F. Cervellati, and E. Grazi.** 1988. On the interaction between xanthine oxidase and actin. *Biochem. Int.* **17**:217–223.
727. **Larsson, H. and U. Lindberg.** 1988. The effect of divalent cations on the interaction between calf spleen profilin and different actins. *Biochim. Biophys. Acta* **953**:95–105.
728. **Lebart, M. C., C. Mejean, M. Boyer, C. Roustan, and Y. Benyamin.** 1990. Localization of a new α-actinin binding site in the COOH-terminal part of actin sequence. *Biochem. Biophys. Res. Commun.* **173**:120–126.
729. **Lebart, M. C., C. Mejean, D. Casanova, E. Audemard, J. Derancourt, C. Roustan, and Y. Benyamin.** 1994. Characterization of the actin binding site on smooth muscle filamin. *J. Biol. Chem.* **269**:4279–4284.
730. **Lebart, M. C., C. Mejean, C. Roustan, and Y. Benyamin.** 1993. Further characterization of the α-actinin–actin interface and comparison with filamin-binding sites on actin. *J. Biol. Chem.* **268**:5642–5648.
730a. **Lees-Miller, J. P., D. M. Helfman and T. A. Schroer.** 1992. A vertebrate actin-related protein is a component of a multisubunit complex involved in microtubule-based vesicle motility. *Nature* **359**:244–246.
731. **Lehman, W., R. Craig, and P. Vibert.** 1994. Ca^{2+}-induced tropomyosin movement in Limulus thin filaments revealed by 3-dimensional reconstruction. *Nature* **368**:65–67.
732. **Levine, B. A., A. J. Moir, E. Audemard, D. Mornet, V. B. Patchell, and S. V. Perry.** 1990. Structural study of gizzard caldesmon and its interaction with actin. Binding involves residues of actin also recognised by myosin subfragment 1. *Eur. J. Biochem.* **193**:687–696.
733. **Levine, B. A., A. J. Moir, V. B. Patchell, and S. V. Perry.** 1990. The interaction of actin with dystrophin. *FEBS Lett.* **263**:159–162.
734. **Levine, B. A., A. J. Moir, V. B. Patchell, and S. V. Perry.** 1992. Binding sites involved in the interaction of actin with the N-terminal region of dystrophin. *FEBS Lett.* **298**:44–48.
735. **Levine, B. A., A. J. Moir, and S. V. Perry.** 1988. The interaction of troponin-I with the N-terminal region of actin. *Eur. J. Biochem.* **172**:389–397.
736. **Lind, S. E. and C. J. Smith.** 1991. Actin is a noncompetitive plasmin inhibitor. *J. Biol. Chem.* **266**:5273–5278.
737. **Llerenas, E. and M. E. Cid.** 1984. The molecular interaction between F-actin and lecithin in a phospholipid monolayer system. *Bol. Estud. Med. Biol.* **33**:33–39.
738. **Lorenz, M., D. Popp, and K. C. Holmes.** 1993. Refinement of the F-actin model against X-ray fiber diffraction data by the use of a directed mutation algorithm. *J. Mol. Biol.* **234**:826–836.
739. **Ma, A. S. P., M. E. Bystol, and A. Tranvan.** 1994. *In vitro* modulation of filament bundling in F-actin and keratins by annexin II and calcium. *In Vitro Cell Dev. Biol.* **30**:329–335.
740. **Mabuchi, I.** 1983. An actin-depolymerizing protein (depactin) from starfish oocytes: properties and interaction with actin. *J. Cell Biol.* **97**:1612–1621.
741. **Maciver, S. K., H. G. Zot, and T. D. Pollard.** 1991. Characterization of actin filament severing by actophorin from *Acanthamoeba castellaniii*. *J. Cell Biol.* **115**:1611–1620.
742. **Maekawa, S., M. Toriyama, S. Hisanaga, N. Yonezawa, S. Endo, N. Hirokawa, and H. Sakai.** 1989. Purification and characterization of a Ca^{2+}-dependent actin filament severing protein from bovine adrenal medulla. *J. Biol. Chem.* **264**:7458–7465.
743. **Makuch, R., J. Kolakowski, and R. Dabrowska.** 1992. The importance of C-terminal amino acid residues of actin to the inhibition of actomyosin ATPase activity by caldesmon and troponin I. *FEBS Lett.* **297**:237–240.
744. **Malm, B.** 1984. Chemical modification of Cys-374 of actin interferes with the formation of the profilactin complex. *FEBS Lett.* **173**:399–402.
745. **Malm, B., H. Larsson, and U. Lindberg.** 1983. The profilin–actin complex: further characterization of profilin and studies on the stability of the complex. *J. Muscle Res. Cell Motil.* **4**:569–588.
746. **Mannherz, H. G., J. Gooch, M. Way, A. G. Weeds, and P. J. McLaughlin.** 1992. Crystallization of the complex of actin with gelsolin segment 1. *J. Mol. Biol.* **226**:899–901.
747. **Maruta, H. and E. D. Korn.** 1977. Purification from *Acanthamoeba castellaniii* of proteins that induce gelation and syneresis of F-actin. *J. Biol. Chem.* **252**:399–402.
748. **McKim, K. S., C. Matheson, M. A. Marra, M. F. Wakarchuk, and D. L. Baillie.** 1994. The *Caenorhabditis elegans* unc-60 gene encodes proteins homologous to a family of actin-binding proteins. *Mol. Gen. Genet.* **242**:346–357.
749. **McLachlan, A. D. and M. Stewart.** 1976. The 14-fold periodicity in α-tropomyosin and the interaction with actin. *J. Mol. Biol.* **103**:271–298.
750.* **McLaughlin, P. J., J. T. Gooch, H. G. Mannherz, and A. G. Weeds.** 1993. Structure of gelsolin segment 1-actin complex and the mechanism of filament severing. *Nature* **364**:685–692.

751. **McWhirter, J. R. and J. Y. Wang.** 1993. An actin-binding function contributes to transformation by the Bcr-Abl oncoprotein of Philadelphia chromosome-positive human leukemias. *EMBO J.* **12**:1533–1546.

752. **Mejean, C., M. Boyer, J. P. Labbe, J. Derancourt, Y. Benyamin, and C. Roustan.** 1986. Antigenic probes locate the myosin subfragment 1 interaction site on the N-terminal part of actin. *Biosci. Rep.* **6**:493–499.

753. **Mejean, C., M. Boyer, J. P. Labbe, L. Marlier, Y. Benyamin, and C. Roustan.** 1987. Anti-actin antibodies. An immunological approach to the myosin–actin and the tropomyosin–actin interfaces. *Biochem. J.* **244**:571–577.

754. **Mejean, C., M. C. Lebart, M. Boyer, C. Roustan, and Y. Benyamin.** 1992. Localization and identification of actin structures involved in the filamin-actin interaction. *Eur. J. Biochem.* **209**:555–562.

755. **Mejean, C., F. Pons, Y. Benyamin, and C. Roustan.** 1989. Antigenic probes locate binding sites for the glycolytic enzymes glyceraldehyde-3-phosphate dehydrogenase, aldolase and phosphofructokinase on the actin monomer in microfilaments. *Biochem. J.* **264**:671–677.

756. **Melki, R., I. E. Vainberg, R. L. Chow, and N. J. Cowan.** 1993. Chaperonin-mediated folding of vertebrate actin-related protein and γ-tubulin. *J. Cell Biol.* **122**:1301–1310.

757. **Mengueld, M., A. Fattoum, J. Derancourt, and R. Kassab.** 1992. Mapping of the functional domains in the amino terminus of calponin. *J. Biol. Chem.* **267**:15943–15951.

758. **Meyer, R. K. and U. Aebi.** 1990. Bundling of actin filaments by α-actinin depends on its molecular length. *J. Cell Biol.* **110**:2013–2024.

759. **Miller, C. A., M. D. Cohen, and M. Costa.** 1991. Complexing of actin and other nuclear proteins to DNA by *cis*-diamminedichloroplatinum(II) and chromium compounds. *Carcinogenesis* **12**:269–276.

760. **Miller, L., M. Kalnoski, Z. Yunossi, J. C. Bulinski, and E. Reisler.** 1987. Antibodies directed against N-terminal residues on actin do not block acto-myosin binding. *Biochemistry* **26**:6064–6070.

761.* **Milligan, R. A., M. Whittaker, and D. Safer.** 1990. Molecular structure of F-actin and location of surface binding sites. *Nature* **348**:217–221.

762. **Mimura, N. and A. Asano.** 1987. Further characterization of a conserved actin-binding 27-kDa fragment of actinogelin and α-actinins and mapping of their binding sites on the actin molecule by chemical cross-linking. *J. Biol. Chem.* **262**:4717–4723.

763. **Miron, T., K. Vancompernolle, J. Vandekerckhove, M. Wilchek, and B. Geiger.** 1991. A 25-kD inhibitor of actin polymerization is a low molecular mass heat shock protein. *J. Cell Biol.* **114**:255–261.

764. **Moir, A. J. and B. A. Levine.** 1986. Protein cognitive sites on the surface of actin. A proton NMR study. *J. Inorg. Biochem.* **28**:271–278.

765. **Moraga, D. M., P. Nunez, J. Garrido, and R. B. Maccioni.** 1993. A tau fragment containing a repetitive sequence induces bundling of actin filaments. *J. Neurochem.* **61**:979–986.

766. **Muguruma, M., S. Matsumura, and T. Fukazawa.** 1992. Augmentation of α-actinin-induced gelation of actin by talin. *J. Biol. Chem.* **267**:5621–5624.

767. **Muneyuki, E., E. Nishida, K. Sutoh, and H. Sakai.** 1985. Purification of cofilin, a 21,000 molecular weight actin-binding protein, from porcine kidney and identification of the cofilin-binding site in the actin sequence. *J. Biochem. Tokyo* **97**:563–568.

768. **Nachmias, V. T.** 1993. Small actin-binding proteins: the β-thymosin family. *Curr. Opin. Cell Biol.* **5**:56–62.

769. **Namba, Y., M. Ito, Y. Zu, K. Shigesada, and K. Maruyama.** 1992. Human T cell L-plastin bundles actin filaments in a calcium-dependent manner. *J. Biochem. Tokyo* **112**:503–507.

770. **Ngai, P. K., U. Groschel Stewart, and M. P. Walsh.** 1986. Comparison of the effects of smooth and skeletal muscle actins on smooth muscle actomyosin Mg^{2+}-ATPase. *Biochem. Int.* **12**:89–93.

771. **Nishida, E.** 1985. Opposite effects of cofilin and profilin from porcine brain on rate of exchange of actin-bound adenosine 5′-triphosphate. *Biochemistry* **24**:1160–1164.

772. **Nishida, E., S. Koyasu, H. Sakai, and I. Yahara.** 1986. Calmodulin-regulated binding of the 90-kDa heat shock protein to actin filaments. *J. Biol. Chem.* **261**:16033–16036.

773. **O'Reilly, G. and F. Clarke.** 1993. Identification of an actin binding region in aldolase. *FEBS Lett.* **321**:69–72.

774. **Ohara, O., S. Takahashi, and T. Ooi.** 1983. Cross-linking study on skeletal muscle actin: interaction of suberimidate-treated actin with deoxyribonuclease I. *J. Biochem. Tokyo* **93**:1547–1556.

775. **Ohnuma, M. and I. Mabuchi.** 1993. 45K actin filament-severing protein from sea urchin eggs: interaction with phosphatidylinositol-4,5-bisphosphate. *J. Biochem. Tokyo* **114**:718–722.

776. **Ohshima, S., H. Abe, and T. Obinata.** 1989. Isolation of profilin from embryonic chicken skeletal muscle and evaluation of its interaction with different actin isoforms. *J. Biochem. Tokyo* **105**:855–857.

777. **Ohtaki, T., S. Tsukita, N. Mimura, and A. Asano.** 1985. Interaction of actinogelin with actin. No nucleation but high gelation activity. *Eur. J. Biochem.* **153**:609–620.

778. **Owen, C. H., D. J. DeRosier, and J. Condeelis.** 1992. Actin crosslinking protein EF-1a of *Dictyostelium discoideum* has a unique bonding rule that allows square-packed bundles. *J. Struct. Biol.* **109**:248–254.

779. **Pacaud, M. and J. Derancourt.** 1993. Purification and further characterization of macrophage 70-kDa protein, a calcium-regulated, actin-binding protein identical to L-plastin. *Biochemistry* **32**:3448–3455.

780. **Paschal, B. M., E. L. Holzbaur, K. K. Pfister, S. Clark, D. I. Meyer, and R. B. Vallee.** 1993. Characterization of a 50-kDa polypeptide in cytoplasmic dynein preparations reveals a complex with p150GLUED and a novel actin. *J. Biol. Chem.* **268**:15318–15323.

781. **Pinder, J. C. and W. B. Gratzer.** 1982. Investigation of the actin-deoxyribonuclease I interaction using a pyrene-conjugated actin derivative. *Biochemistry* **21**:4886–4890.

782. **Plank, L. and B. R. Ware.** 1987. *Acanthamoeba* profilin binding to fluorescein-labeled actins. *Biophys. J.* **51**:985–988.

783. **Pope, B., M. Way, P. T. Matsudaira, and A. Weeds.** 1994. Characterisation of the F-actin binding domains of villin: classification of F-actin binding proteins into two groups according to their binding sites on actin. *FEBS Lett.* **338**:58–62.

784. **Pope, B., M. Way, and A. G. Weeds.** 1991. Two of the three actin-binding domains of gelsolin bind to the same subdomain of actin. Implications of capping and severing mechanisms. *FEBS Lett.* **280**:70–74.

785. **Prendergast, G. C. and E. B. Ziff.** 1991. Mbh 1: a novel gelsolin/severin-related protein which binds actin *in vitro* and exhibits nuclear localization *in vivo*. *EMBO J.* **10**:757–766.

786. **Prochniewicz, E. and T. Yanagida.** 1990. Inhibition of sliding movement of F-actin by crosslinking emphasizes the role of actin structure in the mechanism of motility. *J. Mol. Biol.* **216**:761–772.

787.* **Rayment, I., H. M. Holden, M. Whittaker, C. B. Yohn, M. Lorenz, K. C. Holmes, and R. A. Milligan.** 1993. Structure of the actin-myosin complex and its implications for muscle contraction. *Science* **261**:58–65.

788. **Rayment, I., W. R. Rypniewski, K. Schmidt-Basel, R. Smith, D. R. Tomchick, M. M. Benning, D. A. Winkelmann, G. Wesenberg, and H. M. Holden.** 1993. Three-dimensional structure of myosin subfragment-1: a molecular motor. *Science* **261**:50–58.

789. **Reddy, S. R., A. Houmeida, Y. Benyamin, and C. Roustan.** 1992. Interaction *in vitro* of scallop muscle arginine kinase with filamentous actin. *Eur. J. Biochem.* **206**:251–257.

790. **Reisler, E.** 1993. Actin molecular structure and function. *Curr. Opin. Cell Biol.* **5**:41–47.

791. **Rosenfeld, S. S. and B. Rener.** 1994. The GPQ-rich segment of *Dictyostelium* myosin IB contains an actin binding site. *Biochemistry* **33**:2322–2328.

792. **Roth, G. A., M. D. Gonzalez, C. G. Monferran, M. L. De Santis, and F. A. Cumar.** 1993. Myelin basic protein domains involved in the interaction with actin. *Neurochem. Int.* **23**:459–465.

793. **Rouleau, G. A., P. Merel, M. Lutchman, M. Sanson, J. Zucman, C. Marineau, K. Haong-Xuan, S. Demczuk, S. Desmaze, B. Plougastel, S. M. Pulst, G. Lenoir, R. F. Bijlsma, J. Dumanski, P. de Jong, D. Parry, R. Eldrige, A. Aurias, O. Delattre, and G. Thomas.** 1993. Alteration in a new gene encoding a putative membrane-organising protein causes neuro-fibromatosis type 2. *Nature* **363**:515–521.

793a. **Sanders, C. and L. B. Smillie.** 1984. Chicken gizzard tropomyosin: head-to-tail assembly and interaction with F-actin and troponin. *Can. J. Biochem. Cell Biol.* **62**:443–448.

794. **Sato, N., N. Funayama, A. Nagafuchi, S. Yonemura, and S. Tsukita.** 1992. A gene family consisting of ezrin, radixin and moesin. Its specific localization at actin filament/plasma membrane association sites. *J. Cell Sci.* **103**:131–143.

795. **Sato, N., S. Yonemura, T. Obinata, and S. Tsukita.** 1991. Radixin, a barbed end-capping actin-modulating protein, is concentrated at the cleavage furrow during cytokinesis. *J. Cell Biol.* **113**:321–330.

796. **Sattilaro, R. F.** 1986. Interaction of microtubule-associated protein 2 with actin filaments. *Biochemistry* **25**:2003–2009.

797. **Sattilaro, R. F., W. L. Dentler, and E. L. LeCluyse.** 1981. Microtubule-associated proteins (MAPs) and the organization of actin filaments *in vitro*. *J. Cell Biol.* **90**:467–473.

798. **Sauterer, R. A., R. J. Eddy, A. L. Hall, and J. S. Condeelis.** 1991. Purification and characterization of aginactin, a newly identified agonist-regulated actin-capping protein from *Dictyostelium amoebae*. *J. Biol. Chem.* **266**:24533–24539.

799. **Schliwa, M.** 1981. Proteins associated with cytoplasmic actin. *Cell* **25**:587–590.

800. **Schmid, M. F., J. M. Agris, J. Jakana, P. Matsudaira, and W. Chiu.** 1994. Three-dimensional structure of a single filament in the Limulus acrosomal bundle: scruin binds to homologous helix–loop–β motifs in actin. *J. Cell Biol.* **124**:341–350.

801. **Schmidt, J. M., R. M. Robson, J. Zhang, and M. H. Stromer.** 1993. The marked pH dependence of the talin-actin interaction. *Biochem. Biophys. Res. Commun.* **197**:660–666.

802. **Schoepper, B. and A. Wegner.** 1991. Rate constants and equilibrium constants for binding of actin to the 1:1 gelsolin–actin complex. *Eur. J. Biochem.* **202**:1127–1131.

803. **Schroder, R. R., D. J. Manstein, W. Jahn, H. Holden, I. Rayment, K. C. Holmes, and J. A. Spudich.** 1993. Three-dimensional atomic model of F-actin decorated with *Dictyostelium* myosin. *Nature* **364**:171–174.

804. **Schutt, C. E., U. Lindberg, J. Myslik, and N. Strauss.** 1989. Molecular packing in profilin: actin crystals and its implications. *J. Mol. Biol.* **209**:735–746.

805.* **Schutt, C. E., J. C. Myslik, M. D. Rozycki, N. C. Goonesekere, and U. Lindberg.** 1993. The structure of crystalline profilin-β-actin. *Nature* **365**:810–816.

806. **Selden, S. C. and T. D. Pollard.** 1986. Interaction of actin filaments with microtubules is mediated by microtubule-associated proteins and regulated by phosphorylation. *Ann. N. Y. Acad. Sci.* **466**:803–812.

807. **Selve, N. and A. Wegner.** 1986. Rate constants and equilibrium constants for binding of the gelsolin-actin complex to the barbed ends of actin filaments in the presence and absence of calcium. *Eur. J. Biochem.* **160**:379–387.

808. **Senter, L., M. Luise, C. Presotto, R. Betto, A. Teresi, S. Ceoldo, and G. Salviati.** 1993. Interaction of dystrophin with cytoskeletal proteins: binding to talin and actin. *Biochem. Biophys. Res. Commun.* **192**:899–904.

809. **Shapland, C., J. J. Hsuan, N. F. Totty, and D. Lawson.** 1993. Purification and properties of transgelin: a transformation- and shape-sensitive actin gelling protein. *J. Cell Biol.* **121**:1065–1073.

810. **Shapland, C., P. Lowings, and D. Lawson.** 1988. Identification of new actin-associated polypeptides that are modified by viral transformation and changes in cell shape. *J. Cell Biol.* **107**:153–161.

811. **Shenouda, S. Y. and G. M. Pigott.** 1976. Electron paramagnetic resonance studies of actin-lipid interaction in aqueous media. *J. Agric. Food. Chem.* **24**:11–15.

812. **Shestakova, E. A., L. P. Motuz, A. A. Minin, and L. P. Gavrilova.** 1993. Study of localization of the protein-synthesizing machinery along actin filament bundles. *Cell Biol. Int.* **17**:409–416.

813. **Siegel, D. L. and D. Branton.** 1985. Partial purification and characterization of an actin-bundling protein, band 4.9, from human erythrocytes. *J. Cell Biol.* **100**:775–785.

814. **Snabes, M. C., A. E. Boyd, and J. Bryan.** 1983. Identification of G actin-binding proteins in rat tissues using a gel overlay technique. *Exp. Cell Res.* **146**:63–70.

814a. **Southwick, F. S., N. Tatsumi, and T. P. Stossel.** 1982. Acumentin, an actin-modulating protein of rabbit pulmonary macrophages. *Biochemistry* **21**:6321–6326.

815. **Southwick, F. S. and M. J. DiNubile.** 1986. Rabbit alveolar macrophages contain a Ca^{2+}-sensitive, 41,000-dalton protein which reversibly blocks the "barbed" ends of actin filaments but does not sever them. *J. Biol. Chem.* **261**:14191–14195.

816. **St Onge, D. and C. Gicquaud.** 1989. Evidence of direct interaction between actin and membrane lipids. *Biochem. Cell Biol.* **67**:297–300.

817. **St Onge, D. and C. Gicquaud.** 1990. Research on the mechanism of interaction between actin and membrane lipids. *Biochem. Biophys. Res. Commun.* **167**:40–47.

818. **St Pierre, B., C. Couture, A. Laroche, and D. Pallotta.** 1993. Two developmentally regulated mRNAs encoding actin-binding proteins in *Physarum polycephalum. Biochim. Biophys. Acta* **1173**:107–110.

819. **Sternlicht, H., G. W. Farr, M. L. Sternlicht, J. K. Driscoll, K. Willison, and M. B. Yaffe.** 1993. The t-complex polypeptide 1 complex is a chaperonin for tubulin and actin *in vivo*. *Proc. Natl Acad. Sci. U. S. A.* **90**:9422–9426.

820. **Strasser, P., M. Gimona, M. Herzog, B. Geiger, and J. V. Small.** 1993. Variable and constant regions in the C-terminus of vinculin and metavinculin. Cloning and expression of fragments in *E. coli*. *FEBS Lett.* **317**:189–194.

821. **Sun, H. Q., D. C. Wooten, P. A. Janmey, and H. L. Yin.** 1994. The actin side-binding domain of gelsolin also caps actin filaments – implications for actin filament severing. *J. Biol. Chem.* **269**:9473–9479.

822. **Sutoh, K.** 1982. Identification of myosin-binding sites on the actin sequence. *Biochemistry* **21**:3654–3661.

823. **Sutoh, K.** 1984. Actin–actin and actin–deoxyribonuclease I contact sites in the actin sequence. *Biochemistry* **23**:1942–1946.

824. **Sutoh, K.** 1993. Identification of actin surface interacting with myosin during the actin-myosin sliding. *Adv. Exp. Med. Biol.* **332**:241–244.

825. **Sutoh, K. and S. Hatano.** 1986. Actin–fragmin interactions as revealed by chemical cross-linking. *Biochemistry* **25**:435–440.

826. **Sutoh, K. and I. Mabuchi.** 1986. Improved method for mapping the binding site of an actin-binding protein in the actin sequence. Use of a site-directed antibody against the N-terminal region of actin as a probe of its N-terminus. *Biochemistry* **25**:6186–6192.

827. **Sutoh, K. and I. Mabuchi.** 1989. End-label fingerprintings show that an N-terminal segment of depactin participates in interaction with actin. *Biochemistry* **28**:102–106.

828. **Sutoh, K. and H. L. Yin.** 1989. End-label fingerprintings show that the N- and C-termini of actin are in the contact site with gelsolin. *Biochemistry* **28**:5269–5275.

829. **Szilagyi, L. and R. C. Lu.** 1982. Changes of lysine reactivities of actin in complex with myosin subfragment-1, tropomyosin and troponin. *Biochim. Biophys. Acta* **709**:204–211.

830. **Takahashi, K. and B. Nadal-Ginard.** 1991. Molecular cloning and sequence analysis of smooth muscle calponin. *J. Biol. Chem.* **266**:13284–13288.

831. **Titus, M. A.** 1993. Myosins. *Curr. Opin. Cell Biol.* **5**:77–81.

832. **Tranter, M. P., S. P. Sugrue, and M. A. Schwartz.** 1991. Binding of actin to liver cell membranes: the state of membrane-bound actin. *J. Cell Biol.* **112**:891–901.

833. **Trayer, I. P., H. R. Trayer, and B. A. Levine.** 1987. Evidence that the N-terminal region of A1-light chain of myosin interacts directly with the C-terminal region of actin. A proton magnetic resonance study. *Eur. J. Biochem.* **164**:259–266.

834. **Trifaro, J. M., M. L. Vitale, and A. Rodriguez Del Castillo.** 1993. Scinderin and chromaffin cell actin network dynamics during neurotransmitter release. *J. Physiol. Paris.* **87**:89–106.

835. **Trofatter, J. A., M. M. MacCollin, J. L. Rutter, J. R. Murrell, M. P. Duyao, D. M. Parry, R. Eldridge, N. Kley, A. G. Menon, K. Pulaski, V. H. Haase, C. M. Ambrose, D. Munroe, C. Bove, J. L. Haines, R. L. Martuza, M. E. MacDonald, B. R. Seizinger, M. P. Short, A. J. Buckler, and J. F. Gusella.** 1993. A novel Moesin-, Ezrin-, Radixin-like gene is a candidate for the neurofibromatosis 2 tumor suppressor. *Cell* **72**:791–800.

836.* **Van Etten, R. A., P. K. Jackson, D. Baltimore, M. C. Sanders, P. T. Matsudaira, and P. A. Janmey.** 1994. The COOH terminus of the c-Abl tyrosine kinase contains distinct F- and G-actin binding domains with bundling activity. *J. Cell Biol.* **124**:325–340.

837. **Van Eyk, J. E. and R. S. Hodges.** 1991. A synthetic peptide of the N-terminus of actin interacts with myosin. *Biochemistry* **30**:11676–11682.

838. **Vancompernolle, K., J. Vandekerckhove, M. R. Bubb, and E. D. Korn.** 1991. The interfaces of actin and *Acanthamoeba* actobindin. Identification of a new actin-binding motif. *J. Biol. Chem.* **266**:15427–15431.

839.* **Vandekerckhove, J.** 1990. Actin-binding proteins. *Curr. Opin. Cell Biol.* **2**:41–50.

840. **Vandekerckhove, J. S., D. A. Kaiser, and T. D. Pollard.** 1989. *Acanthamoeba* actin and profilin can be cross-linked between glutamic acid 364 of actin and lysine 115 of profilin. *J. Cell Biol.* **109**:619–626.

841. **Wachsstock, D. H., W. H. Schwartz, and T. D. Pollard.** 1993. Affinity of α-actinin for actin determines the structure and mechanical properties of actin filament gels. *Biophys. J.* **65**:205–214.

842. **Waddle, J. A., J. A. Cooper, and R. H. Waterston.** 1993. The α and β subunits of nematode actin capping protein function in yeast. *Mol. Biol. Cell* **4**:907–917.

843. **Watanabe, S. and K. Maruyama.** 1993. Reannealing of β-actinin-severed actin filaments by troponin and tropomyosin. *J. Biochem. Tokyo* **114**:181–185.

844. **Way, M., B. Pope, R. A. Cross, J. Kendrick Jones, and A. G. Weeds.** 1992. Expression of the N-terminal domain of dystrophin in E. coli and demonstration of binding to F-actin. *FEBS Lett.* **301**:243–245.

845. **Way, M., B. Pope, and A. Weeds.** 1991. Molecular biology of actin binding proteins: evidence for a common structural domain in the F-actin binding sites of gelsolin and α-actinin. *J. Cell Sci. Suppl.* **14**:91–94.

846.* **Way, M., B. Pope, and A. G. Weeds.** 1992. Evidence for functional homology in the F-actin binding domains of gelsolin and α-actinin: implications for the requirements of severing and capping. *J. Cell Biol.* **119**:835–842.

847.* **Weigt, C., A. Gaertner, A. Wegner, H. Korte, and H. E. Meyer.** 1992. Occurrence of an actin-inserting domain in tensin. *J. Mol. Biol.* **227**:593–595.

848.* **Weigt, C., A. Wegner, and M. H. Koch.** 1991. Rate and mechanism of the assembly of tropomyosin with actin filaments. *Biochemistry* **30**:10700–10707.

849.* **Weiner, O. H., J. Murphy, G. Griffiths, M. Schleicher, and A. A. Noegel.** 1993. The actin-binding protein comitin (p24) is a component of the Golgi apparatus. *J. Cell Biol.* **123**:23–34.

850. **Wille, M., I. Just, A. Wegner, and K. Aktories.** 1992. ADP-ribosylation of gelsolin-actin complexes by clostridial toxins. *J. Biol. Chem.* **267**:50–55.

851.* **Winder, J. J., C. Sutherland, and M. P. Walsh.** 1991. Biochemical and functional characterisation of smooth muscle calponin. *Adv. Exp. Med. Biol.* **304**:32–51.

852. **Winder, S. J., G. J. Kargacin, A. A. Bonet-Kerrache, M. D. Pato, and M. P. Walsh.** 1992. Calponin localisation and regulation of smooth muscle actomyosin ATPase. *Jpn. J. Pharmacol.* **58**:29P-34P.

853.* **Wu, H. and J. T. Parsons.** 1993. Cortactin, an 80/85-kilodalton pp60src substrate, is a filamentous actin-binding protein enriched in the cell cortex. *J. Cell Biol.* **120**:1417–1426.

854. **Yamamoto, K.** 1990. Shift of binding site at the interface between actin and myosin. *Biochemistry* **29**:844–848.

855.* **Yamauchi, P. S. and D. L. Purich.** 1993. Microtubule-associated protein interactions with actin filaments: evidence for differential behavior of neuronal MAP-2 and τ in the presence of phosphatidyl-inositol. *Biochem. Biophys. Res. Commun.* **190**:710–715.

856.* **Yang, W. and W. F. Boss.** 1994. Regulation of phosphatidylinositol-4-kinase by the protein activator PIK-A49–activation requires phosphorylation of PIK-A49. *J. Biol. Chem.* **269**:3852–3857.

857. **Yonezawa, N., Y. Homma, I. Yahara, H. Sakai, and E. Nishida.** 1991. A short sequence responsible for both phosphoinositide binding and actin binding activities of cofilin. *J. Biol. Chem.* **266**:17218–17221.

858. **Yonezawa, N., E. Nishida, K. Iida, H. Kumagai, I. Yahara, and H. Sakai.** 1991. Inhibition of actin polymerization by a synthetic dodecapeptide patterned on the sequence around the actin-binding site of cofilin. *J. Biol. Chem.* **266**:10485–10489.

859. **Young, C. L., A. Feierstein, and F. S. Southwick.** 1994. Calcium regulation of actin filament capping and monomer binding by macrophage capping protein. *J. Biol. Chem.* **269**:13997–14002.

860. **Yu, F. X., S. C. Lin, M. Morrison Bogorad, M. A. Atkinson, and H. L. Yin.** 1993. Thymosin β 10 and thymosin β 4 are both actin monomer sequestering proteins. *J. Biol. Chem.* **268**:502–509.

861. **Zalewski, P. D., I. J. Forbes, C. Giannakis, and W. H. Betts.** 1991. Regulation of protein kinase C by Zn^{2+}-dependent interaction with actin. *Biochem. Int.* **24**:1103–1110.

862. **Zechel, K.** 1993. The interaction of 6-propionyl-2-(NN-dimethyl)aminonaphthalene-(PRODAN)-labelled actin with actin-binding proteins and drugs. *Biochem. J.* **290**:411–417.

863. **Zevgolis, V. G., T. G. Sotiroudis, and A. E. Evangelopoulos.** 1991. Phosphorylase kinase from bovine stomach smooth muscle: a Ca^{2+}-dependent protein kinase associated with an actin-like molecule. *Biochim. Biophys. Acta* **1091**:222–230.

SEQUENCE ANALYSIS OF ACTIN

864.* **Adam, L., A. Laroche, A. Barden, G. Lemieux, and D. Pallotta.** 1991. An unusual actin-encoding gene in *Physarum polycephalum*. *Gene* **106**:79–86.

865.* **Adams, A. E. and D. Botstein.** 1989. Dominant suppressors of yeast actin mutations that are reciprocally suppressed. *Genetics* **121**:675–683.

866. **Adams, A. E., J. A. Cooper, and D. G. Drubin.** 1993. Unexpected combinations of null mutations in genes encoding the actin cytoskeleton are lethal in yeast. *Mol. Biol. Cell* **4**:459–468.

867.* **Aerne, B. L., V. Schmid, and P. Schuchert.** 1993. Actin-encoding genes of the hydrozoan *Podocoryne carnea*. *Gene* **131**:183–192.

868.* **Akhurst, R. J., F. J. Calzone, J. J. Lee, R. J. Britten, and E. H. Davidson.** 1987. Structure and organization of the CyIII actin gene subfamily of the sea urchin, *Strongylocentrotus purpuratus*. *J. Mol. Biol.* **194**:193–203.

869. **Akkari, P. A., H. J. Eyre, S. D. Wilton, D. F. Callen, S. A. Lane, C. Meredith, L. Kedes, and N. G. Laing.** 1994. Assignment of the human skeletal muscle α actin gene (ACTA1) to 1q42 by fluorescence *in situ* hybridisation. *Cytogenet. Cell Genet.* **65**:265–267.

870. **Alonso, S.** 1987. Coexpression and evolution of the two sarcomeric actin genes in vertebrates. *Biochimie* **69**:1119–1125.

871. **Alonso, S., A. Minty, Y. Bourlet, and M. Buckingham.** 1986. Comparison of three actin-coding sequences in the mouse; evolutionary relationships between the actin genes of warm-blooded vertebrates. *J. Mol. Evol.* **23**:11–22.

872. **Alving, R. E. and K. Laki.** 1966. N-terminal sequence of actin. *Biochemistry* **5**:2597–2601.

873. **Ampe, C. and J. Vandekerckhove.** 1987. The F-actin capping proteins of Physarum polycephalum: cap42(a) is very similar, if not identical, to fragmin and is structurally and functionally very homologous to gelsolin; cap42(b) is Physarum actin. *EMBO J.* **6**:4149–4157.

874. **Anderson, P. J.** 1976. Chicken muscle and fibroblast actin structure. *Biochem. J.* **159**:185–187.

875.* **Aspenstrom, P., H. Engkvist, U. Lindberg, and R. Karlsson.** 1992. Characterization of yeast-expressed β-actins, site-specifically mutated at the tumor-related residue Gly245. *Eur. J. Biochem.* **207**:315–320.

876.* **Aspenstrom, P. and R. Karlsson.** 1991. Interference with myosin subfragment-1 binding by site-directed mutagenesis of actin. *Eur. J. Biochem.* **200**:35–41.

877. **Aspenstrom, P., U. Lindberg, and R. Karlsson.** 1992. Site-specific amino-terminal mutants of yeast-expressed β-actin. Characterization of the interaction with myosin and tropomyosin. *FEBS Lett.* **303**:59–63.

878.* **Aspenstrom, P., C. E. Schutt, U. Lindberg, and R. Karlsson.** 1993. Mutations in β-actin: influence on polymer formation and on interactions with myosin and profilin. *FEBS Lett.* **329**:163–170.

879. **Beifuss, M. J. and D. S. Durica.** 1992. Sequence analysis of the indirect flight muscle actin-encoding gene of *Drosophila simulans*. *Gene* **118**:163–170.

880.* **Ben Amar, M. F., A. Pays, P. Tebabi, B. Dero, T. Seebeck, M. Steinert, and E. Pays.** 1988. Structure and transcription of the actin gene of *Trypanosoma brucei*. *Mol. Cell Biol.* **8**:2166–2176.

881. **Berger, E. and G. Cox.** 1979. A precursor of cytoplasmic actin in cultured *Drosophila cells*. *J. Cell Biol.* **81**:680–683.

882.* **Bergsma, D. J., K. S. Chang, and R. J. Schwartz.** 1985. Novel chicken actin gene: third cytoplasmic isoform. *Mol. Cell Biol.* **5**:1151–1162.

883.* **Bhattacharya, D., S. K. Stickel, and M. L. Sogin.** 1991. Molecular phylogenetic analysis of actin genic regions from *Achlya bisexualis* (Oomycota) and *Costaria costata* (Chromophyta). *J. Mol. Evol.* **33**:525–536.

884.* **Bhattacharya, D., S. K. Stickel, and M. L. Sogin.** 1993. Isolation and molecular phylogenetic analysis of actin-coding regions from *Emiliania huxleyi*, a Prymnesiophyte alga, by reverse transcriptase and PCR methods. *Mol. Biol. Evol.* **10**:689–703.

885.* **Blum, H., M. Hannappel, and G. J. Arnold.** 1990. Nucleotide and deduced amino acid sequence of an actin cDNA clone from *Plasmodium falciparum*. *Nucl. Acids Res.* **18**:7140.

886. **Bond, B. J. and N. Davidson.** 1986. The *Drosophila melanogaster* actin 5C gene uses two transcription initiation sites and three polyadenylation sites to express multiple mRNA species. *Mol. Cell Biol.* **6**:2080–2088.

886a.* **Boom J.D.G. & Smith M.J.** (1989) Molecular analyses of gene expression during sea star spermatogenesis. *J. Exp. Zool* . **250**: 312–320.

887. **Bridgen, J.** 1971. The amino acid sequence around four cysteine residues in trout actin. *Biochem. J.* **123**:591–600.

888.* **Brown, C. W., K. M. McHugh, and J. L. Lessard.** 1990. A cDNA sequence encoding cytoskeletal γ-actin from rat. *Nucl. Acids Res.* **18**:5312.

889. **Buckingham, M., S. Alonso, G. Bugaisky, P. Barton, A. Cohen, P. Daubas, A. Minty, B. Robert, and A. Weydert.** 1985. The actin and myosin multigene families. *Adv. Exp. Med. Biol.* **182**:333–344.

890. **Cameron, R. A., R. J. Britten, and E. H. Davidson.** 1989. Expression of two actin genes during larval development in the sea urchin *Strongylocentrotus purpuratus*. *Mol. Reprod. Dev.* **1**:149–155.

891.* **Campos, A., P. Bernard, A. Fauconnier, A. Landa, E. Gomez, R. Hernandez, K. Willms, and J. P. Laclette.** 1990. Cloning and sequencing of two actin genes from *Taenia* solium (Cestoda). *Mol. Biochem. Parasitol.* **40**:87–93.

892.* **Carroll, S. L., D. J. Bergsma, and R. J. Schwartz.** 1986. Structure and complete nucleotide sequence of the chicken α-smooth muscle (aortic) actin gene. An actin gene which produces multiple messenger RNAs. *J. Biol. Chem.* **261**:8965–8976.

893.* **Chang, K. S., K. N. Rothblum, and R. J. Schwartz.** 1985. The complete sequence of the chicken α-cardiac actin gene: a highly conserved vertebrate gene. *Nucleic Acids Res.* **13**:1223–1237.

894. **Chang, K. S., W. E. J. Zimmer, D. J. Bergsma, J. B. Dodgson, and R. J. Schwartz.** 1984. Isolation and characterization of six different chicken actin genes. *Mol. Cell Biol.* **4**:2498–2508.

895.* **Chen, X., R. K. Cook, and P. A. Rubenstein.** 1993. Yeast actin with a mutation in the "hydrophobic plug" between subdomains 3 and 4 (L266D) displays a cold-sensitive polymerization defect. *J. Cell Biol.* **123**:1185–1195.

896. **Chou, C. C., R. C. Davis, M. L. Fuller, J. P. Slovin, A. Wong, J. Wright, S. Kania, R. Shaked, R. A. Gatti, and W. A. Salser.** 1987. Γ-actin: unusual mRNA 3′-untranslated sequence conservation and amino acid substitutions that may be cancer related. *Proc. Natl Acad. Sci. U. S. A.* **84**:2575–2579.

897. **Chowdhury, S., K. W. Smith, and M. C. Gustin.** 1992. Osmotic stress and the yeast cytoskeleton: phenotype-specific suppression of an actin mutation. *J. Cell Biol.* **118**:561–571.

898.* **Clark, S. W. and D. I. Meyer.** 1992. Centractin is an actin homologue associated with the centrosome. *Nature* **359**:246–250.

899. **Cleveland, D. W., M. A. Lopata, R. J. MacDonald, N. J. Cowan, W. J. Rutter, and M. W. Kirschner.** 1980. Number and evolutionary conservation of α- and β-tubulin and cytoplasmic β- and γ-actin genes using specific cloned cDNA probes. *Cell* **20**:95–105.

900.* **Collins, J. H. and M. Elzinga.** 1975. The primary structure of actin from rabbit skeletal muscle. Completion and analysis of the amino acid sequence. *J. Biol. Chem.* **250**:5915–5920.

901. **Cook, R. K., W. T. Blake, and P. A. Rubenstein.** 1992. Removal of the amino-terminal acidic residues of yeast actin. Studies *in vitro* and *in vivo*. *J. Biol. Chem.* **267**:9430–9436.

902.* **Cook, R. K., D. Root, C. Miller, E. Reisler, and P. A. Rubenstein.** 1993. Enhanced stimulation of myosin subfragment 1 ATPase activity by addition of negatively charged residues to the yeast actin NH2 terminus. *J. Biol. Chem.* **268**:2410–2415.

903. **Cook, R. K. and P. A. Rubenstein.** 1989. Effects of Asp10 mutations on yeast viability. *J. Cell Biol.* **109**:271a.

904.* **Cook, R. K., D. R. Sheff, and P. A. Rubenstein.** 1991. Unusual metabolism of the yeast actin amino terminus. *J. Biol. Chem.* **266**:16825–16833.

905. **Cooper, A. D. and W. R. J. Crain.** 1982. Complete nucleotide sequence of a sea urchin actin gene. *Nucl. Acids Res.* **10**:4081–4092.

906.* **Crain, W. R. J., M. F. Boshar, A. D. Cooper, D. S. Durica, A. Nagy, and D. Steffen.** 1987. The sequence of a sea urchin muscle actin gene suggests a gene conversion with a cytoskeletal actin gene. *J. Mol. Evol.* **25**:37–45.

907.* **Cresnar, B., W. Mages, K. Muller, J. M. Salbaum, and R. Schmitt.** 1990. Structure and expression of a single actin gene in *Volvox carteri*. *Curr. Genet.* **18**:337–346.

908. **Cripps, R. M., E. Ball, M. Stark, A. Lawn, and J. C. Sparrow.** 1994. Recovery of dominant autosomal flightless mutants of *Drosophila melanogaster* and identification of a new gene required for normal insect structure and function. Genetics **137**:151–164.

909. **Crosbie, R. H., C. Miller, J. M. Chalovich, P. A. Rubenstein, and E. Reisler.** 1994. Caldesmon, N-terminal yeast actin mutants, and the regulation of actomyosin interactions. *Biochemistry* **33**:3210–3216.

910.* **Cross, G. S., C. Wilson, H. P. Erba, and H. R. Woodland.** 1988. Cytoskeletal actin gene families of *Xenopus borealis* and *Xenopus laevis*. *J. Mol. Evol.* **27**:17–28.

911.* **Cupples, C. G. and R. E. Pearlman.** 1986. Isolation and characterisation of the actin gene from *Tetrahymena pyriformis*. *Proc. Natl Acad. Sci. U. S. A.* **83**:5160–5164.

912.* **da Silva, C. M., L. Henrique, B. Ferreira, M. Picon, N. Gorfinkiel, R. Ehrlich, and A. Zaha.** 1993. Molecular cloning and characterization of actin genes from *Echinococcus granulosus*. *Mol. Biochem. Parasitol.* **60**:209–219.

913.* **Dearruda, M. V. and P. Matsudaira.** 1994. Cloning and sequencing of the *Leishmania* major actin-encoding gene. *Gene* **139**:123–125.

914. **Degen, J. L., M. G. Neubauer, S. J. Degen, C. E. Seyfried, and D. R. Morris.** 1983. Regulation of protein synthesis in mitogen-activated bovine lymphocytes. Analysis of actin-specific and total mRNA accumulation and utilization. *J. Biol. Chem.* **258**:12153–12162.

915. **DesGroseillers, L., D. Auclair, and L. Wickham.** 1990. Nucleotide sequence of an actin cDNA gene from *Aplysia californica*. *Nucl. Acids Res.* **18**:3654.

916*. **DesGroseillers, L., D. Auclair, L. Wickham and M. Maalouf.** 1994. A novel actin cDNA is expressed in the neurons of *Aplysia californica*. *Biochim. Biophys. Acta* **1217**:322–324.

917*. **Deshler, J. O., G. P. Larson, and J. J. Rossi. 1989.** *Kluyveromyces lactis* maintains *Saccharomyces cerevisiae* intron-encoded splicing signals. *Mol. Cell Biol.* **9**:2208–2213.

918.* **Drouin, G. and G. A. Dover.** 1990. Independent gene evolution in the potato actin gene family demonstrated by phylogenetic procedures for resolving gene conversions and the phylogeny of angiosperm actin genes. *J. Mol. Evol.* **31**:132–150.

919. **Drummond, D. R., E. S. Hennessey, and J. C. Sparrow.** 1991. Characterisation in the missense mutations in the *Act88F* gene of *Drosophila melanogaster*. *Mol. Gen. Genet.* **226**:70–80.

920. **Drummond, D. R., E. S. Hennessey, and J. C. Sparrow.** 1991. Stability of mutant actins. *Biochem. J.* **274**:301–303.

921.* **Drummond, D. R., E. S. Hennessey, and J. C. Sparrow.** 1992. The binding of mutant actins to profilin, ATP and DNase I. *Eur. J. Biochem.* **209**:171–179.

922.* **Drummond, D. R., M. Peckham, J. C. Sparrow, and D. C. White.** 1990. Alteration in crossbridge kinetics caused by mutations in actin. *Nature* **348**:440–442.

923.* **Dudler, R.** 1990. The single-copy actin gene of *Phytophthora megasperma* encodes a protein considerably diverged from any other known actin. *Plant. Mol. Biol.* **14**:415–422.

924.* **Durica, D. S., D. Garza, M. A. Restrepo, and M. M. Hryniewicz.** 1988. DNA sequence analysis and structural relationships among the cytoskeletal actin genes of the sea urchin *Strongylocentrotus purpuratus. J. Mol. Evol.* **28**:72–86.

925. **Durica, D. S., J. A. Schloss, and W. R. J. Crain.** 1980. Organization of actin gene sequences in the sea urchin: molecular cloning of an intron-containing DNA sequence coding for a cytoplasmic actin. *Proc. Natl Acad. Sci. U. S. A.* **77**:5683–5687.

926.* **Edman, U., I. Meza, and N. Agabian.** 1987. Genomic and cDNA actin sequences from a virulent strain of *Entamoeba histolytica. Proc. Natl Acad. Sci. U. S. A.* **84**:3024–3028.

927. **Elder, P. K., C. L. French, M. Subramaniam, L. J. Schmidt, and M. J. Getz.** 1988. Evidence that the functional β-actin gene is single copy in most mice and is associated with 5′ sequences capable of conferring serum- and cycloheximide-dependent regulation. *Mol. Cell Biol.* **8**:480–485.

928. **Eldridge, J., Z. Zehner, and B. M. Paterson.** 1985. Nucleotide sequence of the chicken cardiac α actin gene: absence of strong homologies in the promoter and 3′-untranslated regions with the skeletal α-actin sequence. *Gene* **36**:55–63.

929. **Elzinga, M.** 1970. Amino acid sequence studies on rabbit skeletal muscle actin. Cyanogen bromide cleavage of the protein and determination of the sequences of seven of the resulting peptides. *Biochemistry* **9**:1365–1374.

930. **Elzinga, M.** 1971. Amino acid sequence around 3-methylhistidine in rabbit skeletal muscle actin. *Biochemistry* **10**:224–229.

931. **Elzinga, M. and J. H. Collins.** 1975. The primary structure of actin from rabbit skeletal muscle. Five cyanogen bromide peptides, including the NH_2 and COOH termini. *J. Biol. Chem.* **250**:5897–5905.

932. **Elzinga, M., J. H. Collins, W. M. Kuehl, and R. S. Adelstein.** 1973. Complete amino-acid sequence of actin of rabbit skeletal muscle. *Proc. Natl Acad. Sci. U. S. A.* **70**:2687–2691.

933. **Elzinga, M. and R. C. Lu.** 1976. Comparative amino acid sequence studies on actins. In: Perry, S. V. and Margreth, A. (eds) *Contractile Systems in Non-muscle Cells*, pp. 29–37. Amsterdam: North-Holland.

934. **Elzinga, M., B. J. Maron, and R. S. Adelstein.** 1976. Human heart and platelet actins are products of different genes. *Science* **191**:94–95.

935. **Engel, J. N., P. W. Gunning, and L. Kedes.** 1981. Isolation and characterization of human actin genes. *Proc. Natl Acad. Sci. U. S. A.* **78**:4674–4678.

936.* **Erba, H. P., R. Eddy, T. Shows, L. Kedes, and P. Gunning.** 1988. Structure, chromosome location, and expression of the human γ-actin gene: differential evolution, location, and expression of the cytoskeletal β- and γ-actin genes. *Mol. Cell Biol.* **8**:1775–1789.

937. **Erba, H. P., P. Gunning, and L. Kedes.** 1986. Nucleotide sequence of the human g cytoskeletal actin mRNA: anomalous evolution of vertebrate non-muscle actin genes. *Nucl. Acids Res.* **14**:5275–5294.

938.* **Fidel, S., J. H. Doonan, and N. R. Morris.** 1988. Aspergillus nidulans contains a single actin gene which has unique intron locations and encodes a γ-actin. *Gene* **70**:283–293.

939.* **Files, J. G., S. Carr, and D. Hirsh.** 1983. Actin gene family of *Caenorhabditis elegans. J. Mol. Biol.* **164**:355–375.

940. **Finlayson, P. J. and C. G. dos Remedios.** 1981. Differences between cardiac and skeletal muscle actins. *J. Mol. Cell Cardiol.* **13**:1081–1086.

941. **Firtel, R. A.** 1981. Multigene families encoding actin and tubulin. *Cell* **24**:6–7.

942. **Firtel, R. A., R. Timm, A. R. Kimmel, and M. McKeown.** 1979. Unusual nucleotide sequences at the 5′ end of actin genes in *Dictyostelium discoideum. Proc. Natl Acad. Sci. U. S. A.* **76**:6206–6210.

943.* **Fisher, D. A. and H. R. Bode.** 1989. Nucleotide sequence of an actin-encoding gene from *Hydra attenuata*: structural characteristics and evolutionary implications. *Gene* **84**:55–64.

943a.* **Fletcher L.D., McDowell J.M., Tidwell R.R., Meagher R.B. and Dykstra C.C.** (1994) Structure, expression and phylogenetic analysis of the gene encoding actin 1 in *Pneumocystis carinii. Genetics* **137**: 743–750.

944. **Flytzanis, C. N., E. A. Bogosian, and C. C. Niemeyer.** 1989. Expression and structure of the CyIIIb actin gene of the sea urchin *Strongylocentrotus purpuratus. Mol. Reprod. Dev.* **1**:208–218.

945.* **Foran, D. R., P. J. Johnson, and G. P. Moore.** 1985. Evolution of two actin genes in the sea urchin *Strongylocentrotus franciscanus. J. Mol. Evol.* **22**:108–116.

946. **Fornwald, J. A., G. Kuncio, I. Peng, and C. P. Ordahl.** 1982. The complete nucleotide sequence of the chick a-actin gene and its evolutionary relationship to the actin gene family. *Nucl. Acids Res.* **10**:3861–3876.

947. **Frankel, S., J. Condeelis, and L. Leinwand.** 1990. Expression of actin in *Escherichia coli*. Aggregation, solubilization, and functional analysis. *J. Biol. Chem.* **265**:17980–17987.

948.* **Frankel, S., M. B. Heintzelman, S. Artavanis Tsakonas, and M. S. Mooseker.** 1994. Identification of a divergent actin-related protein in *Drosophila. J. Mol. Biol.* **235**:1351–1356.

948a*. **Fyrberg C. and Fyrberg E.A.** 1993. A Drosophila homologue of the *Schizosaccharomyces pombe act2* gene. *Bioch. Gen.* **31**: 329–341.

949. **Fyrberg, E. A.** 1984. Structural and functional analyses of *Drosophila melanogaster* actin genes. *Oxf. Surv. Eukaryot. Genes* **1**:61–86.

950.* **Fyrberg, E. A., B. J. Bond, N. D. Hershey, K. S. Mixter, and N. Davidson.** 1981. The actin genes of *Drosophila*: protein coding regions are highly conserved but intron positions are not. *Cell* **24**:107–116.

951. **Fyrberg, E. A., K. L. Kindle, and N. Davidson.** 1980. The actin genes of *Drosophila*: a dispersed multigene family. *Cell* **19**:365–378.

952. **Gallwitz, D. and R. Seidel.** 1980. Molecular cloning of the actin gene from yeast *Saccharomyces cerevisiae. Nucl. Acids Res.* **8**:1043–1059.

953.* **Gallwitz, D. and I. Sures.** 1980. Structure of a split yeast gene: complete nucleotide sequence of the actin gene in *Saccharomyces cerevisiae. Proc. Natl Acad. Sci. U. S. A.* **77**:2546–2550.

954. **Garfinkel, L. I., M. Periasamy, and B. Nadal Ginard.** 1982. Cloning and characterization of cDNA sequences corresponding to myosin light chains 1, 2, and 3, troponin-C, troponin-T, α-tropomyosin, and α-actin. *J. Biol. Chem.* **257**:11078–11086.

955. **Gonzalez-y-Merchand, M. and R. A. Cox.** 1988. Structure and expression of an actin gene of *Physarum polycephalum. J. Mol. Biol.* **202**:161–168.

956.* **Greslin, A. F., S. H. Loukin, Y. Oka, and D. M. Prescott.** 1988. An analysis of the macronuclear actin genes of *Oxytricha. DNA* **7**:529–536.

957. **Greslin, A. F., D. M. Prescott, Y. Oka, S. H. Loukin, and J. C. Chappell.** 1989. Reordering of nine exons is necessary to form a functional actin gene in *Oxytricha nova. Proc. Natl Acad. Sci. U. S. A.* **86**:6264–6268.

958. **Gruenstein, E. and A. Rich.** 1975. Non-identity of muscle and non-muscle actins. *Biochem. Biophys. Res. Commun.* **64**:472–477.

959.* **Gunning, P., P. Ponte, H. Okayama, J. Engel, H. Blau, and L. Kedes.** 1983. Isolation and characterization of full-length cDNA clones for human α-, β-, and γ-actin mRNAs: skeletal but not cytoplasmic actins have an amino-terminal cysteine that is subsequently removed. *Mol. Cell Biol.* **3**:787–795.

960. **Hamada, H., J. Leavitt, and T. Kakunaga.** 1981. Mutated β-actin gene: coexpression with an unmutated allele in a chemically transformed human fibroblast cell line. *Proc. Natl Acad. Sci. U. S. A.* **78**:3634–3638.

961.* **Hamada, H., J. Leavitt, and T. Kakunaga.** 1983. Expression of mutated actin gene associated with malignant transformation. *Prog. Nucl. Acid. Res. Mol. Biol.* **29**:259–262.

962. **Hamada, H., M. G. Petrino, and T. Kakunaga.** 1982. Molecular structure and evolutionary origin of human cardiac muscle actin gene. *Proc. Natl Acad. Sci. U. S. A.* **79**:5901–5905.

963.* **Hamelin, M., L. Adam, G. Lemieux, and D. Pallotta.** 1988. Expression of the three unlinked isocoding actin genes of *Physarum polycephalum. DNA* **7**:317–328.

964. **Hanauer, A., M. Levin, R. Heilig, D. Daegelen, A. Kahn, and J. L. Mandel.** 1983. Isolation and characterization of cDNA clones for human skeletal muscle α actin. *Nucl. Acids Res.* **11**:3503–3516.

965. **Hanukoglu, I., N. Tanese, and E. Fuchs.** 1983. Complementary DNA sequence of a human cytoplasmic actin. Interspecies divergence of 3′ non-coding regions. *J. Mol. Biol.* **163**:673–678.

966. **Harper, D. S. and C. L. Jahn.** 1989. Actin, tubulin and H4 histone genes in three species of hypotrichous ciliated protozoa. *Gene* **75**:93–107.

967. **Harris, D. E. and D. M. Warshaw.** 1993. Smooth and skeletal muscle actin are mechanically indistinguishable in the *in vitro* motility assay. *Circ. Res.* **72**:219–224.

968. **Harris, D. E., D. M. Warshaw, and M. Periasamy.** 1992. Nucleotide sequences of the rabbit α-smooth-muscle and β-non-muscle actin mRNAs. *Gene* **112**:265–266.

969. **Hatanaka, K. and M. Okada.** 1991. Retarded nuclear migration in *Drosophila* embryos with aberrant F-actin reorganization caused by maternal mutations and by cytochalasin treatment. *Development.* **111**:909–920.

970. **Heilig, R., A. Hanauer, K. H. Grzeschik, M. C. Hors Cayla, and J. L. Mandel.** 1984. Actin-like sequences are present on human X and Y chromosomes. *EMBO J.* **3**:1803–1807.

971. **Hennessey, E. S., D. R. Drummond, and J. C. Sparrow.** 1991. Post-translational processing of the amino terminus affects actin function. *Eur. J. Biochem.* **197**:345–352.

972.* **Hennessey, E. S., D. R. Drummond, and J. C. Sparrow.** 1993. Molecular genetics of actin function. *Biochem. J.* **291**:657–671.

973. **Hennessey, E. S., A. Harrison, D. R. Drummond, and J. C. Sparrow.** 1992. Mutant actin: a dead end? *J. Muscle Res. Cell Motil.* **13**:127–131.

974. **Hickey, R. J., M. F. Boshar, and W. R. J. Crain.** 1987. Transcription of three actin genes and a repeated sequence in isolated nuclei of sea urchin embryos. *Dev. Biol.* **124**:215–227.

975.* **Hightower, R. C. and R. B. Meagher.** 1985. Divergence and differential expression of soybean actin genes. *EMBO J.* **4**:1–8.

976. **Hiromi, Y., H. Okamoto, W. J. Gehring, and Y. Hotta.** 1986. Germline transformation with *Drosophila* mutant actin genes induces constitutive expression of heat shock genes. *Cell* **44**:293–301.

977.* **Hirono, M., H. Endoh, N. Okada, O. Numata, and Y. Watanabe.** 1987. *Tetrahymena* actin. Cloning and sequencing of the *Tetrahymena* actin gene and identification of its gene product. *J. Mol. Biol.* **194**:181–192.

978.* **Holmes, K. C., C. Sander, and A. Valencia.** 1993. A new ATP-binding fold in actin, hexokinase and Hsc70. *Trends Cell Biol.* **3**:53–59.

979. **Horovitch, S. J., R. V. Storti, A. Rich, and M. L. Pardue.** 1979. Multiple actins in *Drosophila melanogaster. J. Cell Biol.* **82**:86–92.

980. **Hu, M. C., S. B. Sharp, and N. Davidson.** 1986. The complete sequence of the mouse skeletal α-actin gene reveals several conserved and inverted repeat sequences outside of the protein-coding region. *Mol. Cell Biol.* **6**:15–25.

981.* **Huber, M., L. Garfinkel, C. Gitler, D. Mirelman, M. Revel, and S. Rozenblatt.** 1987. Entamoeba histolytica: cloning and characterization of actin cDNA. *Mol. Biochem. Parasitol.* **24**:227–235.

982. **Hunter, T. and J. I. Garrels.** 1977. Characterization of the mRNAs for α-, β- and γ-actin. *Cell* **12**:767–781.

983.* **Johannes, F. J. and D. Gallwitz.** 1991. Site-directed mutagenesis of the yeast actin gene: a test for actin function *in vivo*. *EMBO J.* **10**:3951–3958.

984.* **Johara, M., Y. Y. Toyoshima, A. Ishijima, H. Kojima, T. Yanagida, and K. Sutoh.** 1993. Charge-reversion mutagenesis of *Dictyostelium* actin to map the surface recognized by myosin during ATP-driven sliding motion. *Proc. Natl Acad. Sci. U. S. A.* **90**:2127–2131.

985. **Johnson, P. J., D. R. Foran, and G. P. Moore.** 1983. Organization and evolution of the actin gene family in sea urchins. *Mol. Cell Biol.* **3**:1824–1833.

986.* **Kabsch, W., H. G. Mannherz, D. Suck, E. F. Pai, and K. C. Holmes.** 1990. Atomic structure of the actin:DNase I complex. *Nature* **347**:37–44.

987. **Kaine, B. P. and B. B. Spear.** 1980. Putative actin genes in the macronucleus of *Oxytricha fallax*. *Proc. Natl Acad. Sci. U. S. A.* **77**:5336–5340.

988.* **Kaine, B. P. and B. B. Spear.** 1982. Nucleotide sequence of a macronuclear gene for actin in *Oxytricha fallax*. *Nature* **295**:430–432.

989. **Kakunaga, T., J. Leavitt, and H. Hamada.** 1984. A mutation in actin associated with neoplastic transformation. *Fed. Proc.* **43**:2275–2279.

990.* **Kamada, S. and T. Kakunaga.** 1989. The nucleotide sequence of a human smooth muscle α-actin (aortic type) cDNA. *Nucl. Acids Res.* **17**:1767.

991. **Kamada, S., Y. Nakano, and T. Kakunaga.** 1989. Structure of 3′-downstream segment of the human smooth muscle (aortic-type) α-actin-encoding gene and isolation of the specific DNA probe. *Gene* **84**:455–462.

992.* **Kang, W. K. and Y. Naya.** 1993. Sequence of the cDNA encoding an actin homolog in the crayfish *Procambarus clarkii*. *Gene* **133**:303–304.

993. **Karlik, C. C., M. D. Coutu, and E. A. Fyrberg.** 1984. A nonsense mutation within the act88F actin gene disrupts myofibril formation in *Drosophila* indirect flight muscles. *Cell* **38**:711–719.

994. **Karlik, C. C., D. L. Saville, and E. A. Fyrberg.** 1987. Two missense alleles of the *Drosophila melanogaster* act88F actin gene are strongly antimorphic but only weakly induce synthesis of heat shock proteins. *Mol. Cell Biol.* **7**:3084–3091.

995.* **Karlsson, R.** 1988. Expression of chicken β-actin in *Saccharomyces cerevisiae*. *Gene* **68**:249–257.

996. **Kawakami, K., K. Omura, F. Ishida, Y. Munemoto, T. Hashimoto, T. Matsu, M. Kawanishi, T. Kawakami, K. Hirano, and Y. Watanabe.** 1993. β-actin mutation in advanced human gastric cancer. *Nippon. Shokakibyo. Gakkai. Zasshi.* **90**:800–803.

997. **Kedes, L., S. Y. Ng, C. S. Lin, P. Gunning, R. Eddy, T. Shows, and J. Leavitt.** 1985. The human β-actin multigene family. *Trans. Assoc. Am. Physicians* **98**:42–46.

998. **Kenny, J. R., B. P. Dancik, L. Z. Florence, and F. E. Nargang.** 1988. Partial sequence of the actin gene from *Pinus contorta* (Shore or Lodgepole pine). *Can. J. For. Res.* **18**:1595–1602.

999. **Khalili, K., C. Salas, and R. Weinmann.** 1983. Isolation and characterization of human actin genes cloned in phage lambda vectors. *Gene* **21**:9–17.

1000. **Khrestchatisky, M. and M. Fontes.** 1987. There is an α-actin skeletal muscle-specific gene in a salamander (*Pleurodeles waltlii*). *J. Mol. Biol.* **193**:409–412.

1001. **Kim, E., S. H. Waters, L. E. Hake, and N. B. Hecht.** 1989. Identification and developmental expression of a smooth-muscle γ-actin in postmeiotic male germ cells of mice. *Mol. Cell Biol.* **9**:1875–1881.

1002.* **Kim, K., L. Gooze, C. Petersen, J. Gut, and R. G. Nelson.** 1992. Isolation, sequence and molecular karyotype analysis of the actin gene of *Cryptosporidium parvum*. *Mol. Biochem. Parasitol.* **50**:105–113.

1003. **Kindle, K. L. and R. A. Firtel.** 1978. Identification and analysis of *Dictyostelium* actin genes, a family of moderately repeated genes. Cell **15**:763–778.

1004. **Knecht, D. A., S. M. Cohen, W. F. Loomis, and H. F. Lodish.** 1986. Developmental regulation of *Dictyostelium discoideum* actin gene fusions carried on low-copy and high-copy transformation vectors. *Mol. Cell Biol.* **6**:3973–3983.

1005.* **Kolattukudy, P. E., L. M. Rogers, A. J. Poulose, S. H. Jang, Y. S. Kim, T. M. Cheesbrough, and D. H. Liggitt.** 1987. Sequence of the β-actin gene from the goose. *Arch. Biochem. Biophys.* **256**:446–454.

1006. **Koncz, C., F. Kreuzaler, Z. Kalman, and J. Schell.** 1984. A simple method to transfer, integrate and study expression of foreign genes, such as chicken ovalbumin and α-actin in plant tumors. *EMBO J.* **3**:1029–1037.

1007.* **Kost, T. A., N. Theodorakis, and S. H. Hughes.** 1983. The nucleotide sequence of the chick cytoplasmic β-actin gene. *Nucl. Acids Res.* **11**:8287–8301.

1008. **Kovacs, A. M. and W. E. Zimmer.** 1993. Molecular cloning and expression of the chicken smooth muscle γ-actin mRNA. *Cell Motil. Cytoskeleton* **24**:67–81.

1009. **Kovesdi, I., F. Preugschat, M. Stuerzl, and M. J. Smith.** 1984. Actin genes from the sea star *Pisaster ochraceus*. *Biochim. Biophys. Acta* **782**:76–86.

1010.* **Kovilur, S., J. W. Jacobson, R. L. Beach, W. R. Jeffery, and C. R. Tomlinson.** 1993. Evolution of the chordate muscle actin gene. *J. Mol. Evol.* **36**:361–368.

1011.* **Kowbel, D. J. and M. J. Smith.** 1989. The genomic nucleotide sequences of two differentially expressed actin-coding genes from the sea star *Pisaster ochraceus*. *Gene* **77**:297–308.

1012. **Krause, M., M. Wild, B. Rosenzweig, and D. Hirsh.** 1989. Wild-type and mutant actin genes in *Caenorhabditis elegans*. *J. Mol. Biol.* **208**:381–392.

1013. **Kubler, E. and H. Riezman.** 1993. Actin and fimbrin are required for the internalization step of endocytosis in yeast. *EMBO J.* **12**:2855–2862.

1014. **Kunieda, T., H. Ikadai, M. Matsui, N. Nomura, R. Ishizaki, and T. Imamichi.** 1990. Polymorphisms detected in actin-related sequences of rats (*Rattus norvegicus*). *Jikken. Dobutsu.* **39**:307–310.

1015.* **Kusakabe, T., J. Suzuki, H. Saiga, W. R. Jeffery, K. W. Makabe, and N. Satoh.** 1991. Sequence of the actin gene from the sea squirt *Holocynthia roretzi*. *Dev. Growth. Differ. Nagoya.* **33**:227–234.

1016. **Landel, C. P., M. Krause, R. H. Waterston, and D. Hirsh.** 1984. DNA rearrangements of the actin gene cluster in *Caenorhabditis elegans* accompany reversion of three muscle mutants. *J. Mol. Biol.* **180**:497–513.

1017. **Larson, G. P., K. Itakura, H. Ito, and J. J. Rossi.** 1983. *Saccharomyces cerevisiae* actin – *Escherichia coli lacZ* gene fusions: synthetic-oligonucleotide-mediated deletion of the 309 base pair intervening sequence in the actin gene. *Gene* **22**:31–39.

1018. **Leader, D. P., I. Gall, P. Campbell, and A. M. Frischauf.** 1986. Isolation and characterization of cDNA clones from mouse skeletal muscle actin mRNA. *DNA* **5**:235–238.

1019. **Leader, D. P., I. Gall, and P. C. Campbell.** 1986. The structure of a cDNA clone corresponding to mouse cardiac muscle actin mRNA. *Biosci. Rep.* **6**:741–747.

1020. **Leavitt, J., G. Bushar, T. Kakunaga, H. Hamada, T. Hirakawa, D. Goldman, and C. Merril.** 1982. Variations in expression of mutant β actin accompanying incremental increases in human fibroblast tumorigenicity. *Cell* **28**:259–268.

1021. **Leavitt, J., D. Goldman, C. Merril, and T. Kakunaga.** 1982. Actin mutations in a human fibroblast model for carcinogenesis. *Clin. Chem.* **28**:850–860.

1022. **Leavitt, J., P. Gunning, P. Porreca, S. Y. Ng, C. S. Lin, and L. Kedes.** 1984. Molecular cloning and characterization of mutant and wild-type human β-actin genes. *Mol. Cell Biol.* **4**:1961–1969.

1023. **Leavitt, J. and T. Kakunaga.** 1980. Expression of a variant form of actin and additional polypeptide changes following chemical-induced *in vitro* neoplastic transformation of human fibroblasts. *J. Biol. Chem.* **255**:1650–1661.

1024.* **Leavitt, J., S. Y. Ng, U. Aebi, M. Varma, G. Latter, S. Burbeck, L. Kedes, and P. Gunning.** 1987. Expression of transfected mutant β-actin genes: alterations of cell morphology and evidence for autoregulation in actin pools. *Mol. Cell Biol.* **7**:2457–2466.

1025. **Leavitt, J., S. Y. Ng, M. Varma, G. Latter, S. Burbeck, P. Gunning, and L. Kedes.** 1987. Expression of transfected mutant β-actin genes: transitions toward the stable tumorigenic state. *Mol. Cell Biol.* **7**:2467–2476.

1026.* **Lees Miller, J. P., G. Henry, and D. M. Helfman.** 1992. Identification of *act2*, an essential gene in the fission yeast *Schizosaccharomyces pombe* that encodes a protein related to actin. *Proc. Natl Acad. Sci. U. S. A.* **89**:80–83.

1028. **Leube, R. and D. Gallwitz.** 1986. Structure of a processed gene of human cytoplasmic γ-actin. *Nucl. Acids Res.* **14**:6339.

1029.* **Lin, C. S., S. Y. Ng, P. Gunning, L. Kedes, and J. Leavitt.** 1985. Identification and order of sequential mutations in β-actin genes isolated from increasingly tumorigenic human fibroblast strains. *Proc. Natl Acad. Sci. U. S. A.* **82**:6995–6999.

1030. **Liu, Z., Z. Zhu, K. Roberg, A. J. Faras, K. S. Guise, A. R. Kapuscinski, and P. B. Hackett.** 1989. The β-actin gene of carp (*Ctenopharyngodon idella*). *Nucl. Acids Res.* **17**:5850.

1031.* **Liu, Z. J., Z. Y. Zhu, K. Roberg, A. Faras, K. Guise, A. R. Kapuscinski, and P. B. Hackett.** 1990. Isolation and characterization of β-actin gene of carp (*Cyprinus carpio*). *DNA Seq.* **1**:125–136.

1032.* **Losberger, C. and J. F. Ernst.** 1989. Sequence of the *Candida albicans* gene encoding actin. *Nucl. Acids Res.* **17**:9488.

1033. **Loukas, M. and F. C. Kafatos.** 1988. Chromosomal locations of actin genes are conserved between the *melanogaster* and *obscura* groups of *Drosophila*. *Genetica.* **76**:33–41.

1034. **Lu, R. C. and M. Elzinga.** 1977. Partial amino acid sequence of brain actin and its homology with muscle actin. *Biochemistry* **16**:5801–5806.

1035.* **Macias, M. T. and L. Sastre.** 1990. Molecular cloning and expression of four actin isoforms during *Artemia* development. *Nucl. Acids Res.* **18**:5219–5225.

1036.* **Manseau, L. J., B. Ganetzky, and E. A. Craig.** 1988. Molecular and genetic characterization of the *Drosophila melanogaster* 87E actin gene region. Genetics **119**:407–420.

1037. **Martin, D. J. and P. A. Rubenstein.** 1987. Alternate pathways for removal of the class II actin initiator methionine. *J. Biol. Chem.* **262**:6350–6356.

1038. **Mayer, Y., H. Czosnek, E. P. Zeelon, D. Yaffe, and U. Nudel.** 1984. Expression of the genes coding for the skeletal muscle and cardiac actins of the heart. *Nucl. Acids Res.* **12**:1087–1100.

1039.* **McElroy, D., M. Rothenberg, K. S. Reece, and R. Wu.** 1990. Characterization of the rice (*Oryza sativa*) actin gene family. *Plant Mol. Biol.* **15**:257–268.

1040. **McElroy, D., M. Rothenberg, and R. Wu.** 1990. Structural characterization of a rice actin gene. *Plant Mol. Biol.* **14**:163–171.

1041.* **McHugh, K. M. and J. L. Lessard.** 1988. The nucleotide sequence of a rat vascular smooth muscle α-actin cDNA. *Nucl. Acids Res.* **16**:4167.

1042. **McHugh, K. M. and J. L. Lessard.** 1988. The development expression of the rat α-vascular and γ-enteric smooth muscle isoactins: isolation and characterization of a rat γ-enteric actin cDNA. *Mol. Cell Biol.* **8**:5224–5231.

1043. **McKeown, M. and R. A. Firtel.** 1981. Differential expression and 5′ end mapping of actin genes in *Dictyostelium*. *Cell* **24**:799–807.

1044. **McKeown, M. and R. A. Firtel.** 1982. Actin multigene family of *Dictyostelium*. *Cold Spring Harb. Symp. Quant. Biol.* 46 Pt **2**:495–505.

1045. **McKeown, M., W. C. Taylor, K. L. Kindle, R. A. Firtel, W. Bender, and N. Davidson.** 1978. Multiple, heterogeneous actin genes in *Dictyostelium*. *Cell* **15**:789–800.

1046.* **McLean, B. G., S. R. Huang, E. C. McKinney, and R. B. Meagher.** 1990. Plants contain highly divergent actin isovariants. *Cell Motil. Cytoskeleton.* **17**:276–290.

1047. **McNally, E., R. Sohn, S. Frankel, and L. Leinwand.** 1991. Expression of myosin and actin in *Escherichia coli*. *Methods Enzymol.* **196**:368–389.

1048.* **Meagher, R. B.** 1991. Divergence and differential expression of actin gene families in higher plants. *Int. Rev. Cytol.* **125**:139–163.

1049. **Merlino, G. T., R. D. Water, J. P. Chamberlain, D. A. Jackson, M. R. El Gewely, and L. J. Kleinsmith.** 1980. Cloning of sea urchin actin gene sequences for use in studying the regulation of actin gene transcription. *Proc. Natl Acad. Sci. U. S. A.* **77**:765–769.

1050.* **Mertins, P. and D. Gallwitz.** 1987. A single intronless actin gene in the fission yeast *Schizosaccharomyces pombe*: nucleotide sequence and transcripts formed in homologous and heterologous yeasts. *Nucl. Acids Res.* **15**:7369–7379.

1051. **Min, B. H., A. R. Strauch, and D. N. Foster.** 1988. Nucleotide sequence of a mouse vascular smooth muscle α-actin cDNA. *Nucl. Acids Res.* **16**:10374.

1052. **Minty, A., M. Caravatti, B. Robert, A. Cohen, P. Daubas, A. Weydert, F. Gros, and M. Buckingham.** 1981. Muscle coding sequences and their regulation during myogenesis: cloning of muscle actin cDNA probes. *Reprod. Nutr. Dev.* **21**:247–255.

1053. **Minty, A. J., S. Alonso, M. Caravatti, and M. E. Buckingham.** 1982. A fetal skeletal muscle actin mRNA in the mouse and its identity with cardiac actin mRNA. *Cell* **30**:185–192.

1054. **Minty, A. J., M. Caravatti, B. Robert, A. Cohen, P. Daubas, A. Weydert, F. Gros, and M. E. Buckingham.** 1981. Mouse actin messenger RNAs. Construction and characterization of a recombinant plasmid molecule containing a complementary DNA transcript of mouse α-actin mRNA. *J. Biol. Chem.* **256**:1008–1014.

1055. **Miwa, T., S. Kamada, and T. Kakunaga.** 1990. The nucleotide sequence of a human smooth muscle (enteric type) γ-actin cDNA. *Nucl. Acids Res.* **18**:4263.

1056.* **Miwa, T., Y. Manabe, K. Kurokawa, S. Kamada, N. Kanda, G. Bruns, H. Ueyama, and T. Kakunaga.** 1991. Structure, chromosome location, and expression of the human smooth muscle (enteric type) γ-actin gene: evolution of six human actin genes. *Mol. Cell Biol.* **11**:3296–3306.

1057. **Mogami, K. and Y. Hotta.** 1981. Isolation of *Drosophila* flightless mutants which affect myofibrillar proteins of indirect flight muscle. *Mol. Gen. Genet.* **183**:409–417.

1058. **Mohun, T., N. Garrett, F. Stutz, and G. Sophr.** 1988. A third striated muscle actin gene is expressed during early development in the amphibian *Xenopus laevis*. *J. Mol. Biol.* **202**:67–76.

1059.* **Mohun, T. J. and N. Garrett.** 1987. An amphibian cytoskeletal-type actin gene is expressed exclusively in muscle tissue. *Development.* **101**:393–402.

1060. **Mohun, T. J., N. Garrett, and J. B. Gurdon.** 1986. Upstream sequences required for tissue-specific activation of the cardiac actin gene in *Xenopus laevis* embryos. *EMBO J.* **5**:3185–3193.

1061.* **Mounier, N., J. Gaillard, and J. C. Prudhomme.** 1987. Nucleotide sequence of the coding region of two actin genes in *Bombyx mori*. *Nucl. Acids Res.* **15**:2781.

1062. **Mounier, N., M. Gouy, D. Mouchiroud, and J. C. Prudhomme.** 1992. Insect muscle actins differ distinctly from invertebrate and vertebrate cytoplasmic actins. *J. Mol. Evol.* **34**:406–415.

1063.* **Mounier, N. and J. C. Prudhomme.** 1986. Isolation of actin genes in *Bombyx mori*: the coding sequence of a cytoplasmic actin gene expressed in the silk gland is interrupted by a single intron in an unusual position. *Biochimie.* **68**:1053–1061.

1064.* **Mounier, N. and J. C. Sparrow.** 1993. Muscle actin genes in insects. *Comp. Biochem. Physiol. B.* **105**:231–238.

1065.* **Nader, W. F., G. Isenberg, and H. W. Sauer.** 1986. Structure of *Physarum* actin gene locus ardA: a non-palindromic sequence causes inviability of phage lambda and recA-independent deletions. *Gene* **48**:133–144.

1066.* **Nagao, R. T., D. M. Shah, V. K. Eckenrode, and R. B. Meagher.** 1981. Multigene family of actin-related sequences isolated from a soybean genomic library. *DNA* **1**:1–9.

1067. **Naharro, G., K. C. Robbins, and E. P. Reddy.** 1984. Gene product of v-fgr onc: hybrid protein containing a portion of actin and a tyrosine-specific protein kinase. *Science* **223**:63–66.

1068.* **Nairn, C. J., L. Winesett, and R. J. Ferl.** 1988. Nucleotide sequence of an actin gene from *Arabidopsis thaliana*. *Gene* **65**:247–257.

1069. **Nakajima Iijima, S., H. Hamada, P. Reddy, and T. Kakunaga.** 1985. Molecular structure of the human cytoplasmic β-actin gene: interspecies homology of sequences in the introns. *Proc. Natl Acad. Sci. U. S. A.* **82**:6133–6137.

1070. **Nau, H., J. A. Kelley, and K. Biemann.** 1973. Determination of the amino acid sequence of the C-terminal cyanogen bromide fragment of actin by computer-assisted gas chromatography–mass spectrometry. *J. Am. Chem. Soc.* **95**:7162–7164.

1071.* **Nellen, W., C. Donath, M. Moos, and D. Gallwitz.** 1981. The nucleotide sequences of the actin genes from *Saccharomyces carlsbergensis* and *Saccharomyces cerevisiae* are identical except for their introns. *J. Mol. Appl. Genet.* **1**:239–244.

1072.* **Nellen, W. and D. Gallwitz.** 1982. Actin genes and actin messenger RNA in *Acanthamoeba castellaniii*. Nucleotide sequence of the split actin gene I. *J. Mol. Biol.* **159**:1–18.

1073.* **Ng, R. and J. Abelson.** 1980. Isolation and sequence of the gene for actin in Saccharomyces cerevisiae. *Proc. Natl Acad. Sci. U. S. A.* **77**:3912–3916.

1074.* **Ng, R., H. Domdey, G. Larson, J. J. Rossi, and J. Abelson.** 1985. A test for intron function in the yeast actin gene. *Nature* **314**:183–184.

1075. **Ng, S. Y., H. Erba, G. Latter, L. Kedes, and J. Leavitt.** 1988. Modulation of microfilament protein composition by transfected cytoskeletal actin genes. *Mol. Cell Biol.* **8**:1790–1794.

1076. **Ng, S. Y., P. Gunning, R. Eddy, P. Ponte, J. Leavitt, T. Shows, and L. Kedes.** 1985. Evolution of the functional human β-actin gene and its multi-pseudogene family: conservation of noncoding regions and chromosomal dispersion of pseudogenes. *Mol. Cell Biol.* **5**:2720–2732.

1077. **Nishizawa, M., K. Semba, T. Yamamoto, and K. Toyoshima.** 1985. Human c-fgr gene does not contain coding sequence for actin-like protein. *Jpn. J. Cancer Res.* **76**:155–159. (See also [1067]).

1078. **Novick, P. and D. Botstein.** 1985. Phenotypic analysis of temperature-sensitive yeast actin mutants. *Cell* **40**:405–416.

1079. **Novick, P., B. C. Osmond, and D. Botstein.** 1989. Suppressors of yeast actin mutations. *Genetics* **121**:659–674.

1080. **Nudel, U., D. Katcoff, R. Zakut, M. Shani, Y. Carmon, M. Finer, H. Czosnek, I. Ginsburg, and D. Yaffe.** 1982. Isolation and characterization of rat skeletal muscle and cytoplasmic actin genes. *Proc. Natl Acad. Sci. U. S. A.* **79**:2763–2767.

1081. **Nudel, U., R. Zakut, M. Shani, S. Neuman, Z. Levy, and D. Yaffe.** 1983. The nucleotide sequence of the rat cytoplasmic β-actin gene. *Nucl. Acids Res.* **11**:1759–1771.

1082. **Okamoto, H., Y. Hiromi, E. Ishikawa, T. Yamada, K. Isoda, H. Maekawa, and Y. Hotta.** 1986. Molecular characterisation of mutant actin genes which induce heat-shock proteins in *Drosophila* flight muscles. *EMBO J.* **5**:589–596.

1083. **Ordahl, C. P., S. M. Tilghman, C. Ovitt, J. Fornwald, and M. T. Largen.** 1980. Structure and developmental expression of the chick α-actin gene. *Nucl. Acids Res.* **8**:4989–5005.

1084. **Ortega, M. A., M. T. Macias, J. L. Martinez, I. Palmero, and L. Sastre.** 1992. Expression of actin isoforms in *Artemia*. *Symp. Soc. Exp. Biol.* **46**:131–137.

1085. **Overbeek, P. A., G. T. Merlino, N. K. Peters, V. H. Cohn, G. P. Moore, and L. J. Kleinsmith.** 1981. Characterization of five members of the actin gene family in the sea urchin. *Biochim. Biophys. Acta* **656**:195–205.

1086. **Pearson, L. and R. B. Meagher.** 1990. Diverse soybean actin transcripts contain a large intron in the 5′ untranslated leader: structural similarity to vertebrate muscle actin genes. *Plant. Mol. Biol.* **14**:513–526.

1087. **Peter, B., Y. M. Man, C. E. Begg, I. Gall, and D. P. Leader.** 1988. Mouse cytoskeletal γ-actin: analysis and implications of the structure of cloned cDNA and processed pseudogenes. *J. Mol. Biol.* **203**:665–675.

1088. **Prakash, K. and V. L. Seligy.** 1985. Oncogene related sequences in fungi: linkage of some to actin. *Biochem. Biophys. Res. Commun.* **133**:293–299.

1089. **Quax, W. J., H. J. Dodemont, J. A. Lenstra, F. C. Ramaekers, P. Soriano, M. E. van Workum, G. Bernardi, and H. Bloemendal.** 1982. Genes coding for vimentin and actin in mammals and birds. *Adv. Exp. Med. Biol.* **158**:349–357.

1090. **Reddy, S., K. Ozgur, M. Lu, W. Chang, S. R. Mohan, C. C. Kumar, and H. E. Ruley.** 1990. Structure of the human smooth muscle α-actin gene. Analysis of a cDNA and 5′ upstream region. *J. Biol. Chem.* **265**:1683–1687.

1091.* **Redman, K. and P. A. Rubenstein.** 1981. NH_2-terminal processing of *Dictyostelium discoideum* actin *in vitro*. *J. Biol. Chem.* **256**:13226–13229.

1092.* **Redman, K. L., D. J. Martin, E. D. Korn, and P. A. Rubenstein.** 1985. Lack of NH_2-terminal processing of actin from *Acanthamoeba castellaniii*. *J. Biol. Chem.* **260**:14857–14861.

1093.* **Reece, K. S., D. McElroy, and R. Wu.** 1990. Genomic nucleotide sequence of four rice (*Oryza sativa*) actin genes. *Plant Mol. Biol.* **14**:621–624.

1094. **Reedy, M. C., C. Beall, and E. Fyrberg.** 1989. Formation of reverse rigor chevrons by myosin heads. *Nature* **339**:481–483.

1095. **Reedy, M. C., C. Beall, and E. Fyrberg.** 1991. Mutations at the N-terminus of *Drosophila* actin. *Biophys. J.* **59**:187a.

1096. **Romans, P. and R. A. Firtel.** 1985. Organization of the actin multigene family of *Dictyostelium discoideum* and analysis of variability in the protein coding regions. *J. Mol. Biol.* **186**:321–335.

1097. **Rubenstein, P. A.** 1990. The functional importance of multiple actin isoforms. *Bioessays* **12**:309–315.

1098.* **Rubenstein, P. A. and D. J. Martin.** 1983. NH_2-terminal processing of *Drosophila melanogaster* actin. Sequential removal of two amino acids. *J. Biol. Chem.* **258**:11354–11360.

1099. **Rubenstein, P. A. and D. J. Martin.** 1983. NH_2-terminal processing of actin in mouse L-cells *in vivo*. *J. Biol. Chem.* **258**:3961–3966.

1100. **Rubenstein, P. A., L. R. Solomon, T. Solomon, and L. Gay.** 1989. Actin structure-function relationships *in vitro* using oligodeoxynucleotide-directed site-specific mutagenesis. *Cell Motil. Cytoskeleton.* **14**:35–39.

1101. **Saborio, J. L. and E. Palmer.** 1981. Brain actin synthesized *in vitro* undergoes two different and sequential posttranslational modifications. *J. Neurochem.* **36**:1659–1669.

1102.* **Sadano, H., S. Taniguchi, and T. Baba.** 1990. Newly identified type of β actin reduces invasiveness of mouse B16-melanoma. *FEBS Lett.* **271**:23–27.

1103. **Sadano, H., S. Taniguchi, T. Kakunaga, and T. Baba.** 1988. cDNA cloning and sequence of a new type of actin in mouse B16 melanoma. *J. Biol. Chem.* **263**:15868–15871.

1104. **Sakai, Y., H. Okamoto, K. Mogami, H. Matsuo, and Y. Hotta.** 1991. Heat shock gene activation by mutant actin is independent of myofibril degeneration in *Drosophila* muscle. *J. Biochem. Tokyo* **109**:670–673.

1105. **Sakai, Y., H. Okamoto, K. Mogami, T. Yamada, and Y. Hotta.** 1990. Actin with tumor-related mutation is antimorphic in *Drosophila* muscle: two distinct modes of myofibrillar disruption by antimorphic actins. *J. Biochem. Tokyo* **107**:499–505.

1106. **Sanchez, F., S. L. Tobin, U. Rdest, E. Zulauf, and B. J. McCarthy.** 1983. Two *Drosophila* actin genes in detail.: gene structure, protein structure and transcription during development. *J. Mol. Biol.* **163**:533–551.

1107. **Sasase, A., Y. Mishima, M. Ichihashi, and S. Taniguchi.** 1989. Biochemical analysis of metastasis-related Ax actin in B16 mouse melanoma cells after chemical reversional modulation and of tumor progression-related A actin in the ontogeny of human malignant melanoma. *Pigment Cell Res.* **2**:493–501.

1108. **Schade, B., W. Richter, P. Spangenberg, and P. Venkov.** 1991. Evidence that the osmotically fragile yeast *S. cerevisiae* VY1160 is an actin mutant. *Acta Histochem. Suppl.* **41**:193–200.

1109. **Schuler, M. A., P. McOsker, and E. B. Keller.** 1983. DNA sequence of two linked actin genes of sea urchin. *Mol. Cell Biol.* **3**:448–456.

1110. **Schwartz, R. J.** 1983. Recombinant DNA studies of muscle and nonmuscle actin genes. *Anat. Rec. Suppl.* **1**:27–50.

1111. **Schwartz, R. J., J. A. Haron, K. N. Rothblum, and A. Dugaiczyk.** 1980. Regulation of muscle differentiation: cloning of sequences from α-actin messenger ribonucleic acid. *Biochemistry* **19**:5883–5890.

1112. **Schwob, E. and R. P. Martin.** 1992. New yeast actin-like gene required late in the cell cycle. *Nature* **355**:179–182.

1113.* **Shah, D. M., R. C. Hightower, and R. B. Meagher.** 1983. Genes encoding actin in higher plants: intron positions are highly conserved but the coding sequences are not. *J. Mol. Appl. Genet.* **2**:111–126.

1114.* **Sheff, D. R. and P. A. Rubenstein.** 1992. Amino-terminal processing of actins mutagenized at the Cys-1 residue. *J. Biol. Chem.* **267**:2671–2678.

1115. **Shimokawa Kuroki, R., H. Sadano, and S. Taniguchi.** 1994. A variant actin (β m) reduces metastasis of mouse B16 melanoma. *Int. J. Cancer* **56**:689–697.

1116. **Shortle, D., P. Novick, and D. Botstein.** 1984. Construction and genetic characterization of temperature-sensitive mutant alleles of the yeast actin gene. *Proc. Natl Acad. Sci. U. S. A.* **81**:4889–4893.

1117. **Solomon, L. R. and P. A. Rubenstein.** 1987. Studies on the role of actin's N τ-methylhistidine using oligodeoxynucleotide-directed site-specific mutagenesis. *J. Biol. Chem.* **262**:11382–11388.

1118.* **Solomon, T. L., L. R. Solomon, L. S. Gay, and P. A. Rubenstein.** 1988. Studies on the role of actin's aspartic acid 3 and aspartic acid 11 using oligodeoxynucleotide-directed site-specific mutagenesis. *J. Biol. Chem.* **263**:19662–19669.

1119. **Sparrow, J., M. Reedy, E. Ball, V. Kyrtatas, J. Molloy, J. Durston, E. Hennessey, and D. White.** 1991. Functional and ultrastructural effects of a missense mutation in the indirect flight muscle-specific actin gene of *Drosophila melanogaster*. *J. Mol. Biol.* **222**:963–982.

1120.* **Sparrow, J. C., D. R. Drummond, E. S. Hennessey, J. D. Clayton, and F. B. Lindegaard.** 1992. *Drosophila* actin mutants and the study of myofibrillar assembly and function. *Symp. Soc. Exp. Biol.* **46**:111–129.

1121. **Storti, R. V. and A. Rich.** 1976. Chick cytoplasmic actin and muscle actin have different structural genes. *Proc. Natl Acad. Sci. U. S. A.* **73**:2346–2350.

1122.* **Stranathan, M., C. Hastings, H. Trinh, and J. L. Zimmerman.** 1989. Molecular evolution of 2 actin genes from carrot. *Plant Mol. Biol.* **13**:375–383.

1123. **Strauch, A. R. and P. A. Rubenstein.** 1984. A vascular smooth muscle α-isoactin biosynthetic intermediate in BC3H1 cells. *J. Biol. Chem.* **259**:7224–7229.

1124. **Stutz, F. and G. Spohr.** 1986. Isolation and characterization of sarcomeric actin genes expressed in *Xenopus laevis* embryos. *J. Mol. Biol.* **187**:349–361.

1125. **Stutz, F. and G. Spohr.** 1987. A processed gene coding for a sarcomeric actin in *Xenopus laevis* and *Xenopus tropicalis*. *EMBO J.* **6**:1989–1995.

1126. **Sutoh, K.** 1993. Identification of actin surface interacting with myosin during the actin-myosin sliding. *Adv. Exp. Med. Biol.* **332**:241–244.

1127.* **Sutoh, K., M. Ando, and Y. Y. Toyoshima.** 1991. Site-directed mutations of *Dictyostelium* actin: disruption of a negative charge cluster at the N terminus. *Proc. Natl Acad. Sci. U. S. A.* **88**:7711–7714.

1128.* **Tanaka, T., F. Shibasaki, M. Ishikawa, N. Hirano, R. Sakai, J. Nishida, T. Takenawa, and H. Hirai.** 1992. Molecular cloning of bovine actin-like protein, actin2. *Biochem. Biophys. Res. Commun.* **187**:1022–1028.

1129. **Taniguchi, S., T. Kawano, T. Kakunaga, and T. Baba.** 1986. Differences in expression of a variant actin between low and high metastatic B16 melanoma. *J. Biol. Chem.* **261**:6100–6106.

1130. **Taniguchi, S. and H. Sadano.** 1992. Biological and biochemical analysis of newly identified actin in mouse B16 melanoma. *Pigment Cell Res.* **2**(Suppl.):185–190.

1131. **Taniguchi, S., H. Sadano, T. Kakunaga, and T. Baba.** 1989. Altered expression of a third actin accompanying malignant progression in mouse B16 melanoma cells. *Jpn. J. Cancer Res.* **80**:31–40.

1132. **Taniguchi, S., J. Sagara, and T. Kakunaga.** 1988. Deficient polymerization *in vitro* of a point-mutated β-actin expressed in a transformed human fibroblast cell line. *J. Biochem. Tokyo* **103**:707–713.

1133. **Taylor, A., H. P. Erba, G. E. Muscat, and L. Kedes.** 1988. Nucleotide sequence and expression of the human skeletal α-actin gene: evolution of functional regulatory domains. *Genomics* **3**:323–336.

1134.* **Thangavelu, M., D. Belostotsky, M. W. Bevan, R. B. Flavell, H. J. Rogers, and D. M. Lonsdale.** 1993. Partial characterization of the Nicotiana tabacum actin gene family: evidence for pollen-specific expression of one of the gene family members. *Mol. Gen. Genet.* **240**:290–295.

1135. **Tobin, S. L., E. Zulauf, F. Sanchez, E. A. Craig, and B. J. McCarthy.** 1980. Multiple actin-related sequences in the *Drosophila melanogaster* genome. *Cell* **19**:121–131.

1136. **Tokunaga, K., K. Takeda, K. Kamiyama, H. Kageyama, K. Takenaga, and S. Sakiyama.** 1988. Isolation of cDNA clones for mouse cytoskeletal γ-actin and differential expression of cytoskeletal actin mRNAs in mouse cells. *Mol. Cell Biol.* **8**:3929–3933.

1137. **Tokunaga, K., H. Taniguchi, K. Yoda, M. Shimizu, and S. Sakiyama.** 1986. Nucleotide sequence of a full-length cDNA for mouse cytoskeletal β-actin mRNA. *Nucl. Acids Res.* **14**:2829.

1138. **Tomlinson, C. R., R. L. Beach, and W. R. Jeffery.** 1987. Differential expression of a muscle actin gene in muscle cell lineages of ascidian embryos. *Development.* **101**:751–765.

1139. **Tsang, A. S., H. Mahbubani, and J. G. Williams.** 1982. Cell-type-specific actin mRNA populations in *Dictyostelium discoideum*. *Cell* **31**:375–382.

1140. **Ueyama, H.** 1991. A *Hin*dIII DNA polymorphism in the human enteric type smooth muscle actin gene (ACTSG). *Nucl. Acids Res.* **19**:411.

1141. **Ueyama, H., H. Hamada, N. Battula, and T. Kakunaga.** 1984. Structure of a human smooth muscle actin gene (aortic type) with a unique intron site. *Mol. Cell Biol.* **4**:1073–1078.

1142.* **Unkles, S. E., R. P. Moon, A. R. Hawkins, J. M. Duncan, and J. R. Kinghorn.** 1991. Actin in the oomycetous fungus *Phytophthora infestans* is the product of several genes. *Gene* **100**:105–112.

1143. **Vandekerckhove, J., A. A. Lal, and E. D. Korn.** 1984. Amino acid sequence of *Acanthamoeba* actin. *J. Mol. Biol.* **172**:141–147.

1144. **Vandekerckhove, J., J. Leavitt, T. Kakunaga, and K. Weber.** 1980. Coexpression of a mutant β-actin and the two normal β- and γ-cytoplasmic actins in a stably transformed human cell line. *Cell* **22**:893–899.

1145. **Vandekerckhove, J. and K. Weber.** 1978. Actin amino-acid sequences. Comparison of actins from calf thymus, bovine brain, and SV40-transformed mouse 3T3 cells with rabbit skeletal muscle actin. *Eur. J. Biochem.* **90**:451–462.

1146.* **Vandekerckhove, J. and K. Weber.** 1978. The amino acid sequence of *Physarum* actin. *Nature* **276**:720–721.

1147. **Vandekerckhove, J. and K. Weber.** 1978. Comparison of the amino acid sequences of three tissue-specific cytoplasmic actins with rabbit skeletal muscle actin. *Arch. Int. Physiol. Biochim.* **86**:891–892.

1148. **Vandekerckhove, J. and K. Weber.** 1978. At least six different actins are expressed in a higher mammal: an analysis based on the amino acid sequence of the amino-terminal tryptic peptide. *J. Mol. Biol.* **126**:783–802.

1149. **Vandekerckhove, J. and K. Weber.** 1978. Mammalian cytoplasmic actins are the products of at least two genes and differ in primary structure in at least 25 identified positions from skeletal muscle actins. *Proc. Natl Acad. Sci. U. S. A.* **75**:1106–1110.

1150.* **Vandekerckhove, J. and K. Weber.** 1979. The complete amino acid sequence of actins from bovine aorta, bovine heart, bovine fast skeletal muscle, and rabbit slow skeletal muscle. A protein-chemical analysis of muscle actin differentiation. *Differentiation.* **14**:123–133.

1151.* **Vandekerckhove, J. and K. Weber.** 1979. The amino acid sequence of actin from chicken skeletal muscle actin and chicken gizzard smooth muscle actin. *FEBS Lett.* **102**:219–222.

1152. **Vandekerckhove, J. and K. Weber.** 1979. Amino-acid sequence analysis of the amino-terminal tryptic peptides of different actins from the same mammal. *Arch. Int. Physiol. Biochim.* **87**:210–212.

1153.* **Vandekerckhove, J. and K. Weber.** 1980. Vegetative *Dictyostelium* cells containing 17 actin genes express a single major actin. *Nature* **284**:475–477.

1154.* **Vandekerckhove, J. and K. Weber.** 1984. Chordate muscle actins differ distinctly from invertebrate muscle actins. The evolution of the different vertebrate muscle actins. *J. Mol. Biol.* **179**:391–413.

1155. **Varma, M., U. Aebi, J. Fleming, and J. Leavitt.** 1987. A 60-kDa polypeptide in mammalian cells with epitopes related to actin. *Exp. Cell Res.* **173**:163–173.

1156. **Waterston, R. H., D. Hirsh, and T. R. Lane.** 1984. Dominant mutations affecting muscle structure in *Caenorhabditis elegans* that map near the actin gene cluster. *J. Mol. Biol.* **180**:473–496.

1157.* **Weber, K. and W. Kabsch.** 1994. Intron positions in actin genes seem unrelated to the secondary structure of the protein. *EMBO J.* **13**:1280–1286.

1158. **Weihing, R. R. and E. D. Korn.** 1972. *Acanthamoeba* actin. Composition of the peptide that contains 3-methylhistidine and a peptide that contains N ϵ-methyllysine. *Biochemistry* **11**:1538–1543.

1159. **Welch, M. D., D. B. Vinh, H. H. Okamura, and D. G. Drubin.** 1993. Screens for extragenic mutations that fail to complement act1 alleles identify genes that are important for actin function in *Saccharomyces cerevisiae. Genetics* **135**:265–274.

1160.* **Wertman, K., D. G. Drubin, and D. Botstein.** 1992. Systematic mutational analysis at the yeast ACT1 gene. *Genetics* **132**:337–350.

1161. **Wertman, K. F. and D. G. Drubin.** 1992. Actin constitution: guaranteeing the right to assemble. *Science* **258**:759–760.

1162.* **Wesseling, J. G., J. M. de Ree, T. Ponnudurai, M. A. Smits, and J. G. Schoenmakers.** 1988. Nucleotide sequence and deduced amino acid sequence of a *Plasmodium falciparum* actin gene. *Mol. Biochem. Parasitol.* **27**:313–320.

1163. **Wesseling, J. G., M. A. Smits, and J. G. Schoenmakers.** 1988. Extremely diverged actin proteins in *Plasmodium falciparum. Mol. Biochem. Parasitol.* **30**:143–153.

1164. **Wesseling, J. G., P. J. Snijders, P. van Someren, J. Jansen, M. A. Smits, and J. G. Schoenmakers.** 1989. Stage-specific expression and genomic organization of the actin genes of the malaria parasite *Plasmodium falciparum. Mol. Biochem. Parasitol.* **35**:167–176.

1165.* **Wildeman, A. G.** 1988. A putative ancestral actin gene present in a thermophilic eukaryote: novel combination of intron positions. *Nucl. Acids Res.* **16**:2553–2564.

1166. **Wilson, C., G. S. Cross, and H. R. Woodland.** 1986. Tissue-specific expression of actin genes injected into *Xenopus* embryos. *Cell* **47**:589–599.

1167.* **Xia, D., B. Peng, D. A. Sesok, and I. Peng.** 1993. Probing actin incorporation into myofibrils using Asp11 and His73 actin mutants. *Cell Motil. Cytoskeleton* **26**:115–124.

1168. **Zafar, R. S. and A. Sodja.** 1983. Homology between actin coding and its adjacent sequences in widely divergent species. *Biochem. Biophys. Res. Commun.* **111**:67–73.

1169. **Zakut, R., M. Shani, D. Givol, S. Neuman, D. Yaffe, and U. Nudel.** 1982. Nucleotide sequence of the rat skeletal muscle actin gene. *Nature* **298**:857–859.

1170.* **Zeng, W. and J. E. Donelson.** 1992. The actin genes of *Onchocerca volvulus. Mol. Biochem. Parasitol.* **55**:207–216.

STRUCTURAL ANALYSIS OF ACTIN

1171. **Aebi, U., W. E. Fowler, G. Isenberg, T. D. Pollard, and P. R. Smith.** 1981. Crystalline actin sheets: their structure and polymorphism. *J. Cell Biol.* **91**: 340–351.

1172.* **Aebi, U., R. Millonig, H. Salvo, and A. Engel.** 1986. The three-dimensional structure of the actin filament revisited. *Ann. N. Y. Acad. Sci.* **483**:100–119.

1173. **Aebi, U., P. R. Smith, G. Isenberg, and T. D. Pollard.** 1980. Structure of crystalline actin sheets. *Nature* **288**:296–298.

1174. **Andersen, P., J. V. Small, H. K. Andersen, and A. Sobieszek.** 1979. Reactivity of smooth-muscle antibodies with F- and G-actin. *Immunology*. **37**:705–709.

1175.* **Ando, T.** 1989. Propagation of Acto-S-1 ATPase reaction-coupled conformational change in actin along the filament. *J. Biochem. Tokyo* **105**:818–822.

1176. **Ando, T. and H. Asai.** 1976. Cooperative conformational change in F-actin filament induced by the binding of heavy meromyosin. *J. Biochem. Tokyo.* **79**:1043–1047.

1177. **Baboonian, C., P. J. Venables, D. G. Williams, R. O. Williams, and R. N. Maini.** 1991. Cross reaction of antibodies to a glycine/alanine repeat sequence of Epstein–Barr virus nuclear antigen-1 with collagen, cytokeratin, and actin. *Ann. Rheum. Dis.* **50**:772–775.

1178. **Barden, J. A.** 1985. Fluorescence energy transfer between Tyr-69 and Cys-374 in actin. *Biochem. Int.* **11**:583–589.

1179. **Barden, J. A., R. Cooke, P. E. Wright, and C. G. dos Remedios.** 1980. Proton nuclear magnetic resonance and electron paramagnetic resonance studies on skeletal muscle actin indicate that the metal and nucleotide binding sites are separate. *Biochemistry* **19**:5912–5916.

1180. **Barden, J. A. and C. G. dos Remedios.** 1978. Evidence for the non-filamentous aggregation of actin induced by lanthanide ions. *Biochim. Biophys. Acta* **537**:417–427.

1181. **Barden, J. A. and C. G. dos Remedios.** 1980. Crystalline actin tubes. I. Is the conformation of the lanthanide-induced actin tube monomer more like F-actin than G-actin? *Biochim. Biophys. Acta* **624**:163–173.

1182. **Barden, J. A. and C. G. dos Remedios.** 1984. The environment of the high-affinity cation binding site on actin and the separation between cation and ATP sites as revealed by proton NMR and fluorescence spectroscopy. *J. Biochem. Tokyo* **96**:913–921.

1183. **Barden, J. A. and C. G. dos Remedios.** 1985. Conformational changes in actin resulting from Ca^{2+}/Mg^{2+} exchange as detected by proton NMR spectroscopy. *Eur. J. Biochem.* **146**:5–8.

1184. **Barden, J. A. and C. G. dos Remedios.** 1987. Fluorescence resonance energy transfer between sites in G-actin. The spatial relationship between Cys-10, Tyr-69, Cys-374, the high-affinity metal and the nucleotide binding sites. *Eur. J. Biochem.* **168**:103–109.

1185. **Barden, J. A. and M. Miki.** 1986. The distances separating Tyr-69 from the high-affinity nucleotide and metal binding sites in actin. *Biochem. Int.* **12**:321–329.

1186. **Barden, J. A., P. A. Tulloch, and C. G. dos Remedios.** 1981. Crystalline actin tubes. IV. Structural information on actin monomers obtained from computer-averaged lattice images. *J. Biochem. Tokyo* **90**:287–290.

1187. **Belagyi, J., P. Grof, G. Pallai, and J. Tigyi.** 1979. The study of nitroxide radical active esters as spin labels on muscle protein actin. *Acta Biochim. Biophys. Acad. Sci. Hung.* **14**:183–188.

1188. **Bork, P., C. Sander, and A. Valencia.** 1992. An ATPase domain common to prokaryotic cell cycle proteins, sugar kinases, actin, and hsp70 heat shock proteins. *Proc. Natl Acad. Sci. U. S. A.* **89**:7290–7294.

1189. **Brauer, M. and B. D. Sykes.** 1981. Phosphorus-31 nuclear magnetic resonance studies of adenosine 5′-triphosphate bound to a nitrated derivative of G-actin. *Biochemistry* **20**:6767–6775.

1190. **Brauer, M. and B. D. Sykes.** 1981. Phosphorus-31 nuclear magnetic resonance studies of the adenosine 5′-triphosphate–calcium–G-actin complex. *Biochemistry* **20**:2060–2064.

1191. **Brauer, M. and B. D. Sykes.** 1982. Effects of manganous ion on the phosphorus-31 nuclear magnetic resonance spectrum of adenosine triphosphate bound to nitrated G-actin: proximity of divalent metal ion and nucleotide binding sites. *Biochemistry* **21**:5934–5939.

1192. **Brauer, M. and B. D. Sykes.** 1986. 19F nuclear magnetic resonance studies of selectively fluorinated derivatives of G- and F-actin. *Biochemistry* **25**:2187–2191.

1193.* **Bremer, A. and U. Aebi.** 1992. The structure of the F-actin filament and the actin molecule. *Curr. Opin. Cell Biol.* **4**:20–26.

1194.* **Bremer, A., R. C. Millonig, R. Sutterlin, A. Engel, T. D. Pollard, and U. Aebi.** 1991. The structural basis for the intrinsic disorder of the actin filament: the "lateral slipping" model. *J. Cell Biol.* **115**:689–703.

1195.* **Bullitt, E. S., D. J. DeRosier, L. M. Coluccio, and L. G. Tilney.** 1988. Three-dimensional reconstruction of an actin bundle. *J. Cell Biol.* **107**:597–611.

1196. **Burley, R. W., J. C. Seidel, and J. Gergely.** 1971. The stoichiometry of the reaction of the spin labeling of F-actin and the effect of orientation of spin-labeled F-actin filaments. *Arch. Biochem. Biophys.* **146**:597–602.

1197. **Burley, R. W., J. C. Seidel, and J. Gergely.** 1972. The effect of divalent metal binding on the electron spin resonance spectra of spin-labeled actin. Evidence for spin–spin interactions involving manganese. II. *Arch. Biochem. Biophys.* **150**:792–796.

1198. **Callebaut, I., M. G. Catelli, D. Portetelle, A. Burny, E. E. Baulieu, and J. P. Mornon.** 1994. Structural similarities between chaperone molecules of the Hsp60 and Hsp70 families deduced from hydrophobic cluster analysis. *FEBS Lett.* **342**:242–248.

1199. **Carlsson, L., L. E. Nystrom, U. Lindberg, K. K. Kannan, H. Cid Dresdner, and S. Lovgren.** 1976. Crystallization of a non-muscle actin. *J. Mol. Biol.* **105**:353–366.

1200. **Casey, R., G. Creissen, C. Domoney, C. Forster, and N. Matta.** 1994. An intron-containing nuclear stress-70 gene from *Pisum sativum. Mol. Biol. Evol.* **11**:325–326.

1201. **Censullo, R. and H. C. Cheung.** 1993. A rotational offset model for two-stranded F-actin. *J. Struct. Biol.* **110**:75–83.

1202. **Cheung, H. C., R. Cooke, and L. Smith.** 1971. The G-actin is greater than F-actin transformation as studied by the fluorescence of bound dansyl cystine. *Arch. Biochem. Biophys.* **142**:333–339.

1203. **Cheung, H. C. and B. M. Liu.** 1984. Distance between nucleotide site and cysteine-373 of G-actin by resonance energy transfer measurements. *J. Muscle Res. Cell Motil.* **5**:65–80.

1204. **Chu, M. L., R. Z. Zhang, T. C. Pan, D. Stokes, D. Conway, H. J. Kuo, R. Glanville, U. Mayer, K. Mann, R. Deutzmann, and R. Timpl.** 1990. Mosaic structure of globular domains in the human type VI collagen α3 chain: similarity to von Willebrand factor, fibronectin, actin, salivary proteins and aprotinin type protease inhibitors. *EMBO J.* **9**:385–393.

1205. **Contaxis, C. C., C. C. Bigelow, and C. G. Zarkadas.** 1977. The thermal denaturation of bovine cardiac G-actin. *Can. J. Biochem.* **55**:325–331.

1206. **Coppin, C. M. and P. C. Leavis.** 1992. Quantitation of liquid-crystalline ordering in F-actin solutions. *Biophys. J.* **63**:794–807.

1207. **Cozzone, P. J., D. J. Nelson, and O. Jardetzky.** 1974. Fourier transform phosphorus magnetic resonance study of ATP–calcium–G-actin complex. *Biochem. Biophys. Res. Commun.* **60**:341–347.

1208. **Curmi, P. M., J. A. Barden, and C. G. dos Remedios.** 1984. Actin tube formation: effects of variations in commonly used solvent conditions. *J. Muscle Res. Cell Motil.* **5**:423–430.

1209.* **De Rosier, D. J.** 1990. Protein structure. The changing shape of actin. *Nature* **347**:21–22.

1210. **DeRosier, D., E. Mandelkow, and A. Silliman.** 1977. Structure of actin-containing filaments from two types of non-muscle cells. *J. Mol. Biol.* **113**:679–695.

1211. **DeRosier, D., L. Tilney, and P. Flicker.** 1980. A change in the twist of the actin-containing filaments occurs during the extension of the acrosomal process in *Limulus* sperm. *J. Mol. Biol.* **137**:375–389.

1212.* **DeRosier, D. J. and L. G. Tilney.** 1982. How actin filaments pack into bundles. *Cold Spring Harb. Symp. Quant. Biol.* 46 Pt **2**:525–540.

1213. **DeRosier, D. J. and L. G. Tilney.** 1984. The form and function of actin. A product of its unique design. *Cell Muscle Motil.* **5**:139–169.

1214. **DeRosier, D. J. and L. G. Tilney.** 1984. How to build a bend into an actin bundle. *J. Mol. Biol.* **175**:57–73.

1215. **dos Remedios, C. G., J. A. Barden, and A. A. Valois.** 1980. Crystalline actin tubes. II. The effect of various lanthanide ions on actin tube formation. *Biochim. Biophys. Acta* **624**:174–186.

1216. **dos Remedios, C. G. and M. J. Dickens.** 1978. Actin microcrystals and tubes formed in the presence of gadolinium ions. *Nature* **276**:731–733.

1217. **dos Remedios, C. G., M. Miki, and J. A. Barden.** 1987. Fluorescence resonance energy transfer measurements of distances in actin and myosin. A critical evaluation. *J. Muscle Res. Cell Motil.* **8**:97–117.

1218. **Drewes, G. and H. Faulstich.** 1991. A reversible conformational transition in muscle actin is caused by nucleotide exchange and uncovers cysteine in position 10. *J. Biol. Chem.* **266**:5508–5513.

1219. **Egelman, E. H.** 1985. The structure of F-actin. *J. Muscle Res. Cell Motil.* **6**:129–151.

1220. **Egelman, E. H.** 1994. The ghost of ribbons past. *Curr. Biol.* **4**:79–81.

1221. **Egelman, E. H. and D. J. DeRosier.** 1991. Angular disorder in actin: is it consistent with general principles of protein structure? *J. Mol. Biol.* **217**:405–408.

1222.* **Egelman, E. H. and D. J. DeRosier.** 1992. Image analysis shows that variations in actin crossover spacings are random, not compensatory. *Biophys. J.* **63**:1299–1305.

1223. **Egelman, E. H., N. Francis, and D. J. DeRosier.** 1982. F-actin is a helix with a random variable twist. *Nature* **298**:131–135.

1224. **Egelman, E. H., N. Francis, and D. J. DeRosier.** 1983. Helical disorder and the filament structure of F-actin are elucidated by the angle-layered aggregate. *J. Mol. Biol.* **166**:605–629.

1225. **Erickson, H. P.** 1989. Co-operativity in protein-protein association. The structure and stability of the actin filament. *J. Mol. Biol.* **206**:465–474.

1226. **Fausnaugh, J. L., J. F. Blazyk, S. C. el Saleh, and P. Johnson.** 1984. Calorimetric studies on monomeric and polymeric actin. *Experientia* **40**:83–84.

1227. **Fievez, S. and M. F. Carlier.** 1993. Conformational changes in subdomain-2 of G-actin upon polymerization into F-actin and upon binding myosin subfragment-1. *FEBS Lett.* **316**:186–190.

1228. **Fisher, A. J., P. M. Curmi, J. A. Barden, and C. G. dos Remedios.** 1983. A re-investigation of actin monomer conformation under polymerizing conditions based on rates of enzymatic digestion and ultraviolet difference spectroscopy. *Biochim. Biophys. Acta* **748**:220–229.

1229.* **Flaherty, K. M., D. B. McKay, W. Kabsch, and K. C. Holmes.** 1991. Similarity of the three-dimensional structures of actin and the ATPase fragment of a 70-kDa heat shock cognate protein. *Proc. Natl Acad. Sci. U. S. A.* **88**:5041–5045.

1230.* **Flicker, P. F., R. A. Milligan, and D. Applegate.** 1991. Cryo-electron microscopy of S1-decorated actin filaments. *Adv. Biophys.* **27**:185–196.

1231. **Fowler, W. E. and U. Aebi.** 1982. Polymorphism of actin paracrystals induced by polylysine. *J. Cell Biol.* **93**:452–458.

1232. **Fowler, W. E. and U. Aebi.** 1983. A consistent picture of the actin filament related to the orientation of the actin molecule. *J. Cell Biol.* **97**:264–269.

1233. **Fowler, W. E., E. L. Buhle, and U. Aebi.** 1984. Tubular arrays of the actin–DNase I complex induced by gadolinium. *Proc. Natl Acad. Sci. U. S. A.* **81**:1669–1673.

1234. **Francis, N. R. and D. J. DeRosier.** 1990. A polymorphism peculiar to bipolar actin bundles. *Biophys. J.* **58**:771–776.

1235. **Fujime, S., S. Ishiwata, and T. Maeda.** 1984. Dynamic light scattering study of muscle F-actin. *Biophys. Chem.* **20**:1–21.

1236. **Furukawa, R., R. Kundra, and M. Fechheimer.** 1993. Formation of liquid crystals from actin filaments. *Biochemistry* **32**:12346–12352.

1237. **Gao, B. C., L. Green, and E. Eisenberg.** 1994. Characterisation of nucleotide-free uncoating ATPase and its binding to ATP, ADP and ATP analogues. *Biochemistry* **33**:2048–2054.

1238. **Gao, Y., J. O. Thomas, R. L. Chow, G. H. Lee, and N. J. Cowan.** 1992. A cytoplasmic chaperonin that catalyzes β-actin folding. *Cell* **69**:1043–1050.

1239. **Garcia, C. R., J. A. Amaral Junior, P. Abrahamsohn, and S. Verjovski Almeida.** 1992. Dissociation of F-actin induced by hydrostatic pressure. *Eur. J. Biochem.* **209**:1005–1011.

1240. **Graham, G. N. and D. B. Sellen.** 1971. Light-scattering Raleigh linewidth measurements on G-actin. *Biochem. J.* **125**:104P-105P.

1241. **Grazi, E., E. Magri, C. Schwienbacher, and G. Trombetta.** 1994. Actin may contribute to the power stroke in the binary actomyosin system. *Biochem. Biophys. Res. Commun.* **200**:59–64.

1242. **Grazi, E., C. Schwienbacher, and E. Magri.** 1993. Osmotic stress is the main determinant of the diameter of the actin filament. *Biochem. Biophys. Res. Commun.* **197**:1377–1381.

1243. **Greene, P. R.** 1985. Thermal torsion of F-actin. *J. Theor. Biol.* **117**:489–492.

1244. **Hanson, J.** 1973. Evidence from electron microscope studies on actin paracrystals concerning the origin of the cross-striation in the thin filaments of vertebrate skeletal muscle. *Proc. R. Soc. Lond. Biol.* **183**:39–58.

1245. **Harvey, S. C., H. C. Cheung, and K. E. Thames.** 1977. Cooperativity in F-actin filaments on binding of myosin subfragments, demonstrated by fluorescence of 1, N6-ethenoadenosine diphosphate. *Arch. Biochem. Biophys.* **179**:391–396.

1246. **Harwell, O. D., M. L. Sweeney, and F. H. Kirkpatrick.** 1980. Conformation changes of actin during formation of filaments and paracrystals and upon interaction with DNase I, cytochalasin B, and phalloidin. *J. Biol. Chem.* **255**:1210–1220.

1247. **Haus, U., P. Trommler, P. R. Fisher, H. Hartmann, F. Lottspeich, A. A. Noegel, and M. Schleicher.** 1993. The heat shock cognate protein from *Dictyostelium* affects actin polymerization through interaction with the actin-binding protein cap32/34. EMBO J. **12**:3763–3771.

1248.* **Heuser, J. E. and R. Cooke.** 1983. Actin–myosin interactions visualized by the quick-freeze, deep-etch replica technique. *J. Mol. Biol.* **169**:97–122.

1249. **Hirokawa, N., L. G. Tilney, K. Fujiwara, and J. E. Heuser.** 1982. Organization of actin, myosin, and intermediate filaments in the brush border of intestinal epithelial cells. *J. Cell Biol.* **94**:425–443.

1250. **Holmes, K. C., R. S. Goody, and L. A. Amos.** 1982. The structure of S1-decorated actin filaments calculated from x-ray diffraction data with phases derived from electron micrographs. *Ultramicroscopy* **9**:37–44.

1251.* **Holmes, K. C., D. Popp, W. Gebhard, and W. Kabsch.** 1990. Atomic model of the actin filament. *Nature* **347**:44–49.

1252.* **Holmes, K. C., C. Sander, and A. Valencia.** 1993. A new ATP-binding fold in actin, hexokinase and Hsc70. *Trends Cell Biol.* **3**:53–59.

1253.* **Holmes, K. C., M. Tirion, D. Popp, M. Lorenz, W. Kabsch, and R. A. Milligan.** 1993. A comparison of the atomic model of F-actin with cryo-electron micrographs of actin and decorated actin. *Adv. Exp. Med. Biol.* **332**:15–22.

1254. **Horie, T. and J. M. Vanderkooi.** 1982. Phosphorescence of tryptophan from parvalbumin and actin in liquid solution. *FEBS Lett.* **147**:69–73.

1255. **Hue, H. K., Y. Benyamin, and C. Roustan.** 1989. Comparative study of invertebrate actins: antigenic cross-reactivity versus sequence variability. *J. Muscle Res. Cell Motil.* **10**:135–142.

1256. **Iio, T., S. Takahashi, and S. Sawada.** 1993. Fluorescent molecular rotor binding to actin. *J. Biochem. Tokyo* **113**:196–199.

1257. **Ikeuchi, Y., T. Ito, and T. Fukazawa.** 1981. A kinetic analysis of thermal denaturation of F-actin. *Int. J. Biochem.* **13**:1065–1069.

1258. **Ikkai, T., P. Wahl, and J. C. Auchet.** 1979. Anisotropy decay of labelled actin. Evidence of the flexibility of the peptide chain in F-actin molecules. *Eur. J. Biochem.* **93**:397–408.

1259. **Ishiwata, S.** 1976. Freezing of actin. Reversible oxidation of a sulfhydryl group and structural change. *J. Biochem. Tokyo* **80**:595–609.

1260. **Kabsch, W., H. G. Mannherz, and D. Suck.** 1985. Three-dimensional structure of the complex of actin and DNase I at 4.5 A resolution. *EMBO J.* **4**:2113–2118.

1261.* **Kabsch, W., H. G. Mannherz, D. Suck, E. F. Pai, and K. C. Holmes.** 1990. Atomic structure of the actin:DNase I complex. *Nature* **347**:37–44.

1262.* **Kabsch, W. and J. Vandekerckhove.** 1992. Structure and function of actin. *Annu. Rev. Biophys. Biomol. Struct.* **21**:49–76.

1263. **Kakar, S. and F. A. Bettelheim.** 1991. The hydration of actin. *Invest. Ophthalmol. Vis. Sci.* **32**:562–566.

1264. **Kasai, M. and H. Hama.** 1969. Natural F-actin. 3. Natural F-actin as inactive polymer. *Biochim. Biophys. Acta* **180**:550–561.

1265. **Kasprzak, A. A., R. Takashi, and M. F. Morales.** 1988. Orientation of actin monomer in the F-actin filament: radial coordinate of glutamine-41 and effect of myosin subfragment 1 binding on the monomer orientation. *Biochemistry* **27**:4512–4522.

1266. **Katayama, E. and T. Wakabayashi.** 1981. Three-dimensional image analysis of the complex of thin filaments and myosin molecules from skeletal muscle. III. The multi-domain structure of actin-heavy meromyosin complex. *J. Biochem. Tokyo* **90**:703–714.

1267. **Kawamura, M. and K. Maruyama.** 1970. Polymorphism of F-actin. I. Three forms of paracrystals. *J. Biochem. Tokyo* **68**:885–899.

1268. **Kinosita, K. J., N. Suzuki, S. Ishiwata, T. Nishizaka, H. Itoh, H. Hakozaki, G. Marriott, and H. Miyata.** 1993. Orientation of actin monomers in moving actin filaments. *Adv. Exp. Med. Biol.* **332**:321–328.

1269. **Konno, K.** 1988. G-actin structure revealed by chymotryptic digestion. *J. Biochem. Tokyo* **103**:386–392.

1270. **Kreuder, V., J. Dieckhoff, M. Sittig, and H. G. Mannherz.** 1984. Isolation, characterisation and crystallization of deoxyribonuclease I from bovine and rat parotid gland and its interaction with rabbit skeletal muscle actin. *Eur. J. Biochem.* **139**:389–400.

1271. **Kuroda, M. and K. Maruyama.** 1972. Polymorphism of F-actin. II. ATPase activity at acid pH. *J. Biochem. Tokyo* **71**:39–45.

1272. **Le Bihan, T. and C. Gicquaud.** 1993. Kinetic study of the thermal denaturation of G actin using differential scanning calorimetry and intrinsic fluorescence spectroscopy. *Biochem. Biophys. Res. Commun.* **194**:1065–1073.

1273. **Lehrer, S. S. and G. Kerwar.** 1972. Intrinsic fluorescence of actin. *Biochemistry* **11**:1211–1217.

1274. **Levine, B. A., A. J. Moir, A. J. Goodearl, and I. P. Trayer.** 1991. Characterization of the nucleotide-modulated interaction of myth actin by n.m.r. spectroscopy. *Biochem. Soc. Trans.* **19**:423–425.

1275. **Lin, E. C. and H. F. Cantiello.** 1993. A novel method to study the electrodynamic behavior of actin filaments. Evidence for cable-like properties of actin. *Biophys. J.* **65**:1371–1378.

1276. **Lorenz, M., D. Popp, and K. C. Holmes.** 1993. Refinement of the F-actin model against X-ray fiber diffraction data by the use of a directed mutation algorithm. *J. Mol. Biol.* **234**:826–836.

1277. **Loscalzo, J. and G. H. Reed.** 1976. Spectroscopic studies of actin–metal–nucleotide complexes. *Biochemistry* **15**:5407–5413.

1278. **Loscalzo, J., G. H. Reed, and A. Weber.** 1975. Conformational change and cooperativity in actin filaments free of tropomyosin. *Proc. Natl Acad. Sci. U. S. A.* **72**:3412–3415.

1279. **Maeda, Y., C. Boulin, A. Gabriel, I. Sumner, and M. H. Koch.** 1986. Intensity increases of actin layer-lines on activation of the *Limulus* muscle. *Biophys. J.* **50**:1035–1042.
1280. **Mannherz, H. G.** 1992. Crystallization of actin in complex with actin-binding proteins. *J. Biol. Chem.* **267**:11661–11664.
1281. **Mannherz, H. G., W. Kabsch, and R. Leverman.** 1977. Crystals of skeletal muscle actin: pancreatic DNAase I complex. *FEBS Lett.* **73**:141–143.
1282. **McCubbin, W. D. and C. M. Kay.** 1970. Physicochemical studies on bovine cardiac globular actin. *Biochim. Biophys. Acta* **214**:272–281.
1283. **McCubbin, W. D., K. Oikawa, and C. M. Kay.** 1981. The effect of terbium on the structure of actin and myosin subfragment 1 as measured by circular dichroism. *FEBS Lett.* **127**:245–249.
1284.* **McLaughlin, P. J., J. T. Gooch, H. G. Mannherz, and A. G. Weeds.** 1993. Structure of gelsolin segment 1-actin complex and the mechanism of filament severing. *Nature* **364**:685–692.
1285. **Melki, R. and N. J. Cowan.** 1994. Facilitated folding of actins and tubulins occurs via a nucleotide-dependent interaction between cytoplasmic chaperonin and distinctive folding intermediates. *Mol. Cell Biol.* **14**:2895–2904.
1286. **Mihashi, K.** 1972. Alkylation of urea-denatured actin with iodoacetate: functional reorganization of actin. *Biochim. Biophys. Acta* **267**:409–421.
1287. **Mihashi, K., H. Yoshimura, T. Nishio, A. Ikegami, and K. J. Kinosita.** 1983. Internal motion of F-actin in 10^{-6}–10^{-3} s time range studied by transient absorption anisotropy: detection of torsional motion. *J. Biochem. Tokyo* **93**:1705–1707.
1288. **Miki, M.** 1991. Detection of conformational changes in actin by fluorescence resonance energy transfer between tyrosine-69 and cysteine-374. *Biochemistry* **30**:10878–10884.
1289. **Miki, M., J. A. Barden, B. D. Hambly, and C. G. dos Remedios.** 1986. Fluorescence energy transfer between Cys-10 residues in F-actin filaments. *Biochem. Int.* **12**:725–731.
1290. **Miki, M. and C. G. dos Remedios.** 1990. A determination of the radial coordinate of Tyr-69 in F-actin using fluorescence energy transfer. *Biochem. Int.* **22**:125–132.
1291. **Miki, M., C. G. dos Remedios, and J. A. Barden.** 1987. Spatial relationship between the nucleotide-binding site, Lys-61 and Cys-374 in actin and a conformational change induced by myosin subfragment-1 binding. *Eur. J. Biochem.* **168**:339–345.
1292. **Miki, M. and K. Mihashi.** 1978. Fluorescence energy transfer between epsilon-ATP at the nucleotide binding site and N-(4-dimethylamino-3,5-dinitrophenyl)-maleimide at Cys-373 of G-actin. *Biochim. Biophys. Acta* **533**:163–172.
1293.* **Miki, M., S. I. O'Donoghue, and C. G. dos Remedios.** 1992. Structure of actin observed by fluorescence resonance energy transfer spectroscopy. *J. Muscle Res. Cell Motil.* **13**:132–145.
1294. **Miki, M., P. Wahl, and J. C. Auchet.** 1982. Fluorescence anisotropy of labelled F-actin. Influence of Ca^{2+} on the flexibility of F-actin. *Biophys. Chem.* **16**:165–172.
1295. **Milligan, R. A. and P. F. Flicker.** 1987. Structural relationships of actin, myosin, and tropomyosin revealed by cryo-electron microscopy. *J. Cell Biol.* **105**:29–39.
1296.* **Milligan, R. A., M. Whittaker, and D. Safer.** 1990. Molecular structure of F-actin and location of surface binding sites. *Nature* **348**:217–221.
1297. **Moncman, C. L., I. Peng, and D. A. Winkelmann.** 1993. Actin filament structure probed with monoclonal antibodies. *Cell Motil. Cytoskeleton* **25**:73–86.
1298. **Montague, C., K. W. Rhee, and F. D. Carlson.** 1983. Measurement of the translational diffusion constant of G-actin by photon correlation spectroscopy. *J. Muscle Res. Cell Motil.* **4**:95–101.
1299.* **Moore, P. B., H. E. Huxley, and D. J. DeRosier.** 1970. Three-dimensional reconstruction of F-actin, thin filaments and decorated thin filaments. *J. Mol. Biol.* **50**:279–295.
1300. **Murphy, A. J.** 1971. Circular dichroism of the adenine and 6-mercaptopurine nucleotide complexes of actin. *Biochemistry* **10**:3723–3728.
1301. **Naber, N. and R. Cooke.** 1994. Mobility and orientation of spin probes attached to nucleotides incorporated into actin. *Biochemistry* **33**:3855–3861.
1302. **Naber, N., M. Lorenz, and R. Cooke.** 1994. The orientation of spin-probes attached to Cys374 on actin in oriented gels. *J. Mol. Biol.* **236**:703–709.
1303. **Nagashima, H. and S. Asakura.** 1980. Dark-field light microscopic study of the flexibility of F-actin complexes. *J. Mol. Biol.* **136**:169–182.
1304. **Nagy, B.** 1966. Optical rotatory dispersion of G-actin-adenosine-5′-diphosphate. *Biochim. Biophys. Acta* **115**:498–500.
1305. **Noda, S., M. Ito, S. Watanabe, K. Takahashi, and K. Maruyama.** 1992. Conformational changes of actin induced by calponin. *Biochem. Biophys. Res. Commun.* **185**:481–487.
1306. **O'Donoghue, S. I., B. D. Hambly, and C. G. dos Remedios.** 1992. Models of the actin monomer and filament from fluorescence resonance-energy transfer. *Eur. J. Biochem.* **205**:591–601.
1307. **O'Donoghue, S. I., M. Miki, and C. G. dos Remedios.** 1992. Removing the two C-terminal residues of actin affects the filament structure. *Arch. Biochem. Biophys.* **293**:110–116.
1308. **Oosawa, F.** 1980. The flexibility of F-actin. *Biophys. Chem.* **11**:443–446.
1309. **Oosawa, F., Y. Maeda, S. Fujime, S. Ishiwata, T. Yanagida, and M. Taniguchi.** 1977. Dynamic characteristics of F-actin and thin filaments *in vivo and in vitro. J. Mechanochem. Cell Motil.* **4**:63–78.
1310. **Oriol, C., C. Dubord, and F. Landon.** 1977. Crystallization of native striated-muscle actin. *FEBS Lett.* **73**:89–91.
1311.* **Orlova, A. and E. H. Egelman.** 1992. Structural basis for the destabilization of F-actin by phosphate release following ATP hydrolysis. *J. Mol. Biol.* **227**:1043–1053.
1312.* **Orlova, A. and E. H. Egelman.** 1993. A conformational change in the actin subunit can change the flexibility of the actin filament. *J. Mol. Biol.* **232**:334–341.
1313. **Orlova, A., X. Yu, and E. H. Egelman.** 1994. Three-dimensional reconstruction of a Co-complex of F-actin with antibody Fab fragments to actin's NH_2 terminus. *Biophys. J.* **66**:276–285.
1314. **Owen, C. and D. DeRosier.** 1993. A 13-A map of the actin-scruin filament from the limulus acrosomal process. *J. Cell Biol.* **123**:337–344.
1315. **Pajot Augy, E. and M. A. Axelos.** 1992. The effect of organic cryosolvents on actin structure: studies by small angle X-ray scattering. *Eur. Biophys. J.* **21**:179–184.
1316. **Phillips, L., F. Separovic, B. A. Cornell, J. A. Barden, and C. G. dos Remedios.** 1991. Actin dynamics studied by solid-state NMR spectroscopy. *Eur. Biophys. J.* **19**:147–155.
1317. **Pollard, T. D., E. Shelton, R. R. Weihing, and E. D. Korn.** 1970. Ultrastructural characterization of F-actin isolated from *Acanthamoeba castellaniii* and identification of cytoplasmic filaments as F-actin by reaction with rabbit heavy meromyosin. *J. Mol. Biol.* **50**:91–97.
1318. **Porter, M. and A. Weber.** 1979. Non-cooperative response of actin-cysteine 373 in cooperatively behaving regulated actin filaments. *FEBS Lett.* **105**:259–262.
1319. **Prochniewicz, E., E. Katayama, T. Yanagida, and D. D. Thomas.** 1993. Cooperativity in F-actin: chemical modifications of actin monomers affect the functional interactions of myosin with unmodified monomers in the same actin filament. *Biophys. J.* **65**:113–123.
1320.* **Rayment, I., H. M. Holden, M. Whittaker, C. B. Yohn, M. Lorenz, K. C. Holmes, and R. A. Milligan.** 1993. Structure of the actin–myosin complex and its implications for muscle contraction. *Science* **261**:58–65.
1321. **Rayment, I., W. R. Rypniewski, K. Schmidt-Basel, R. Smith, D. R. Tomchick, M. M. Benning, D. A. Winkelmann, G. Wesenberg, and H. M. Holden.** 1993. Three-dimensional structure of myosin subfragment-**1**: a molecular motor. *Science* **261**:50–58.
1322.* **Reisler, E.** 1993. Actin molecular structure and function. *Curr. Opin. Cell Biol.* **5**:41–47.
1323. **Reizer, J., A. Reizer, M. H. J. Saier, P. Bork, and C. Sander.** 1993. Exopolyphosphate phosphatase and guanosine pentaphosphate phosphatase belong to the sugar kinase/actin/hsp 70 superfamily. *Trends Biochem. Sci.* **18**:247–248.
1324. **Rich, S. A. and J. E. Estes.** 1976. Detection of conformational changes in actin by proteolytic digestion: evidence for a new monomeric species. *J. Mol. Biol.* **104**:777–792.
1325. **Rioux, L. and C. Gicquaud.** 1985. Actin paracrystalline sheets formed at the surface of positively charged liposomes. *J. Ultrastruct. Res.* **93**:42–49.
1326. **Roustan, C., Y. Benyamin, M. Boyer, and J. C. Cavadore.** 1986. Structural variations in actins. A study of the immunological reactivity of the N-terminal region. *Biochem. J.* **233**:193–197.
1327. **Rubenstein, P. A., R. K. Cook, and G. Babcock.** 1991. Do residues 265–268 in yeast actin stabilise the F-actin helix? *J. Cell Biol.* **115**:160a.
1328. **Safer, D., L. Bolinger, and J. S. Leigh.** 1986. Location of cys-374 labelled with undeca-gold. *J. Inorg. Biochem.* **26**:77–91.
1329. **Sakabe, N., K. Sakabe, K. Sasaki, H. Kondo, T. Ema, N. Kamiya, and M. Matsushima.** 1983. Crystallographic studies of the chicken gizzard G-actin X DNase I complex at 5A resolution. *J. Biochem. Tokyo.* **93**:299–302.
1330. **Sawyer, W. H., A. G. Woodhouse, J. J. Czarnecki, and E. Blatt.** 1988. Rotational dynamics of actin. *Biochemistry* **27**:7733–7740.
1331. **Schmid, M. F., J. M. Agris, J. Jakana, P. Matsudaira, and W. Chiu.** 1994. Three-dimensional structure of a single filament in the *Limulus* acrosomal bundle: scruin binds to homologous helix–loop–β motifs in actin. *J. Cell Biol.* **124**:341–350.
1332. **Schmid, M. F., J. Jakana, P. Matsudaira, and W. Chiu.** 1992. Effects of radiation damage with 400-kV electrons on frozen, hydrated actin bundles. *J. Struct. Biol.* **108**:62–68.
1333.* **Schrader, M., C. J. Temmgrove, J. L. Lessard, and B. M. Jockusch.** 1994. Chicken antibodies to rabbit muscle actin with a restricted repertoire of F-actin recognition. *Eur. J. Cell Biol.* **63**:326–335.
1334.* **Schroder, R. R., D. J. Manstein, W. Jahn, H. Holden, I. Rayment, K. C. Holmes, and J. A. Spudich.** 1993. Three-dimensional atomic model of F-actin decorated with *Dictyostelium* myosin. *Nature* **364**:171–174.
1335. **Schutt, C., U. Lindberg, and J. Myslik.** 1991. Actin in ribbons. *Nature* **353**:508.
1336. **Schutt, C. E., U. Lindberg, J. Myslik, and N. Strauss.** 1989. Molecular packing in profilin: actin crystals and its implications. *J. Mol. Biol.* **209**:735–746.
1337.* **Schutt, C. E., J. C. Myslik, M. D. Rozycki, N. C. Goonesekere, and U. Lindberg.** 1993. The structure of crystalline profilin-β-actin. *Nature* **365**:810–816.
1338. **Schutt, C. E., M. D. Rozycki, and U. Lindberg.** 1994. What's the matter with the ribbon? *Curr. Biol.* **4**:185–186.
1339. **Shu, W. P., D. Wang, and A. Stracher.** 1992. Chemical evidence for the existence of activated G-actin. *Biochem. J.* **283**:567–573.
1340. **Smith, P. R., W. E. Fowler, and U. Aebi.** 1984. Towards an alignment of the actin molecule within the actin filament. *Ultramicroscopy.* **13**:113–123.
1341. **Smith, P. R., W. E. Fowler, T. D. Pollard, and U. Aebi.** 1983. Structure of the actin molecule determined from electron micrographs of crystalline actin sheets with a tentative alignment of the molecule in the actin filament. *J. Mol. Biol.* **167**:641–660.

1342. **Spencer, M.** 1969. Low-angle X-ray diffraction from concentrated sols of F-actin. *Nature* **223**:1361–1362.
1343. **Stokes, D. L. and D. J. DeRosier.** 1987. The variable twist of actin and its modulation by actin-binding proteins. *J. Cell Biol.* **104**:1005–1017.
1344. **Stone, D. B., S. C. Prevost, and J. Botts.** 1970. Studies on spin-labeled actin. *Biochemistry* **9**:3937–3947.
1345. **Stournaras, C.** 1990. Exposure of thiol groups and bound nucleotide in G-actin: thiols as an indicator for the native state of actin. *Anticancer. Res.* **10**:1651–1659.
1346. **Strambini, G. B. and S. S. Lehrer.** 1991. Tryptophan phosphorescence of G-actin and F-actin. *Eur. J. Biochem.* **195**:645–651.
1347. **Strzelecka Golaszewska, H., J. Moraczewska, S. Y. Khaitlina, and M. Mossakowska.** 1993. Localization of the tightly bound divalent-cation-dependent and nucleotide-dependent conformation changes in G-actin using limited proteolytic digestion. *Eur. J. Biochem.* **211**:731–742.
1348. **Strzelecka Golaszewska, H., B. Nagy, and J. Gergely.** 1974. Changes in conformation and nucleotide binding of Ca, Mn, or MgG-actin upon removal of the bound divalent cation. Studies of ultraviolet difference spectra and optical rotation. *Arch. Biochem. Biophys.* **161**:559–569.
1349. **Strzelecka Golaszewska, H., S. Y. Venyaminov, S. Zmorzynski, and M. Mossakowska.** 1985. Effects of various amino acid replacements on the conformational stability of G-actin. *Eur. J. Biochem.* **147**:331–342.
1350. **Suck, D., W. Kabsch, and H. G. Mannherz.** 1981. Three-dimensional structure of the complex of skeletal muscle actin and bovine pancreatic DNAse I at 6-A resolution. *Proc. Natl Acad. Sci. U. S. A.* **78**:4319–4323.
1351. **Sugino, H., N. Sakabe, K. Sakabe, S. Hatano, F. Oosawa, T. Mikawa, and S. Ebashi.** 1979. Crystallization and preliminary crystallographic data of chicken gizzard G-actin. DNase I complex and *Physarum* G-actin. DNase I complex. *J. Biochem. Tokyo* **86**:257–260.
1352. **Suzuki, A., T. Maeda, and T. Ito.** 1991. Formation of liquid crystalline phase of actin filament solutions and its dependence on filament length as studied by optical birefringence. *Biophys. J.* **59**:25–30.
1353. **Suzuki, S., H. Noda, and K. Maruyama.** 1973. Natural F-actin. VI. Degeneration of natural F-actin and its protection by ATP. *J. Biochem. Tokyo* **73**:695–703.
1354. **Swezey, R. R. and G. N. Somero.** 1982. Polymerization thermodynamics and structural stabilities of skeletal muscle actins from vertebrates adapted to different temperatures and hydrostatic pressures. *Biochemistry* **21**:4496–4503.
1355. **Takebayashi, T., Y. Morita, and F. Oosawa.** 1977. Electronmicroscopic investigation of the flexibility of F-actin. *Biochim. Biophys. Acta* **492**:357–363.
1356. **Taniguchi, M.** 1976. Diphasic transformations of F-actin. Effects of urea and $MgCl_2$ on F-actin. *Biochim. Biophys. Acta* **427**:126–140.
1357. **Tao, T.** 1978. Nanosecond fluorescence depolarization studies on actin labeled with 1,5-IAEDANS and dansyl chloride. Evidence for label flexibility. *FEBS Lett.* **93**:146–150.
1358. **Tao, T. and J. Cho.** 1979. Fluorescence lifetime quenching studies on the accessibilities of actin sulfhydryl sites. *Biochemistry* **18**:2759–2765.
1359. **Tao, T., M. Lamkin, and C. J. Scheiner.** 1985. The conformation of the C-terminal region of actin: a site-specific photocrosslinking study using benzophenone-4-maleimide. *Arch. Biochem. Biophys.* **240**:627–634.
1360. **Taylor, K. A. and D. W. Taylor.** 1992. Formation of 2-D paracrystals of F-actin on phospholipid layers mixed with quaternary ammonium surfactants. *J. Struct. Biol.* **108**:140–147.
1361. **Thomas, D. D., J. C. Seidel, and J. Gergely.** 1979. Rotational dynamics of spin-labeled F-actin in the sub-millisecond time range. *J. Mol. Biol.* **132**:257–273.
1362. **Tirion, M. M. and D. ben Avraham.** 1993. Normal mode analysis of G-actin. *J. Mol. Biol.* **230**:186–195.
1363. **Toyoshima, C. and T. Wakabayashi.** 1985. Three-dimensional image analysis of the complex of thin filaments and myosin molecules from skeletal muscle. IV. Reconstitution from minimal- and high-dose images of the actin-tropomyosin-myosin subfragment-1 complex. *J. Biochem. Tokyo* **97**:219–243.
1364. **Toyoshima, C. and T. Wakabayashi.** 1985. Three-dimensional image analysis of the complex of thin filaments and myosin molecules from skeletal muscle. V. Assignment of actin in the actin–tropomyosin–myosin subfragment-1 complex. *J. Biochem Tokyo* **97**:245–263.
1365. **Trachtenberg, S., D. Stokes, E. Bullitt, and D. DeRosier.** 1986. Actin and flagellar filaments: two helical structures with variable twist. *Ann. N. Y. Acad. Sci.* **483**:88–99.
1366.* **Trinick, J., J. Cooper, J. Seymour, and E. H. Egelman.** 1986. Cryo-electron microscopy and three-dimensional reconstruction of actin filaments. *J. Microsc.* **141**:349–360.
1367. **Vazina, A. A., I. A. Bolotina, M. V. Vol'kenshtein, I. Lyasotskaya, and G. M. Frank.** 1966. Configuration of polypeptide chain in G- and F-actin. *Fed. Proc. Transl. Suppl.* **25**:476–478.
1368. **Wahl, P., K. Mihashi, and J. C. Auchet.** 1975. Nanosecond pulse fluorometry in polarized light of dansyl-L-cysteine linked to a unique SH group of F-actin; the influence of regulatory proteins and myosin moiety. *FEBS Lett.* **60**:164–167.
1369.* **Wakabayashi, K., H. Tanaka, H. Saito, N. Moriwaki, Y. Ueno, and Y. Amemiya.** 1991. Dynamic X-ray diffraction of skeletal muscle contraction: structural change of actin filaments. *Adv. Biophys.* **27**:3–13.
1370. **Wakabayashi, T., H. E. Huxley, L. A. Amos, and A. Klug.** 1975. Three-dimensional image reconstruction of actin–tropomyosin complex and actin–tropomyosin–troponin T–troponin I complex. *J. Mol. Biol.* **93**:477–497.
1371. **Wakabayashi, T., C. Toyoshima, and E. Katayama.** 1984. Image analysis of the complex of actin-tropomyosin and myosin subfragment 1. *Adv. Exp. Med. Biol.* **170**:21–28.
1372. **Waring, A. J. and R. Cooke.** 1987. The molecular dynamics of actin measured by a spin probe attached to lysine. *Arch. Biochem. Biophys.* **252**:197–205.
1373. **West, J. J.** 1970. Adenosine triphosphate and inosine triphosphate dependent conformational changes of adenosine diphosphate-G-actin. *Biochemistry* **9**:3847–3853.
1374. **Yamamoto, K., M. Yanagida, M. Kawamura, K. Maruyama, and H. Noda.** 1975. A study on the structure of paracrystals of F-actin. *J. Mol. Biol.* **91**:463–469.
1375. **Yoshimura, H., T. Nishio, K. Mihashi, K. J. Kinosita, and A. Ikegami.** 1984. Torsional motion of eosin-labeled F-actin as detected in the time-resolved anisotropy decay of the probe in the sub-millisecond time range. *J. Mol. Biol.* **179**:453–467.
1376. **Zechel, K.** 1991. The formation of a paracrystalline structure from muscle F-actin and oligolysine. *Biol. Chem. Hoppe Seyler* **372**:331–335.
1377. **Zheleznaya, L. A., N. G. Mevkh, and V. S. Gherasimov.** 1980. X-ray diffraction study of F-actin induced by phalloidin. *Naturwissenschaften* **67**:363–364.
1378. **Aktories, K., U. Geipel, M. Wille, and I. Just.** 1990. Characterization of the ADP-ribosylation of actin by *Clostridium botulinum* C2 toxin and *Clostridium perfringens* iota toxin. *J. Physiol. Paris.* **84**:262–266.

COVALENT DERIVATIZATION OF ACTIN

1379. **Amato, P. A. and D. L. Taylor.** 1986. Probing the mechanism of incorporation of fluorescently labeled actin into stress fibers. *J. Cell Biol.* **102**:1074–1084.
1380. **Ando, T., N. Kobayashi, and E. Munekata.** 1993. Electrostatic potential around actin. *Adv. Exp. Med. Biol.* **332**:361–376.
1381. **Asatoor, A. M. and M. D. Armstrong.** 1967. 3-methylhistidine, a component of actin. *Biochem. Biophys. Res. Commun.* **26**:168–174.
1382.* **Ball, E., C. C. Karlik, C. J. Beall, D. L. Saville, J. C. Sparrow, B. Bullard, and E. A. Fyrberg.** 1987. Arthrin, a myofibrillar protein of insect flight muscle, is an actin–ubiquitin conjugate. *Cell* **51**:221–228.
1383. **Barden, J. A. and C. G. dos Remedios.** 1992. The environment of the high-affinity cation-binding site on actin and the separation between cation and ATP sites as revealed by proton NMR and fluorescence spectroscopy. *J. Biochem. Tokyo* **96**:913–921.
1384.* **Barden, J. A., M. Miki, and C. G. dos Remedios.** 1986. Selective labelling of Cys-10 on actin. *Biochem. Int.* **12**:95–101.
1385. **Bates, P. C., G. K. Grimble, M. P. Sparrow, and D. J. Millward.** 1983. Myofibrillar protein turnover. Synthesis of protein-bound 3-methylhistidine, actin, myosin heavy chain and aldolase in rat skeletal muscle in the fed and starved states. *Biochem. J.* **214**:593–605.
1386.* **Bender, N., H. Fasold, A. Kenmoku, G. Middelhoff, and K. E. Volk.** 1976. The selective blocking of the polymerization reaction of striated muscle actin leading to a derivative suitable for crystallization. Modification of Tyr-53 by 5-diazonium-(1H)tetrazole. *Eur. J. Biochem.* **64**:215–218.
1387. **Berger, E. M., G. Cox, L. Weber, and J. S. Kenney.** 1981. Actin acetylation in *Drosophila* tissue culture cells. *Biochem. Genet.* **19**:321–331.
1388. **Bertrand, R., P. Chaussepied, E. Audemard, and R. Kassab.** 1989. Functional characterization of skeletal F-actin labeled on the NH_2-terminal segment of residues 1–28. *Eur. J. Biochem.* **181**:747–754.
1389. **Bettache, N., R. Bertrand, and R. Kassab.** 1990. Maleimidobenzoyl-G-actin: structural properties and interaction with skeletal myosin subfragment-1. *Biochemistry* **29**:9085–9091.
1390. **Borovikov, Y. S., E. Nowak, M. I. Khoroshev, and R. Dabrowska.** 1993. The effect of Ca^{2+} on the conformation of tropomyosin and actin in regulated actin filaments with or without bound myosin subfragment 1. *Biochim. Biophys. Acta* **1163**:280–286.
1391. **Bridgen, J.** 1972. The reactivity and function of thiol groups in trout actin. *Biochem. J.* **126**:21–25.
1392. **Bright, G. R. and B. S. Spooner.** 1983. Preparation and reactions of an iodinated imidoester reagent with actin and α-actinin. *Anal. Biochem.* **131**:301–311.
1393. **Bullard, B., J. Bell, R. Craig, and K. Leonard.** 1985. Arthrin: a new actin-like protein in insect flight muscle. *J. Mol. Biol.* **182**:443–454.
1394. **Burtnick, L. D.** 1984. Modification of actin with fluorescein isothiocyanate. *Biochim. Biophys. Acta* **791**:57–62.
1395. **Burtnick, L. D. and C. D. Isaac.** 1984. Modification of actin with 2-(N-methylanilino)-naphthalene-6-sulfonyl chloride. *Can. J. Biochem. Cell Biol.* **62**:191–195.
1396. **Carrascosa, J. M. and O. H. Wieland.** 1986. Evidence that (a) serine specific protein kinase(s) different from protein kinase C is responsible for the insulin-stimulated actin phosphorylation by placental membrane. *FEBS Lett.* **201**:81–86.
1397. **Carsten, M. E.** 1966. Actin. Its thiol groups. *Biochemistry* **5**:297–300.

1398. **Cass, K. A., E. B. Clark, and P. A. Rubenstein.** 1983. Is the onset of actin histidine methylation under development control in the chick embryo. *Arch. Biochem. Biophys.* **225**:731–739.

1399. **Chai, Y. C., S. S. Ashraf, K. Rokutan, R. B. Johnston, and J. A. Thomas.** 1994. S-thiolation of individual human neutrophil proteins, including actin, by stimulation of the respiratory burst: evidence against a role for glututhione disulphide. *Arch. Biochem. Biophys.* **310**:273–281.

1400. **Chantler, P. D. and W. B. Gratzer.** 1975. Effects of specific chemical modification of actin. *Eur. J. Biochem.* **60**:67–72.

1401.* **Clancy, R. M., J. Leszczynska Piziak, and S. B. Abramson.** 1993. Nitric oxide stimulates the ADP-ribosylation of actin in human neutrophils. *Biochem. Biophys. Res. Commun.* **191**:847–852.

1402. **Clark, S. J., M. E. O'Brien, and G. B. Ralston.** 1988. The effects of p-mercuribenzenesulfonate on purified spectrin and actin. *Biochim. Biophys. Acta* **957**:243–253.

1403. **Combeau, C. and M. F. Carlier.** 1992. Covalent modification of G-actin by pyridoxal 5′-phosphate: polymerization properties and interaction with DNase I and myosin subfragment 1. *Biochemistry* **31**:300–309.

1404. **DeBiasio, R. L., L. L. Wang, G. W. Fisher, and D. L. Taylor.** 1988. The dynamic distribution of fluorescent analogues of actin and myosin in protrusions at the leading edge of migrating Swiss 3T3 fibroblasts. *J. Cell Biol.* **107**:2631–2645.

1405. **Derrick, N. and K. Laki.** 1966. Enzymatic labelling of actin and tropomyosin with 14 C-labelled putrescine. *Biochem. Biophys. Res. Commun.* **22**:82–88.

1406. **Drewes, G. and H. Faulstich.** 1990. The enhanced ATPase activity of glutathione-substituted actin provides a quantitative approach to filament stabilization. *J. Biol. Chem.* **265**:3017–3021.

1407. **Dunkley, P. R. and P. J. Robinson.** 1983. The *in vitro* phosphorylation of actin from rat cerebral cortex. *Neurochem. Res.* **8**:865–871.

1408. **Elzinga, M.** 1971. Amino acid sequence aroung 3-methylhistidine in rabbit skeletal muscle actin. *Biochemistry* **10**:224–229.

1409. **Elzinga, M. and J. J. Phelan.** 1984. F-actin is intermolecularly crosslinked by N,N′-p-phenylenedimaleimide through lysine-191 and cysteine-374. *Proc. Natl Acad. Sci. U. S. A.* **81**:6599–6602.

1410. **Faulstich, H., D. Heintz, and G. Drewes.** 1992. Interchain and intrachain crosslinking of actin thiols by a bifunctional thiol reagent. *FEBS Lett.* **302**:201–205.

1411. **Fazekas, S., I. Ovary, and V. Szekessy Hermann.** 1985. Composition of amino acid phosphates in phosphorylated G-actin of rabbit skeletal muscle. *Acta Physiol. Hung.* **66**:69–76.

1412. **Field, S. J., J. C. Pinder, B. Clough, A. R. Dluzewski, R. J. Wilson, and W. B. Gratzer.** 1993. Actin in the merozoite of the malaria parasite, *Plasmodium falciparum*. *Cell Motil. Cytoskeleton.* **25**:43–48.

1413.* **Furuhashi, K. and S. Hatano.** 1992. Actin kinase: a protein kinase that phosphorylates actin of fragmin–actin complex. *J. Biochem. Tokyo* **111**:366–370.

1414. **Furuhashi, K., S. Hatano, S. Ando, K. Nishizawa, and M. Inagaki.** 1992. Phosphorylation by actin kinase of the pointed end domain on the actin molecule. *J. Biol. Chem.* **267**:9326–9330.

1415. **Gaetjens, E. and M. Barany.** 1966. N-acetyl-aspartic acid in G-actin. *Biochim. Biophys. Acta* **117**:176–183.

1416. **Garrels, J. I. and T. Hunter.** 1979. Post-translational modification of actins synthesized *in vitro*. *Biochim. Biophys. Acta* **564**:517–525.

1417. **Gerber, B. R. and T. Ooi.** 1968. Effect of dinitrophenylation of the properties of G- and F-actin. *Biochim. Biophys. Acta* **154**:162–174.

1418.* **Gettemans, J., Y. De Ville, J. Vandekerckhove, and E. Waelkens.** 1992. Physarum actin is phosphorylated as the actin-fragmin complex at residues Thr203 and Thr202 by a specific 80 kDa kinase. *EMBO J.* **11**:3185–3191.

1419. **Gettemans, J., Y. De Ville, J. Vandekerckhove, and E. Waelkens.** 1993. Purification and partial amino acid sequence of the actin-fragmin kinase from *Physarum polycephalum*. *Eur. J. Biochem.* **214**:111–119.

1420. **Gilbert, H. R. and C. Frieden.** 1983. Preparation, purification and properties of a crosslinked trimer of G-actin. *Biochem. Biophys. Res. Commun.* **111**:404–408.

1421. **Haverberg, L. N., H. N. Munro, and V. R. Young.** 1974. Isolation and quantitation of N τ-methylhistidine in actin and myosin of rat skeletal muscle: use of pyridine elution of protein hydrolysates on ion-exchange resins. *Biochim. Biophys. Acta* **371**:226–237.

1422. **Hegyi, G., H. Michel, J. Shabanowitz, D. F. Hunt, N. Chatterjie, G. Healy Louie, and M. Elzinga.** 1992. Gln-41 is intermolecularly cross-linked to Lys-113 in F-actin by N-(4-azidobenzoyl)-putrescine. *Protein. Sci.* **1**:132–144.

1423. **Hegyi, G., G. Premecz, B. Sain, and A. Muhlrad.** 1974. Selective carbethoxylation of the histidine residues of actin by diethylpyrocarbonate. *Eur. J. Biochem.* **44**:7–12.

1424. **Hofstein, R., M. Hershkowitz, I. Gozes, and D. Samuel.** 1980. The characterization and phosphorylation of an actin-like protein in synaptosomal membranes. *Biochim. Biophys. Acta* **624**:153–162.

1425. **Houk, T. W., M. Ovnic, and S. Karipides.** 1983. pH and polymerization dependence of the site of labeling of actin by 7-chloro-4-nitrobenzo-2-oxa-1,3-diazole. *J. Biol. Chem.* **258**:5419–5423.

1426. **Howard, P. K., B. M. Sefton, and R. A. Firtel.** 1993. Tyrosine phosphorylation of actin in *Dictyostelium* associated with cell-shape changes. *Science* **259**:241–244.

1427. **Hozumi, T.** 1991. The interaction of maleimidobenzoyl G-actin with myosin subfragment 1 in solution: characterization of the MgATPase activity after chemical crosslinking. *Biochem. Int.* **23**:835–843.

1428. **Johnson, P. and J. M. Blazyk.** 1978. Involvement of an arginine residue of actin in tropomyosin binding. *Biochem. Biophys. Res. Commun.* **82**:1013–1018.

1429. **Johnson, P., C. I. Harris, and S. V. Perry.** 1967. 3-methylhistidine in actin and other muscle proteins. *Biochem. J.* **105**:361–370.

1430. **Johnson, P., G. E. Lobley, and S. V. Perry.** 1969. Distribution and biological role of 3-methyl-histidine in actin and myosin. *Biochem. J.* **114**:34P.

1431. **Johnson, P. and S. V. Perry.** 1968. Chemical studies on the cysteine and terminal peptides in tryptic digests of actin. *Biochem. J.* **110**:207–216.

1432. **Johnson, P. and S. V. Perry.** 1970. Biological activity and the 3-methylhistidine content of actin and myosin. *Biochem. J.* **119**:293–298.

1433. **Jungbluth, A., V. Vonarnim, E. Biegelmann, B. Humbel, A. Schweiger, and G. Gerisch.** 1994. Strong increase in the tyrosine phosphorylation of actin upon inhibition of oxidative phosphorylation – correlation with reversible rearrangements in the actin skeleton of *Dictyostelium* cells. *J. Cell Sci.* **107**:117–125.

1434. **Just, I., U. Geipel, A. Wegner, and K. Aktories.** 1990. De-ADP-ribosylation actin by *Clostridium perfringens* iota-toxin and *Clostridium botulinum* C2 toxin. *Eur. J. Biochem.* **192**:723–727.

1435.* **Just, I., P. Wollenberg, J. Moss, and K. Aktories.** 1994. Cysteine-specific ADP-ribosylation of actin. *Eur. J. Biochem.* **221**:1047–1054.

1436. **Kagawa, H.** 1981. The reaction of a positively-charged N-ethylmaleimide derivative with actin and myosin subfragment-1. *Int. J. Biochem.* **13**:871–873.

1437. **Katoh, S., T. Inoue, H. Kohno, and Y. Ohkubo.** 1994. Transglutaminase-modified actin decreases epidermal growth factor binding to its receptor in rat liver membrane. *Biomed. Res.* **15**:1–8.

1438. **Kawasaki, Y., K. Mihashi, H. Tanaka, and H. Ohnuma.** 1976. Fluorescence study of N-(3-pyrene)maleimide conjugated to rabbit skeletal F-actin and plasmodium actin polymers. *Biochim. Biophys. Acta* **446**:166–178.

1439. **Khalili, A. D. and C. G. Zarkadas.** 1988. Determination of myofibrillar and connective tissue protein contents of young and adult avian (Gallus domesticus) skeletal muscles and the N τ-methylhistidine content of avian actins. *Poult. Sci.* **67**:1593–1614.

1440.* **Knight, P. and G. Offer.** 1978. p-NN′-phenylenebismaleimide, a specific cross-linking agent for F-actin. *Biochem. J.* **175**:1023–1032.

1441. **Lehrer, S. S. and M. Elzinga.** 1992. Fluorescence studies on nitrated actin. *Fed. Proc.* **31**:502.

1442. **Lin, T. I.** 1978. Fluorimetric studies of actin labeled with dansyl aziridine. *Arch. Biochem. Biophys.* **185**:285–299.

1443. **Lin, T. I. and R. M. Dowben.** 1982. Fluorescence spectroscopic studies of pyrene-actin adducts. *Biophys. Chem.* **15**:289–298.

1444. **Lin, T. I., M. Kim, and R. M. Dowben.** 1990. Accessibility of thiols in actin – a kinetic study with fluorescent maleimide probes. *Prog. Clin. Biol. Res.* **327**:771–778.

1445. **Liu, D. F., D. Wang, and A. Stracher.** 1990. The accessibility of the thiol groups on G- and F-actin of rabbit muscle. *Biochem. J.* **266**:453–459.

1446. **Ludescher, R. D. and Z. Liu.** 1993. Characterization of skeletal muscle actin labeled with the triplet probe erythrosin-5-iodoacetamide. *Photochem. Photobiol.* **58**:858–866.

1447. **Lusty, C. J. and H. Fasold.** 1969. Characterization of sulfhydryl groups of actin. *Biochemistry* **8**:2933–2939.

1448. **Machicao, F., T. Urumow, and O. H. Wieland.** 1983. Evidence for phosphorylation of actin by the insulin receptor-associated protein kinase from human placenta. *FEBS Lett.* **163**:76–80.

1449. **Machicao, F. and O. H. Wieland.** 1985. Insulin-stimulated phosphorylation of actin by human placental insulin receptor preparations. *Curr. Top. Cell Regul.* **27**:95–105.

1450. **Marriott, G., K. Zechel, and T. M. Jovin.** 1988. Spectroscopic and functional characterization of an environmentally sensitive fluorescent actin conjugate. *Biochemistry* **27**:6214–6220.

1451. **Martonosi, A.** 1968. The sulfhydryl groups of actin. *Arch. Biochem. Biophys.* **123**:29–40.

1451a. **Maruta, H. and G. Isenberg.** 1983. Ca^{2+}-dependent actin-binding phosphoprotein in *Physarum polycephalum*. II. Ca^{2+}-dependent F-actin-capping activity of subunit a and its regulation by phosphorylation of subunit b. *J. Biol. Chem.* **258**:10151–10158.

1452.* **Matsuyama, S. and S. Tsuyama.** 1991. Mono-ADP-ribosylation in brain: purification and characterization of ADP-ribosyltransferases affecting actin from rat brain. *J. Neurochem.* **57**:1380–1387.

1453. **Mauss, S., C. Chaponnier, I. Just, K. Aktories, and G. Gabbiani.** 1990. ADP-ribosylation of actin isoforms by *Clostridium botulinum* C2 toxin and *Clostridium perfringens* iota toxin. *Eur. J. Biochem.* **194**:237–241.

1454. **Miki, M.** 1988. Characterization of carbethoxylated actin. *J. Biochem. Tokyo* **104**:312–315.

1455. **Miki, M.** 1989. Interaction of Lys-61 labeled actin with myosin subfragment-1 and the regulatory proteins. *J. Biochem. Tokyo* **106**:651–655.

1456. **Miki, M. and C. G. dos Remedios.** 1988. Fluorescence quenching studies of fluorescein attached to Lys-61 or Cys-374 in actin: effects of polymerization, myosin subfragment-1 binding, and tropomyosin-troponin binding. *J. Biochem. Tokyo* **104**:232–235.

1457. **Miki, M. and T. Hozumi.** 1991. Interaction of maleimidobenzoyl actin with myosin subfragment 1 and tropomyosin-troponin. *Biochemistry* **30**:5625–5630.

1458. **Miki, M., S. I. O'Donoghue, and C. G. dos Remedios.** 1992. Structure of actin observed by fluorescence resonance energy transfer spectroscopy. *J. Muscle Res. Cell Motil.* **13**:132–145.

1459. **Miki, M. and P. Wahl.** 1984. Fluorescence energy transfers in labelled G-actin and F-actin. *Biochim. Biophys. Acta* **786**:188–196.

1460. **Miki, M. and P. Wahl.** 1985. Fluorescence energy transfer between points in G-actin: the nucleotide-binding site, the metal-binding site and Cys-373 residue. *Biochim. Biophys. Acta* **828**:188–195.
1461. **Miller, L., M. Phillips, and E. Reisler.** 1988. Polymerization of actin modified with fluorescein isothiocyanate. *Eur. J. Biochem.* **174**:23–29.
1462. **Millonig, R., H. Salvo, and U. Aebi.** 1988. Probing actin polymerization by intermolecular cross-linking. *J. Cell Biol.* **106**:785–796.
1463. **Mockrin, S. C. and E. D. Korn.** 1981. Isolation and characterization of covalently cross-linked actin dimer. *J. Biol. Chem.* **256**:8228–8233.
1464. **Muhlrad, A.** 1968. Studies on the properties of chemically modified actin. II. Trinitrophenylation. *Biochim. Biophys. Acta* **162**:444–451.
1465. **Muhlrad, A., A. Corsi, and A. L. Granata.** 1968. Studies on the properties of chemically modified actin. I. Photooxidation, succinylation, nitration. *Biochim. Biophys. Acta* **162**:435–443.
1466. **Muhlrad, A. and K. Ferencz.** 1973. Studies on the properties of chemically modified actin IV: activation of myosin ATPase by actin and trinitrophenylated actin. *Physiol. Chem. Phys.* **5**:13–26.
1467. **Muhlrad, A., G. Hegyi, and M. Horanyi.** 1969. Studies on the properties of chemically modified actin. 3. Carbethoxylation. *Biochim. Biophys. Acta* **181**:184–190.
1468. **Ohara, O., S. Takahashi, T. Ooi, and Y. Fujiyoshi.** 1981. Fixation of skeletal muscle actin in F-state by chemical cross-linking with bis-imidoesters. *Biochem. Biophys. Res. Commun.* **100**:988–994.
1469. **Ohara, O., S. Takahashi, T. Ooi, and Y. Fujiyoshi.** 1982. Cross-linking study on skeletal muscle actin: properties of suberimidate-treated actin. *J. Biochem. Tokyo* **91**:1999–2012.
1470. **Ohmi, K. and S. Nakamura.** 1991. Analysis of the interaction of reserpine with actin by the photoaffinity labelling method. *Eur. J. Pharmacol.* **207**:305–310.
1471. **Ohta, Y., T. Akiyama, E. Nishida, and H. Sakai.** 1987. Protein kinase C and cAMP-dependent protein kinase induce opposite effects on actin polymerizability. *FEBS Lett.* **222**:305–310.
1472. **Palmer, E., H. de la Vega, D. Grana, and J. L. Saborio.** 1980. Posttranslational processing of brain actin. *J. Neurochem.* **34**:911–915.
1472a. **Pekiner, C., N. A. Cullum, J. N. Hughes, A. J. Hargreaves, J. Mahon, I. F. Casson, and W. G. McLean.** 1993. Glycation of brain actin in experimental diabetes. *J. Neurochem.* **61**:436–442.
1473. **Prat, A. G., A. M. Bertorello, D. A. Ausiello, and H. F. Cantiello.** 1993. Activation of epithelial Na^+ channels by protein kinase A requires actin filaments. *Am. J. Physiol.* **265**:C224-C233.
1474. **Prochniewicz, E.** 1979. Effect of crosslinking by glutaraldehyde on interaction of F-actin with heavy meromyosin. *Biochim. Biophys. Acta* **579**:346–358.
1475. **Raghavan, M., U. Lindberg, and C. Schutt.** 1992. The use of alternative substrates in the characterization of actin-methylating and carnosine-methylating enzymes. *Eur. J. Biochem.* **210**:311–318.
1476. **Raghavan, M., C. K. Smith, and C. E. Schutt.** 1989. Analytical determination of methylated histidine in proteins: actin methylation. *Anal. Biochem.* **178**:194–197.
1477. **Redman, K. L. and P. A. Rubenstein.** 1984. Actin amino-terminal acetylation and processing in a rabbit reticulocyte lysate. *Methods Enzymol.* **106**:179–192.
1478. **Rokutan, K., R. B. Johnston, and K. Kawai.** 1994. Oxidative stress induces S-thiolation of specific proteins in cultured gastric mucosal cells. *Am. J. Physiol.* **266**:G247-G254.
1479. **Rubenstein, P. and J. Deuchler.** 1979. Acetylated and nonacetylated actins in *Dictyostelium discoideum*. *J. Biol. Chem.* **254**:11142–11147.
1480. **Rubenstein, P., P. Smith, J. Deuchler, and K. Redman.** 1981. NH_2-terminal acetylation of *Dictyostelium discoideum* actin in a cell-free protein-synthesizing system. *J. Biol. Chem.* **256**:8149–8155.
1481. **Rubenstein, P. A., K. L. Redman, L. R. Solomon, and D. J. Martin.** 1987. Amino-terminal processing of *Dictyostelium discoideum* actin. *Methods Cell Biol.* **28**:231–243.
1482. **Schick, B. P. and J. A. Jacoby.** 1994. Sulfation of guinea pig megakaryocyte and platelet proteins. *J. Cell Physiol.* **159**:356–364.
1483. **Sheff, D. R. and P. A. Rubenstein.** 1989. Identification of N-acetylmethionine as the product released during the NH_2-terminal processing of a pseudo-class I actin. *J. Biol. Chem.* **264**:11491–11496.
1484. **Shibayama, T., K. Shinkawa, S. Nakajo, K. Nakaya, and Y. Nakamura.** 1986. Phosphorylation of muscle and non-muscle actins by casein kinase 1 *in vitro*. *Biochem. Int.* **13**:367–373.
1485. **Simon, J. R., A. Gough, E. Urbanik, F. Wang, F. Lanni, B. R. Ware, and D. L. Taylor.** 1988. Analysis of rhodamine and fluorescein-labeled F-actin diffusion *in vitro* by fluorescence photobleaching recovery. *Biophys. J.* **54**:801–815.
1486. **Sonobe, S., S. Takahashi, S. Hatano, and K. Kuroda.** 1986. Phosphorylation of Amoeba G-actin and its effect on actin polymerization. *J. Biol. Chem.* **261**:14837–14843.
1487. **Steinberg, R. A.** 1980. Actin nascent chains are substrates for cyclic AMP-dependent phosphorylation *in vivo*. *Proc. Natl Acad. Sci. U. S. A.* **77**:910–914.
1488. **Stournaras, C., G. Drewes, H. Blackholm, I. Merkler, and H. Faulstich.** 1990. Glutathionyl (cysteine-374) actin forms filaments of low mechanical stability. *Biochim. Biophys. Acta* **1037**:86–91.
1489. **Sussman, D. J., J. R. Sellers, P. Flicker, E. Y. Lai, L. E. Cannon, A. G. Szent Gyorgyi, and C. Fulton.** 1984. Actin of Naegleria gruberi. Absence of N τ-methylhistidine. *J. Biol. Chem.* **259**:7349–7354.
1490. **Takashi, R.** 1988. A novel actin label: a fluorescent probe at glutamine-41 and its consequences. *Biochemistry* **27**:938–943.
1491. **Tawada, K., H. Asai, and B. R. Gerber.** 1969. Spectral changes accompanying the polymerization of the dinitrophenyl derivative of G-actin. *Biochim. Biophys. Acta* **194**:486–497.
1492. **Tawada, K. and F. Oosawa.** 1969. Activation of H-meromyosin ATPase by polymers of actin and carboxymethylated actin. *J. Mol. Biol.* **44**:309–317.
1493. **Tawada, K., P. Wahl, and J. C. Auchet.** 1978. Study of actin and its interactions with heavy meromyosin and the regulatory proteins by the pulse fluorimetry in polarized light of a fluorescent probe attached to an actin cysteine. *Eur. J. Biochem.* **88**:411–419.
1494. **Walsh, M. P., S. Hinkins, and D. J. Hartshorne.** 1981. Phosphorylation of smooth muscle actin by the catalytic subunit of the cAMP-dependent protein kinase. *Biochem. Biophys. Res. Commun.* **102**:149–157.
1495. **Wang, Y. L. and D. L. Taylor.** 1979. Distribution of fluorescently labeled actin in living sea urchin eggs during early development. *J. Cell Biol.* **81**:672–679.
1496. **Wang, Y. L. and D. L. Taylor.** 1980. Preparation and characterization of a new molecular cytochemical probe: 5-iodoacetamidofluorescein-labeled actin. *J. Histochem. Cytochem.* **28**:1198–1206.
1497.* **Wegner, A. and K. Aktories.** 1988. ADP-ribosylated actin caps the barbed ends of actin filaments. *J. Biol. Chem.* **263**:13739–13742.
1498. **Weigt, C., I. Just, A. Wegner, and K. Aktories.** 1989. Nonmuscle actin ADP-ribosylated by botulinum C2 toxin caps actin filaments. *FEBS Lett.* **246**:181–184.
1499. **Weihing, R. R. and E. D. Korn.** 1969. Ameba actin: the presence of 3-methylhistidine. *Biochem. Biophys. Res. Commun.* **35**:906–912.
1500. **Weihing, R. R. and E. D. Korn.** 1970. Epsilon-N-dimethyllysine in amoeba actin. *Nature* **227**:1263–1264.
1501. **Weltman, J. K., A. R. J. Frackelton, R. P. Szaro, and R. M. Dowben.** 1971. Synthesis and fluorescence polarization analysis of an organomercurial derivative of actin. *J. Lab. Clin. Med.* **78**:808.
1502. **Xu, D. S., R. B. Jennett, S. L. Smith, M. F. Sorrell, and D. J. Tuma.** 1989. Covalent interactions of acetaldehyde with the actin/microfilament system. *Alcohol Alcohol.* **24**:281–289.

PROTEOLYTIC CLEAVAGE OF ACTIN

1503. **Adams, L. D., A. G. Tomasselli, P. Robbins, B. Moss, and R. L. Heinrikson.** 1992. HIV-1 protease cleaves actin during acute infection of human T-lymphocytes. *AIDS Res. Hum. Retroviruses.* **8**:291–295.
1504.* **Adelstein, R. S. and W. M. Kuehl.** 1970. Structural studies on rabbit skeletal actin. I. Isolation and characterization of the peptides produced by cyanogen bromide cleavage. *Biochemistry* **9**:1355–1364.
1505. **Barber, B. H. and T. L. Delovitch.** 1979. The identification of actin as a major lymphocyte component. *J. Immunol.* **122**:320–325.
1506. **Billeter, R., C. W. Heizmann, U. Reist, H. Howald, and E. Jenny.** 1982. Two-dimensional peptide analysis of myosin heavy chains and actin from single-typed human skeletal muscle fibers. *FEBS Lett.* **139**:45–48.
1507. **Bochkov, A. F., V. V. Sova, and S. Kirkwood.** 1972. The study of the actin pattern of an exo -(1–3)-D-glucanase. *Biochim. Biophys. Acta* **258**:531–540.
1508. **Burtnick, L. D.** 1987. Quenching of the fluorescence of actin modified at lysine-61. *Biochem. Int.* **14**:547–551.
1509. **Burtnick, L. D. and K. W. Chan.** 1980. Protection of actin against proteolysis by complex formation with deoxyribonuclease I. *Can. J. Biochem.* **58**:1348–1354.
1510. **Cavadore, J. C., C. Roustan, Y. Benyamin, M. Boyer, and J. Haiech.** 1985. Structural variations in actins. Biochemical and immunological tools for probing the structure of rabbit skeletal-muscle and bovine aortic actins. *Biochem. J.* **231**:363–368.
1511. **Collins, J. H. and M. Elzinga.** 1975. The primary structure of actin from rabbit skeletal muscle. Three cyanogen bromide peptides that are insoluble at neutral pH. *J. Biol. Chem.* **250**:5906–5914.
1512. **DasGupta, B. R. and W. Tepp.** 1993. Protease activity of botulinum neurotoxin type E and its light chain: cleavage of actin. *Biochem. Biophys. Res. Commun.* **190**:470–474.
1513. **de Vries, J. and T. Wieland.** 1978. Influence of phallotoxins and metal ions on the rate of proteolysis of actin. *Biochemistry* **17**:1965–1968.
1514.* **Frenette, G., J. Y. Dube, and R. R. Tremblay.** 1986. Proteolytic activity of arginine esterase from dog seminal plasma towards actin and other structural proteins. Comparison with trypsin and kallikrein. *Int. J. Biochem.* **18**:697–703.
1515. **Gerday, C., E. Robyns, and C. Gosselin Rey.** 1968. High resolution techniques of peptide mapping. Separation of bovine carotid actin peptides on cellulose thin layers and of the corresponding dansyl-peptides on polyamide thin layers. *J. Chromatogr.* **38**:408–411.
1516. **Goldschmidt Clermont, P. J., E. L. Van Alstyne, J. R. Day, D. L. Emerson, A. E. Nel, J. Lazarchick, and R. M. Galbraith.** 1986. Group-specific component (vitamin D binding protein) prevents the interaction between G-actin and profilin. *Biochemistry* **25**:6467–6472.
1517. **Gosselin Rey, C., C. Gerday, A. Gaspar Godfroid, and M. E. Carsten.** 1969. Amino acid analysis and peptide mapping of bovine carotid actin. *Biochim. Biophys. Acta* **175**:165–173.
1518. **Higashi Fujime, S., M. Suzuki, K. Titani, and T. Hozumi.** 1992. Muscle actin cleaved by proteinase K: its polymerization and *in vitro* motility. *J. Biochem. Tokyo* **112**:568–572.
1519. **Hozumi, T.** 1988. Structural aspects of skeletal muscle F-actin as studied by tryptic digestion: evidence for a second nucleotide interacting site. *J. Biochem. Tokyo* **104**:285–288.

1520. **Hozumi, T.** 1988. Structural aspects of skeletal muscle G-actin molecule as studied by proteolytic digestion: effect of nucleotide. *Biochem. Int.* **17**:171–178.

1521. **Hozumi, T.** 1988. Effect of divalent cation on the structure of skeletal muscle G-actin molecule. *Biochem. Int.* **16**:59–67.

1522. **Johnson, P., P. J. Wester, and R. S. Hikida.** 1979. Protein–protein interactions of proteolytic fragments of actin. *Biochim. Biophys. Acta* **578**:253–257.

1523. **Khaitlina, S. Y., J. H. Collins, I. M. Kuznetsova, V. P. Pershina, I. G. Synakevich, K. K. Turoverov, and A. M. Usmanova.** 1991. Physico-chemical properties of actin cleaved with bacterial protease from E. coli A2 strain. *FEBS Lett.* **279**:49–51.

1524. **Khaitlina, S. Y., T. D. Smirnova, and A. M. Usmanova.** 1988. Limited proteolysis of actin by a specific bacterial protease. *FEBS Lett.* **228**:172–174.

1525. **Konno, K.** 1988. G-actin structure revealed by chymotryptic digestion. *J. Biochem. Tokyo* **103**:386–392.

1526. **Kuehl, W. M., M. A. Conti, and R. S. Adelstein.** 1975. Structural studies on rabbit skeletal muscle actin. Ordering of the peptides produced by cleavage with cyanogen bromide. *J. Biol. Chem.* **250**:5890–5896.

1527.* **Mornet, D. and K. Ue.** 1984. Proteolysis and structure of skeletal muscle actin. *Proc. Natl Acad. Sci. U. S. A.* **81**:3680–3684.

1528. **Muszbek, L., L. Fesus, E. Olveti, and T. Szabo.** 1976. Cleavage of thrombosthenin A by thrombin. Evidence for the existence of two types of bovine platelet actin. *Biochim. Biophys. Acta* **427**:171–177.

1529. **Muszbek, L., J. A. Gladner, and K. Laki.** 1975. The fragmentation of actin by thrombin. Isolation and characterization of the split products. *Arch. Biochem. Biophys.* **167**:99–103.

1530. **Muszbek, L. and M. Hauck.** 1979. Fragmentation of actin by thrombin-like snake venom proteases. *Biochim. Biophys. Acta* **577**:34–43.

1531. **Muszbek, L. and K. Laki.** 1974. Cleavage of actin by thrombin. *Proc. Natl Acad. Sci. U. S. A.* **71**:2208–2211.

1532. **Rosenbusch, J. P., G. R. Jacobson, and J. C. Jaton.** 1976. Does a bacterial elongation factor share a common evolutionary ancestor with actin? *J. Supramol. Struct.* **5**:391–396.

1533.* **Schwyter, D. H., S. J. Kron, Y. Y. Toyoshima, J. A. Spudich, and E. Reisler.** 1990. Subtilisin cleavage of actin inhibits *in vitro* sliding movement of actin filaments over myosin. *J. Cell Biol.* **111**:465–470.

1534. **Shainoff, J. R.** 1973. Studies on reaction between thrombin and actin. *Ser. Haematol.* **6**:392–402.

1535. **Sheff, D. R. and P. A. Rubenstein.** 1992. Isolation and characterization of the rat liver actin N-acetylaminopeptidase. *J. Biol. Chem.* **267**:20217–20224.

1536. **Tomasselli, A. G., J. O. Hui, L. Adams, J. Chosay, D. Lowery, B. Greenberg, A. Yem, M. R. Deibel, H. Zurcher Neely, and R. L. Heinrikson.** 1991. Actin, troponin C, Alzheimer amyloid precursor protein and pro-interleukin 1 β as substrates of the protease from human immunodeficiency virus. *J. Biol. Chem.* **266**:14548–14553.

1537. **Vandekerckhove, J. and K. Weber.** 1981. Actin typing on total cellular extracts: a highly sensitive protein-chemical procedure able to distinguish different actins. *Eur. J. Biochem.* **113**:595–603.

1538. **Yoshida, H., T. Murachi, and I. Tsukahara.** 1984. Degradation of actin and vimentin by calpain II, a Ca^{2+}-dependent cysteine proteinase, in bovine lens. *FEBS Lett.* **170**:259–262.

1539. **Zechel, K.** 1979. Localization of the charge differences in the actins of rabbit skeletal muscle and chicken gizzard by two-dimensional gel electrophoretic analysis of tryptic fragments. *Hoppe Seylers Z. Physiol. Chem.* **360**:777–782.

EXPRESSION AND LOCALIZATION OF ACTIN

1540. **Abbas, M. K. and G. D. Cain.** 1989. Analysis of isoforms of actin from *Schistosoma mansoni* by two-dimensional gel electrophoresis. *Parasitol. Res.* **76**:178–180.

1541. **Abraham, E. G., N. Mounier, and G. Bosquet.** 1993. Expression of a Bombyx cytoplasmic actin gene in cultured *Drosophila* cells: influence of 20-hydroxyecdysone and interference with expression of endogenous cytoplasmic actin genes. *Insect. Biochem. Mol. Biol.* **23**:905–912.

1542. **Adams, A. E. and J. R. Pringle.** 1984. Relationship of actin and tubulin distribution to bud growth in wild-type and morphogenetic-mutant *Saccharomyces cerevisiae*. *J. Cell Biol.* **98**:934–945.

1543. **Amankwah, K. S. and U. De Boni.** 1994. Ultrastructural localization of filamentous actin within neuronal interphase nuclei in situ. *Exp. Cell Res.* **210**:315–325.

1544.* **Anderson, J. M. and D. R. Soll.** 1986. Differences in actin localization during bud and hypha formation in the yeast *Candida albicans*. *J. Gen. Microbiol.* **132**:2035–2047.

1545. **Andrawis, N. S., E. H. Ruley, and D. R. Abernethy.** 1993. Angiotensin II regulates human vascular smooth muscle α-actin gene expression. *Biochem. Biophys. Res. Commun.* **196**:962–968.

1546. **Antecol, M. H., A. Darveau, N. Sonenberg, and B. B. Mukherjee.** 1986. Altered biochemical properties of actin in normal skin fibroblasts from individuals predisposed to dominantly inherited cancers. *Cancer Res.* **46**:1867–1873.

1547. **Asante, E. A., J. M. Boswell, D. W. Burt, and G. Bulfield.** 1994. Tissue specific expression of an α-skeletal actin-lacZ fusion gene during development in transgenic mice. *Transgenic Res.* **3**:59–66.

1548. **Bach, M. A., D. E. Lewis, J. E. McClure, N. Parikh, H. M. Rosenblatt, and W. T. Shearer.** 1986. Monoclonal anti-actin antibody recognizes a surface molecule on normal and transformed human B lymphocytes: expression varies with phase of cell cycle. *Cell Immunol.* **98**:364–374.

1549. **Baines, I. and C. A. King.** 1989. Demonstration of actin in the protozoon; *Gregarina*. *Cell Biol. Int. Rep.* **13**:679–686.

1550. **Baines, I. and C. A. King.** 1989. Demonstration of actin in sporozoites of the protozoon *Eimeria*. *Cell Biol. Int. Rep.* **13**:639–641.

1551. **Baird, W. V. and R. B. Meagher.** 1987. A complex gene superfamily encodes actin in petunia. *EMBO J.* **6**:3223–3231.

1552.* **Barja, F., B. N. Thi, and G. Turian.** 1990. Localization of actin and characterization of its isoforms in the hyphae of *Neurospora crassa*. *FEMS. Microbiol. Lett.* **61**:19–24.

1553. **Beach, R. L. and W. R. Jeffery.** 1990. Temporal and spatial expression of a cytoskeletal actin gene in the ascidian *Styela clava*. *Dev. Genet.* **11**:2–14.

1554. **Beach, R. L. and W. R. Jeffery.** 1992. Multiple actin genes encoding the same α-muscle isoform are expressed during ascidian development. *Dev. Biol.* **151**:55–66.

1555. **Bearer, E. L.** 1991. Actin in the *Drosophila* embryo: is there a relationship to developmental cue localization? *Bioessays* **13**:199–204.

1556. **Benkoel, L., J. Brisse, C. Capo, A. M. Benoliel, P. Bongrand, T. Garcia, and A. Chamlian.** 1992. Localization of actin in normal human hepatocytes using fluorescent phallotoxins and immunohistochemical amplification. *Cell Mol. Biol.* **38**:377–383.

1557. **Bernheim, A., R. Berger, and P. Szabo.** 1984. Localization of actin-related sequences by in situ hybridization to R-banded human chromosomes. *Chromosoma.* **89**:163–167.

1558. **Bremer, J. W., H. Busch, and L. C. Yeoman.** 1981. Evidence for a species of nuclear actin distinct from cytoplasmic and muscles actins. *Biochemistry* **20**:2013–2017.

1559.* **Brett, J. G. and J. Tannenbaum.** 1985. Cytochalasin D-induced increase in actin synthesis and content in a variety of cell types. *Cell Biol. Int. Rep.* **9**:723–730.

1560. **Buckingham, M. E.** 1985. Actin and myosin multigene families: their expression during the formation of skeletal muscle. *Essays. Biochem.* **20**:77–109.

1561. **Buckingham, M. E., M. Carvatti, A. Minty, B. Robert, S. Alonso, A. Cohen, P. Daubas, and A. Weydert.** 1982. Skeletal muscle myogenesis: the expression of actin and myosin mRNAs. *Adv. Exp. Med. Biol.* **158**:331–347.

1562. **Bullock, B. P., P. E. Nisson, and W. R. J. Crain.** 1988. The timing of expression of four actin genes and an RNA polymerase III-transcribed repeated sequence is correct in hybrid embryos of the sea urchin species *Strongylocentrotus purpuratus* and *Lytechinus pictus*. *Dev. Biol.* **130**:335–347.

1563. **Burn, T. C., J. O. Vigoreaux, and S. L. Tobin.** 1989. Alternative 5C actin transcripts are localized in different patterns during *Drosophila* embryogenesis. *Dev. Biol.* **131**:345–355.

1564. **Cao, L. G., D. J. Fishkind, and Y. L. Wang.** 1993. Localization and dynamics of non-filamentous actin in cultured cells. *J. Cell Biol.* **123**:173–181.

1565. **Carey, A. V., R. M. Carey, and R. A. Gomez.** 1992. Expression of α-smooth muscle actin in the developing kidney vasculature. *Hypertension* **19**:II168–II175.

1566. **Carrier, L., K. R. Boheler, C. Chassagne, D. de la Bastie, C. Wisnewsky, E. G. Lakatta, and K. Schwartz.** 1992. Expression of the sarcomeric actin isogenes in the rat heart with development and senescence. *Circ. Res.* **70**:999–1005.

1567. **Charlton, C. A. and L. E. Volkman.** 1991. Sequential rearrangement and nuclear polymerization of actin in baculovirus-infected *Spodoptera frugiperda* cells. *J. Virol.* **65**:1219–1227.

1568. **Cheng, H. and M. Bjerknes.** 1989. Asymmetric distribution of actin mRNA and cytoskeletal pattern generation in polarized epithelial cells. *J. Mol. Biol.* **210**:541–549.

1569. **Clowes, A. W., M. M. Clowes, O. Kocher, P. Ropraz, C. Chaponnier, and G. Gabbiani.** 1988. Arterial smooth muscle cells *in vivo*: relationship between actin isoform expression and mitogenesis and their modulation by heparin. *J. Cell Biol.* **107**:1939–1945.

1570. **Courchesne-Smith, C. L. and S. L. Tobin.** 1989. Tissue-specific expression of the 79B actin gene during *Drosophila* development. *Dev. Biol.* **133**:313–321.

1571.* **Cox, K. H., L. M. Angerer, J. J. Lee, E. H. Davidson, and R. C. Angerer.** 1986. Cell lineage-specific programs of expression of multiple actin genes during sea urchin embryogenesis. *J. Mol. Biol.* **188**:159–172.

1572. **Crosby, J. L., S. J. Phillips, and J. H. Nadeau.** 1989. The cardiac actin locus (Actc-1) is not on mouse chromosome 17 but is linked to β_2-microglobulin on chromosome 2. *Genomics.* **5**:19–23.

1573. **Czosnek, H., U. Nudel, Y. Mayer, P. E. Barker, D. D. Pravtcheva, F. H. Ruddle, and D. Yaffe.** 1983. The genes coding for the cardiac muscle actin, the skeletal muscle actin and the cytoplasmic β-actin are located on three different mouse chromosomes. *EMBO J.* **2**:1977–1979.

1574. **Davis, A. H., R. Blanton, and P. Klich.** 1985. Stage and sex specific differences in actin gene expression in *Schistosoma mansoni*. *Mol. Biochem. Parasitol.* **17**:289–298.

1575. **Denning, G. M., I. S. Kim, and A. B. Fulton.** 1988. Shedding of cytoplasmic actins by developing muscle cells. *J. Cell Sci.* **89**:273–282.

1576.* **DeNofrio, D., T. C. Hoock, and I. M. Herman.** 1989. Functional sorting of actin isoforms in microvascular pericytes. *J. Cell Biol.* **109**:191–202.

1577. **Dueland, S., M. S. Nenseter, and C. A. Drevon.** 1991. Uptake and degradation of filamentous actin and vitamin D-binding protein in the rat. *Biochem. J.* **274**:237–241.

1578. **Dunwoodie, S. L., J. E. Joya, R. M. Arkell, and E. C. Hardeman.** 1994. Multiple regions of the human cardiac actin gene are necessary for maturation-based expression in striated muscle. *J. Biol. Chem.* **269**:12212–12219.

1579. **Durica, D. S. and W. R. J. Crain.** 1982. Analysis of actin synthesis in early sea urchin development. *Dev. Biol.* **92**:428–439.

1580. **Etoh, S., H. Matsui, M. Tokuda, T. Itano, M. Nakamura, and O. Hatase.** 1990. Purification and immunohistochemical study of actin in mitochondrial matrix. *Biochem. Int.* **20**:599–606.

1581. **Fahrni, J. F.** 1992. Actin in the ciliated protozoan *Climacostomum virens*: purification by DNAse I affinity chromatography, electrophoretic characterization, and immunological analysis. *Cell Motil. Cytoskeleton* **22**:62–71.

1582. **Friedman, E., M. Verderame, S. Winawer, and R. Pollack.** 1984. Actin cytoskeletal organization loss in the benign-to-malignant tumor transition in cultured human colonic epithelial cells. *Cancer Res.* **44**:3040–3050.

1583. **Fukuda, Y., S. Uchiyama, Y. Masuda, and Y. Masugi.** 1987. Intranuclear rod-shaped actin filament bundles in poorly differentiated axillary adenosquamous cell carcinoma. *Cancer* **60**:2979–2984.

1584. **Funabiki, R. and M. Kandatsu.** 1968. An attempt to estimate the type of metabolic turnover of actin by continuous administration of acid hydrolysate of 14-C-labelled *Chlorella* protein. *J. Biochem. Tokyo* **64**:717–718.

1585. **Fyrberg, E. A. and J. J. Donady.** 1979. Actin Heterogeneity in primary embryonic culture cells from *Drosophila melanogaster*. *Dev. Biol.* **68**:487–502.

1586. **Fyrberg, E. A., J. W. Mahaffey, B. J. Bond, and N. Davidson.** 1983. Transcripts of the six *Drosophila* actin genes accumulate in a stage- and tissue-specific manner. *Cell* **33**:115–123.

1587. **Garcia, R., B. Paz Aliaga, S. G. Ernst, and W. R. J. Crain.** 1984. Three sea urchin actin genes show different patterns of expression: muscle specific, embryo specific, and constitutive. *Mol. Cell Biol.* **4**:840–845.

1588. **Ghosh, G., J. Mukherjee, S. Biswas, and A. Pal.** 1987. Actin-like protein from *Mimosa pudica* L. *Indian J. Biochem. Biophys.* **24**:336–339.

1589. **Goldstein, D. and J. Leavitt.** 1985. Expression of neoplasia-related proteins of chemically transformed HuT fibroblasts in human osteosarcoma HOS fibroblasts and modulation of actin expression upon elevation of tumorigenic potential. *Cancer Res.* **45**:3256–3261.

1590. **Goodyer, I. D., P. Towner, and D. J. Hayes.** 1991. The use of nested PCR and Southern blotting to confirm the presence of actin in *Plasmodium yoelli*. *Biochem. Soc. Trans.* **19**:419S.

1591. **Gunning, P., P. Ponte, L. Kedes, R. Eddy, and T. Shows.** 1984. Chromosomal location of the co-expressed human skeletal and cardiac actin genes. *Proc. Natl Acad. Sci. U. S. A.* **81**:1813–1817.

1592. **Gunning, P., P. Ponte, L. Kedes, R. J. Hickey, and A. I. Skoultchi.** 1984. Expression of human cardiac actin in mouse L cells: a sarcomeric actin associates with a nonmuscle cytoskeleton. *Cell* **36**:709–715.

1593.* **Gurdon, J. B., T. J. Mohun, S. Brennan, and S. Cascio.** 1985. Actin genes in *Xenopus* and their developmental control. *J. Embryol. Exp. Morphol.* 89 Suppl:125–136.

1594.* **Handel, S. E., S. M. Wang, M. L. Greaser, E. Schultz, J. C. Bulinski, and J. L. Lessard.** 1989. Skeletal muscle myofibrillogenesis as revealed with a monoclonal antibody to titin in combination with detection of the α- and γ-isoforms of actin. *Dev. Biol.* **132**:35–44.

1595. **Hanstein, C., U. Lange, H. A. Schneider Poetsch, F. Grolig, and G. Wagner.** 1991. Detection of actin and localization of phytochrome in the green alga *Mougeotia* by monoclonal antibodies. *Acta Histochem. Suppl.* **41**:223–230.

1596. **Hasek, J., I. Rupes, J. Svobodova, and E. Streiblova.** 1987. Tubulin and actin topology during zygote formation of *Saccharomyces cerevisiae*. *J. Gen. Microbiol.* **133**:3355–3363.

1597.* **Haugland, R. P., W. You, V. B. Paragas, K. S. Wells, and D. A. Dubose.** 1994. Simultaneous visualization of G- and F-actin in endothelial cells. *J. Histochem. Cytochem.* **42**:345–350.

1598. **Hayward, L. J. and R. J. Schwartz.** 1986. Sequential expression of chicken actin genes during myogenesis. *J. Cell Biol.* **102**:1485–1493.

1599. **Hayward, L. J., Y. Y. Zhu, and R. J. Schwartz.** 1988. Cellular localization of muscle and nonmuscle actin mRNAs in chicken primary myogenic cultures: the induction of α-skeletal actin mRNA is regulated independently of α-cardiac actin gene expression. *J. Cell Biol.* **106**:2077–2086.

1600. **Hemminki, K.** 1973. Turnover of actin in rat brain. *Brain Res.* **57**:259–260.

1601. **Hill, M. A. and P. Gunning.** 1993. β and γ actin mRNAs are differentially located within myoblasts. *J. Cell Biol.* **122**:825–832.

1602. **Hirono, M., M. Nakamura, M. Tsunemoto, T. Yasuda, H. Ohba, O. Numata, and Y. Watanabe.** 1987. *Tetrahymena* actin: localization and possible biological roles of actin in *Tetrahymena* cells. *J. Biochem. Tokyo* **102**:537–545.

1603. **Hirono, M., K. Saijo Kurita, J. Kurashima, Y. Watanabe, and T. Ohno.** 1992. Expression of *Tetrahymena* actin in mammalian cells. *Cell Biol. Int. Rep.* **16**:645–651.

1604. **Hoch, H. C. and R. C. Staples.** 1983. Visualization of actin *in situ* by rhodamine-conjugated phalloin in the fungus *Uromyces phaseoli*. *Eur. J. Cell Biol.* **32**:52–58.

1605. **Hoock, T. C., P. M. Newcomb, and I. M. Herman.** 1991. B actin and its mRNA are localized at the plasma membrane and the regions of moving cytoplasm during the cellular response to injury. *J. Cell Biol.* **112**:653–664.

1606. **Hough-Evans, B. R., R. R. Franks, R. W. Zeller, R. J. Britten, and E. H. Davidson.** 1990. Negative spatial regulation of the lineage specific CyIIIa actin gene in the sea urchin embryo. *Development.* **110**:41–50.

1607. **Humphries, S. E., R. Whittall, A. Minty, M. Buckingham, and R. Williamson.** 1981. There are approximately 20 actin genes in the human genome. *Nucl. Acids Res.* **9**:4895–4908.

1608. **Iida, K. and I. Yahara.** 1986. Reversible induction of actin rods in mouse C3H-2K cells by incubation in salt buffers and by treatment with non-ionic detergents. *Exp. Cell Res.* **164**:492–506.

1609. **Infante, A. A. and L. J. Heilmann.** 1981. Distribution of messenger ribonucleic acid in polysomes and nonpolysomal particles of sea urchin embryos: translational control of actin synthesis. *Biochemistry* **20**:1–8.

1610. **Intres, R. and J. J. Donady.** 1985. A constitutively transcribed actin gene is associated with the nuclear matrix in a *Drosophila* cell line. *In Vitro Cell Dev. Biol.* **21**:641–648.

1611. **Jahoda, C. A., A. J. Reynolds, C. Chaponnier, J. C. Forester, and G. Gabbiani.** 1991. Smooth muscle α-actin is a marker for hair follicle dermis *in vivo* and *in vitro*. *J. Cell Sci.* **99**:627–636.

1612. **Jantzen, H.** 1981. Control of actin synthesis during the development of *Acanthamoeba castellaniii*. *Dev. Biol.* **82**:113–126.

1613. **Jochova, J., I. Rupes, and E. Streiblova.** 1991. F-actin contractile rings in protoplasts of the yeast *Schizosaccharomyces*. *Cell Biol. Int. Rep.* **15**:607–610.

1614. **Jockusch, B. M., D. F. Brown, and H. P. Rusch.** 1971. Synthesis and some properties of an actin-like nuclear protein in the slime mold *Physarum polycephalum*. *J. Bacteriol.* **108**:705–714.

1615. **Jung, G. and W. Wernicke.** 1991. Patterns of actin filaments during cell shaping in developing mesophyll of wheat (*Triticum aestivum L.*). *Eur. J. Cell Biol.* **56**:139–146.

1616. **Karlsson, R., P. Aspenstrom, and A. S. Bystrom.** 1991. A chicken β-actin gene can complement a disruption of the *Saccharomyces cerevisiae* ACT1 gene. *Mol. Cell Biol.* **11**:213–217.

1617. **Kedersha, N. L., D. Broek, and R. A. Berg.** 1986. A novel isoform of cytoplasmic actin that binds poly-L-proline. *Biochem. J.* **238**:561–570.

1618. **Kersken, H., J. Vilmart Seuwen, M. Momayezi, and H. Plattner.** 1986. Filamentous actin in *Paramecium* cells: mapping by phalloidin affinity labeling *in vivo* and *in vitro*. *J. Histochem. Cytochem.* **34**:443–454.

1619. **Kilmartin, J. V. and A. E. Adams.** 1984. Structural rearrangements of tubulin and actin during the cell cycle of the yeast *Saccharomyces*. *J. Cell Biol.* **98**:922–933.

1620. **Kim, E., Y. K. Kwon, J. M. Trasler, C. A. Kozak, and N. B. Hecht.** 1990. The mouse smooth muscle γ actin gene is on chromosome 6. *Somat. Cell Mol. Genet.* **16**:287–291.

1621. **Kimura, M., M. Hirono, T. Takemasa, and Y. Watanabe.** 1991. Drastic change in the level of actin mRNA in the course of synchronous division in *Tetrahymena*. *J. Biochem. Tokyo* **109**:399–403.

1622. **Kislauskis, E. H., Z. Li, R. H. Singer, and K. L. Taneja.** 1993. Isoform-specific 3′-untranslated sequences sort α-cardiac and β-cytoplasmic actin messenger RNAs to different cytoplasmic compartments. *J. Cell Biol.* **123**:165–172.

1623. **Kuroda, M.** 1985. Change of actin isomers during differentiation of smooth muscle. *Biochim. Biophys. Acta* **843**:208–213.

1624.* **Kuznetsov, S. A., G. M. Langford, and D. G. Weiss.** 1992. Actin-dependent organelle movement in squid axoplasm. *Nature* **356**:722–725.

1625. **Leavitt, J., A. Leavitt, and A. M. Attallah.** 1980. Dissimilar modes of expression of β- and γ-actin in normal and leukemic human T lymphocytes. *J. Biol. Chem.* **255**:4984–4987.

1626. **Lee, J. J., F. J. Calzone, R. J. Britten, R. C. Angerer, and E. H. Davidson.** 1986. Activation of sea urchin actin genes during embryogenesis. Measurement of transcript accumulation from five different genes in *Strongylocentrotus purpuratus*. *J. Mol. Biol.* **188**:173–183.

1627. **Lee, J. J., F. J. Calzone, and E. H. Davidson.** 1992. Modulation of sea urchin actin mRNA prevalence during embryogenesis: nuclear synthesis and decay rate measurements of transcripts from five different genes. *Dev. Biol.* **149**:415–431.

1628. **Lee, J. J., R. J. Shott, S. J. Rose, T. L. Thomas, R. J. Britten, and E. H. Davidson.** 1984. Sea urchin actin gene subtypes. Gene number, linkage and evolution. *J. Mol. Biol.* **172**:149–176.

1629. **Lloyd, C., G. Schevzov, and P. Gunning.** 1992. Transfection of nonmuscle β- and γ-actin genes into myoblasts elicits different feedback regulatory responses from endogenous actin genes. *J. Cell Biol.* **117**:787–797.

1630.* **Lyons, G. E., M. E. Buckingham, and H. G. Mannherz.** 1991. α-Actin proteins and gene transcripts are colocalized in embryonic mouse muscle. *Development.* **111**:451–454.

1631. **MacLeod, C., R. A. Firtel, and J. Papkoff.** 1980. Regulation of actin gene expression during spore germination in *Dictyostelium discoideum*. *Dev. Biol.* **76**:263–274.

1632. **Masibay, A. S., P. K. Qasba, D. N. Sengupta, G. P. Damewood, and T. Sreevalsan.** 1988. Cell-cycle-specific and serum-dependent expression of γ-actin mRNA in Swiss mouse 3T3 cells. *Mol. Cell Biol.* **8**:2288–2294.

1633.* **McCurdy, D. W. and R. E. Williamson.** 1987. An actin-related protein inside pea chloroplasts. *J. Cell Sci.* **87**:449–456.

1634. **McDonald, A. R., D. J. Garbary, and J. G. Duckett.** 1993. Rhodamine-phalloidin staining of F-actin in rhodophyta. *Biotech. Histochem.* **68**:91–98.

1635. **McKenna, N., J. B. Meigs, and Y. L. Wang.** 1985. Identical distribution of fluorescently labeled brain and muscle actins in living cardiac fibroblasts and myocytes. *J. Cell Biol.* **100**:292–296.

1636. **McKenna, N. M. and Y. L. Wang.** 1986. Possible translocation of actin and α-actinin along stress fibers. *Exp. Cell Res.* **167**:95–105.

1637. **McKeown, M. and R. A. Firtel.** 1981. Evidence for sub-families of actin genes in *Dictyostelium* as determined by comparisons of 3′ end sequences. *J. Mol. Biol.* **151**:593–606.

1638.* **McKeown, M., K. P. Hirth, C. Edwards, and R. A. Firtel.** 1982. Examination of the regulation of the actin multigene family in *Dictyostelium discoideum*. *Prog. Clin. Biol. Res.* 85 Pt A:51–78.

1639.* **McLean, B. G., S. Eubanks, and R. B. Meagher.** 1990. Tissue-specific expression of divergent actins in soybean root. *Plant. Cell* **2**:335–344.

1640. **McLean, M., A. G. Gerats, W. V. Baird, and R. B. Meagher.** 1990. Six actin gene subfamilies map to five chromosomes of *Petunia hybrida*. *J. Hered.* **81**:341–346.

1641. **Merlino, G. T., R. D. Water, G. P. Moore, and L. J. Kleinsmith.** 1981. Change in expression of the actin gene family during early sea urchin development. *Dev. Biol.* **85**:505–508.

1642. **Milankov, K. and U. De Boni.** 1993. Cytochemical localization of actin and myosin aggregates in interphase nuclei *in situ*. *Exp. Cell Res.* **209**:189–199.

1643. **Moalic, J. M., J. Bercovici, and B. Swynghedauw.** 1984. Myosin heavy chain and actin fractional rates of synthesis in normal and overload rat heart ventricles. *J. Mol. Cell Cardiol.* **16**:875–884.

1644. **Mohun, T., N. Garrett, F. Stutz, and G. Sophr.** 1988. A third striated muscle actin gene is expressed during early development in the amphibian *Xenopus laevis*. *J. Mol. Biol.* **202**:67–76.

1645. **Mohun, T. J., S. Brennan, N. Dathan, S. Fairman, and J. B. Gurdon.** 1984. Cell type-specific activation of actin genes in the early amphibian embryo. *Nature* **311**:716–721.

1646. **Mohun, T. J., S. Brennan, and J. B. Gurdon.** 1984. Region-specific regulation of the actin multi-gene family in early amphibian embryos. *Phil. Trans. R. Soc. Lond. Biol.* **307**:337–342.

1647. **Moroianu, J., J. W. Fett, J. F. Riordan, and B. L. Vallee.** 1993. Actin is a surface component of calf pulmonary artery endothelial cells in culture. *Proc. Natl Acad. Sci. U. S. A.* **90**:3815–3819.

1648. **Mortara, R. A.** 1989. Studies on trypanosomatid actin. I. Immunochemical and biochemical identification. *J. Protozool.* **36**:8–13.

1649. **Mossakowska, M. and H. Strzelecka Golaszewska.** 1985. Identification of amino acid substitutions differentiating actin isoforms in their interaction with myosin. *Eur. J. Biochem.* **153**:373–381.

1650. **Mostow, W. R., A. G. Ferguson, M. Lesch, R. S. Decker, and A. M. Samarel.** 1988. Nonrandom turnover of actin and tubulin in cultured rabbit cardiac fibroblasts. *Am. J. Physiol.* **255**:C202-C213.

1651. **Mulholland, J., D. Preuss, A. Moon, A. Wong, D. Drubin, and D. Botstein.** 1994. Ultrastructure of the yeast actin cytoskeleton and its association with the plasma membrane. *J. Cell Biol.* **125**:381–391.

1652. **Mulholland, J. M. and A. G. Wildeman.** 1991. Differential modulation of yeast actin, tubulin, and YPT1 mRNA levels by cycloheximide. *Gene* **101**:81–87.

1653. **Murphy, R. A.** 1992. Do the cytoplasmic and muscle-specific isoforms of actin and myosin heavy and light chains serve different functions in smooth muscle? *Jpn. J. Pharmacol.* 58 Suppl **2**:67P-74P.

1654. **Nakayasu, H. and K. Ueda.** 1983. Association of actin with the nuclear matrix from bovine lymphocytes. *Exp. Cell Res.* **143**:55–62.

1655. **Nakayasu, H. and K. Ueda.** 1984. Small nuclear RNA-protein complex anchors on the actin filaments in bovine lymphocyte nuclear matrix. *Cell Struct. Funct.* **9**:317–325.

1656. **Nakayasu, H. and K. Ueda.** 1986. Preferential association of acidic actin with nuclei and nuclear matrix from mouse leukemia L5178Y cells. *Exp. Cell Res.* **163**:327–336.

1657. **Nishida, E., K. Iida, N. Yonezawa, S. Koyasu, I. Yahara, and H. Sakai.** 1987. Cofilin is a component of intranuclear and cytoplasmic actin rods induced in cultured cells. *Proc. Natl Acad. Sci. U. S. A.* **84**:5262–5266.

1658. **Nisson, P. E., L. E. Dike, and W. R. Crain.** 1989. Three Strongylocentrotus purpuratus actin genes show correct cell-specific expression in hybrid embryos of *S. purpuratus* and *Lytechinus pictus*. *Development*. **105**:407–413.

1659.* **North, A. J., M. Gimona, Z. Lando, and J. V. Small.** 1994. Actin isoform compartments in chicken gizzard smooth muscle cells. *J. Cell Sci.* **107**:445–455.

1660. **Ordahl, C. P.** 1986. The skeletal and cardiac α-actin genes are coexpressed in early embryonic striated muscle. *Dev. Biol.* **117**:488–492.

1661. **Otey, C. A., M. H. Kalnoski, and J. C. Bulinski.** 1987. Identification and quantification of actin isoforms in vertebrate cells and tissues. *J. Cell Biochem.* **34**:113–124.

1662.* **Otey, C. A., M. H. Kalnoski, and J. C. Bulinski.** 1988. Immunolocalization of muscle and nonmuscle isoforms of actin in myogenic cells and adult skeletal muscle. *Cell Motil. Cytoskeleton.* **9**:337–348.

1663.* **Otey, C. A., M. H. Kalnoski, J. L. Lessard, and J. C. Bulinski.** 1986. Immunolocalization of the γ isoform of nonmuscle actin in cultured cells. *J. Cell Biol.* **102**:1726–1737.

1664. **Pahlic, M.** 1985. Multiple forms of actin in *Physarum polycephalum*. *Eur. J. Cell Biol.* **36**:169–175.

1665. **Palevitz, B. A.** 1987. Actin in the preprophase band of *Allium cepa*. *J. Cell Biol.* **104**:1515–1519.

1666. **Palevitz, B. A. and P. K. Hepler.** 1975. Identification of actin in situ at the ectoplasm-endoplasm interface of *Nitella*. Microfilament–chloroplast association. *J. Cell Biol.* **65**:29–38.

1667.* **Pardo, J. V., M. F. Pittenger, and S. W. Craig.** 1983. Subcellular sorting of isoactins: selective association of γ actin with skeletal muscle mitochondria. *Cell* **32**:1093–1103.

1668. **Peng, I. and D. A. Fischman.** 1991. Post-translational incorporation of actin into myofibrils *in vitro*: evidence for isoform specificity. *Cell Motil. Cytoskeleton.* **20**:158–168.

1669. **Peres, H. E., N. Sanchez, L. Vidali, J. M. Hernandez, M. Lara, and F. Sanchez.** 1994. Actin isoforms in non-infected roots and symbiotic root nodules of *Phaseolus vulgaris* L. *Planta* **193**:51–56.

1670. **Pierron, G., D. S. Durica, and H. W. Sauer.** 1984. Invariant temporal order of replication of the four actin gene loci during the naturally synchronous mitotic cycles of *Physarum polycephalum*. *Proc. Natl Acad. Sci. U. S. A.* **81**:6393–6397.

1671. **Prime, S. S. and B. H. Toh.** 1980. Co-capping of insulin, epidermal growth factor and concanavalin A surface receptors with each other and with cytoplasmic actin. *Cytobios.* **28**:151–160.

1672.* **Reuner, K. H., K. Schlegel, I. Just, K. Aktories, and N. Katz.** 1991. Autoregulatory control of actin synthesis in cultured rat hepatocytes. *FEBS Lett.* **286**:100–104.

1673. **Roberts, S. G., G. H. Cope, and C. J. McDonald.** 1991. β-Adrenergic regulation of β-actin mRNA abundance in mouse parotid glands by a post-transcriptional mechanism. *J. Mol. Endocrinol.* **6**:79–86.

1674.* **Romans, P., R. A. Firtel, and C. L. Saxe.** 1985. Gene-specific expression of the actin multigene family of *Dictyostelium discoideum*. *J. Mol. Biol.* **186**:337–355.

1675. **Rubin, R. W. and M. Maher.** 1976. Actin turnover during encystation in *Acanthamoeba*. *Exp. Cell Res.* **103**:159–168.

1676. **Rubinstein, N., J. Chi, and H. Holtzer.** 1976. Coordinated synthesis and degradation of actin and myosin in a variety of myogenic and non-myogenic cells. *Exp. Cell Res.* **97**:387–393.

1677. **Ruzicka, D. L. and R. J. Schwartz.** 1988. Sequential activation of α-actin genes during avian cardiogenesis: vascular smooth muscle α-actin gene transcripts mark the onset of cardiomyocyte differentiation. *J. Cell Biol.* **107**:2575–2586.

1678. **Sabanero, M. and R. Zazueta.** 1989. Actin in *Mucor rouxii*. *FEMS Microbiol. Lett.* **51**:227–231.

1679. **Saborio, J. L., M. Segura, M. Flores, R. Garcia, and E. Palmer.** 1979. Differential expression of gizzard actin genes during chick embryogenesis. *J. Biol. Chem.* **254**:11119–11125.

1680.* **Sahlas, D. J., K. Milankov, P. C. Park, and U. De Boni.** 1993. Distribution of snRNPs, splicing factor SC-35 and actin in interphase nuclei: immunocytochemical evidence for differential distribution during changes in functional states. *J. Cell Sci.* **105**:347–357.

1681. **Sanchez, F., S. L. Tobin, U. Rdest, E. Zulauf, and B. J. McCarthy.** 1983. Two *Drosophila* actin genes in detail. Gene structure, protein structure and transcription during development. *J. Mol. Biol.* **163**:533–551.

1682. **Sanders, S. K. and S. W. Craig.** 1983. A lymphocyte cell surface molecule that is antigenically related to actin. *J. Immunol.* **131**:370–377.

1683. **Sanger, J. W., J. M. Sanger, T. E. Kreis, and B. M. Jockusch.** 1980. Reversible translocation of cytoplasmic actin into the nucleus caused by dimethyl sulfoxide. *Proc. Natl Acad. Sci. U. S. A.* **77**:5268–5272.

1684. **Schedl, T. and W. F. Dove.** 1982. Mendelian analysis of the organization of actin sequences in *Physarum polycephalum*. *J. Mol. Biol.* **160**:41–57.

1685. **Schevzov, G., C. Lloyd, and P. Gunning.** 1992. High level expression of transfected β- and γ-actin genes differentially impacts on myoblast cytoarchitecture. *J. Cell Biol.* **117**:775–785.

1686. **Schmit, A. C. and A. M. Lambert.** 1987. Characterization and dynamics of cytoplasmic F-actin in higher plant endosperm cells during interphase, mitosis, and cytokinesis. *J. Cell Biol.* **105**:2157–2166.

1687. **Schuler, M. A. and E. B. Keller.** 1981. The chromosomal arrangement of two linked actin genes in the sea urchin *S. purpuratus*. *Nucl. Acids Res.* **9**:591–604.

1688. **Seiler Tuyns, A., J. D. Eldridge, and B. M. Paterson.** 1984. Expression and regulation of chicken actin genes introduced into mouse myogenic and nonmyogenic cells. *Proc. Natl Acad. Sci. U. S. A.* **81**:2980–2984.

1689.* **Serpinskaya, A. S., O. N. Denisenko, V. I. Gelfand, and A. D. Bershadsky.** 1990. Stimulation of actin synthesis in phalloidin-treated cells. Evidence for autoregulatory control. *FEBS Lett.* **277**:11–14.

1690. **Shaw, E. M., K. S. Guise, and R. N. Shoffner.** 1989. Chromosomal localization of chicken sequences homologous to the β-actin gene by *in situ* hybridization. *J. Hered.* **80**:475–478.

1691. **Shimizu, N. and T. Obinata.** 1986. Actin concentration and monomer-polymer ratio in developing chicken skeletal muscle. *J. Biochem. Tokyo* **99**:751–759.

1692. **Shortle, D., J. E. Haber, and D. Botstein.** 1982. Lethal disruption of the yeast actin gene by integrative DNA transformation. *Science* **217**:371–373.

1693. **Shott, R. J., J. J. Lee, R. J. Britten, and E. H. Davidson.** 1984. Differential expression of the actin gene family of *Strongylocentrotus purpuratus*. *Dev. Biol.* **101**:295–306.

1694. **Skalli, O., F. Gabbiani, and G. Gabbiani.** 1990. Action of general and α-smooth muscle-specific actin antibody microinjection on stress fibers of cultured smooth muscle cells. *Exp. Cell Res.* **187**:119–125.

1695. **Skalli, O., J. Vandekerckhove, and G. Gabbiani.** 1987. Actin-isoform pattern as a marker of normal or pathological smooth-muscle and fibroblastic tissues. *Differentiation.* **33**:232–238.

1696. **Sodja, A., R. Arking, and R. S. Zafar.** 1982. Actin gene expression during embryogenesis of *Drosophila melanogaster*. *Dev. Biol.* **90**:363–368.

1697. **Sodja, A., R. M. Rizki, T. M. Rizki, and R. S. Zafar.** 1982. Overlapping deficiencies refine the map position of the sex-linked actin gene of *Drosophila melanogaster*. *Chromosoma.* **86**:293–298.

1698. **Soriano, P., P. Szabo, and G. Bernardi.** 1982. The scattered distribution of actin genes in the mouse and human genomes. *EMBO J.* **1**:579–583.

1699. **Storti, R. V., D. M. Coen, and A. Rich.** 1976. Tissue-specific forms of actin in the developing chick. *Cell* **8**:521–527.

1700. **Stozharov, A. N.** 1986. Synthesis of actin-like protein in rat liver mitochondria. *Gen. Physiol. Biophys.* **5**:285–295.
1701. **Sugi, Y. and J. Lough.** 1992. Onset of expression and regional deposition of α-smooth and sarcomeric actin during avian heart development. *Dev. Dyn.* **193**:116–124.
1702. **Sundell, C. L. and R. H. Singer.** 1990. Actin mRNA localizes in the absence of protein synthesis. *J. Cell Biol.* **111**:2397–2403.
1703. **Sundell, C. L. and R. H. Singer.** 1991. Requirement of microfilaments in sorting of actin messenger RNA. *Science* **253**:1275–1277.
1704. **Sympson, C. J. and T. E. Geoghegan.** 1990. Actin gene expression in murine erythroleukemia cells treated with cytochalasin D. *Exp. Cell Res.* **189**:28–32.
1705. **Sympson, C. J., D. Singleton, and T. E. Geoghegan.** 1993. Cytochalasin D-induced actin gene expression in murine erythroleukemia cells. *Exp. Cell Res.* **205**:225–231.
1706. **Takiya, S., T. Tabata, M. Iwabuchi, S. Hirose, and Y. Suzuki.** 1984. Transcription of cloned actin genes of *Dictyostelium discoideum* in the cell-free system. *J. Biochem. Tokyo* **95**:1367–1377.
1707. **Tannenbaum, J. and G. C. Godman.** 1983. Cytochalasin D induces increased actin synthesis in HEp-2 cells. *Mol. Cell Biol.* **3**:132–142.
1708. **Tannenbaum, J. and A. F. Miranda.** 1986. Stimulation of actin synthesis by cytochalasin D is specific for the isoactins normally expressed in muscle or non-muscle cells. *J. Cell Sci.* **84**:253–262.
1709. **Temperli, E., U. P. Roos, and H. R. Hohl.** 1990. Actin and tubulin cytoskeletons in germlings of the oomycete fungus *Phytophthora infestans*. *Eur. J. Cell Biol.* **53**:75–88.
1710. **Tobin, S. L., P. J. Cook, and T. C. Burn.** 1990. Transcripts of individual *Drosophila* actin genes are differentially distributed during embryogenesis. *Dev. Genet.* **11**:15–26.
1711. **Tomlinson, C. R., W. R. Bates, and W. R. Jeffery.** 1987. Development of a muscle actin specified by maternal and zygotic mRNA in ascidian embryos. *Dev. Biol.* **123**:470–482.
1712. **Traas, J. A., J. H. Doonan, D. J. Rawlins, P. J. Shaw, J. Watts, and C. W. Lloyd.** 1987. An actin network is present in the cytoplasm throughout the cell cycle of carrot cells and associates with the dividing nucleus. *J. Cell Biol.* **105**:387–395.
1713. **Ueyama, H., G. Bruns, and N. Kanda.** 1990. Assignment of the vascular smooth muscle actin gene ACTSA to human chromosome 10. *Jinrui. Idengaku. Zasshi.* **35**:145–150.
1714. **Ueyama, H., K. Kurokawa, I. Sasaki, and K. Ueda.** 1987. Characterization of acidic actin in mouse sarcoma 180 cells. *Cell Struct. Funct.* **12**:463–470.
1715. **Ueyama, H., H. Nakayasu, and K. Ueda.** 1987. Nuclear actin and transport of RNA. *Cell Biol. Int. Rep.* **11**:671–677.
1716. **Vandekerckhove, J., G. Bugaisky, and M. Buckingham.** 1986. Simultaneous expression of skeletal muscle and heart actin proteins in various striated muscle tissues and cells. A quantitative determination of the two actin isoforms. *J. Biol. Chem.* **261**:1838–1843.
1717. **Vandekerckhove, J., M. Osborn, M. Altmannsberger, and K. Weber.** 1987. Actin typing of rhabdomyosarcomas shows the presence of the fetal and adult forms of sarcomeric muscle actin. *Differentiation.* **35**:126–131.
1718. **Vigoreaux, J. O. and S. L. Tobin.** 1987. Stage-specific selection of alternative transcriptional initiation sites from the 5C actin gene of *Drosophila melanogaster*. *Genes. Dev.* **1**:1161–1171.
1719. **Watson, P. A., J. P. Stein, and F. W. Booth.** 1984. Changes in actin synthesis and α-actin-mRNA content in rat muscle during immobilization. *Am. J. Physiol.* **247**:C39-C44.
1720. **Williamson, R. E.** 1974. Actin in the alga, *Chara corallina*. *Nature* **248**:801–802.
1721. **Winrow, M. A. and A. Sodja.** 1991. Initial characterization of several actin genes in the parasitic nematode, *Ascaris suum*. *Biochem. Biophys. Res. Commun.* **178**:578–585.
1722.* **Yeh, B. and K. K. H. Svoboda.** 1993. Intracellular labelling of β-actin messenger RNA using reverse transcriptase incorporated biotin dUTP into the actin cortical mat of corneal epithelial cells. *Micron* **24**:595–602.
1723.* **Yeh, B. and K. K. H. Svoboda.** 1994. Intracellular distribution of β-actin mRNA is polarised in embryonic corneal epithelium. *J. Cell Sci.* **107**:105–115.
1724. **Zimmerman, A. M., S. Zimmerman, J. Thomas, and I. Ginzburg.** 1983. Control of tubulin and actin gene expression in *Tetrahymena pyriformis* during the cell cycle. *FEBS Lett.* **164**:318–321.

METHODS FOR ANALYSING ACTIN

1725. Wilson, L. 1982. *Methods in Cell Biology: The Cytoskeleton Part B*; Vol. 24. New York, Academic Press.
1726. Wilson, L. 1982. *Methods in Cell Biology: the Cytoskeleton Part A*; Vol. 24. New York, Academic Press.
1727. **Aakhus, A. M., M. Wilkinson, T. M. Pedersen, and N. O. Solum.** 1989. The use of PhastSystem crossed immunoelectrophoresis with immunoblotting to demonstrate a complex between glycoprotein Ib and the actin-binding protein (ABP) of human platelets. *Electrophoresis* **10**:758–761.
1728. **Abandowitz, H. M. and P. K. Basrur.** 1973. Isolation and immunofluorescent localization of actin derived from rat skeletal muscle. *Histochemie.* **36**:313–320.
1729. **Adams, A. E. and J. R. Pringle.** 1991. Staining of actin with fluorochrome-conjugated phalloidin. *Methods Enzymol.* **194**:729–731.
1730. **Adelman, M. R. and E. W. Taylor.** 1969. Further purification and characterization of slime mold myosin and slime mold actin. *Biochemistry* **8**:4976–4988.
1731. **Adelman, M. R. and E. W. Taylor.** 1969. Isolation of an actomyosin-like protein complex from slime mold plasmodium and the separation of the complex into actin- and myosin-like fractions. *Biochemistry* **8**:4964–4975.
1732. **Ahn, T. I. and K. W. Jeon.** 1989. Antigenic reactivity of a monoclonal antibody against *Amoeba proteus* actin. *J. Protozool.* **36**:560–562.
1733. **Ajtai, K., S. Danko, V. Harsanyi, and E. N. Biro.** 1979. Thymus actin: preparation and characterization. *Acta Biochim. Biophys. Acad. Sci. Hung.* **14**:43–52.
1734. **Amato, P. A. and D. L. Taylor.** 1986. Probing the mechanism of incorporation of fluorescently labeled actin into stress fibers. *J. Cell Biol.* **102**:1074–1084.
1735. **Bailin, G. and M. Barany.** 1972. A simple procedure for the preparation of tropomyosin-free F-actin. *J. Mechanochem. Cell Motil.* **1**:189–190.
1736. **Baumert, H. G., A. Kenmoku, G. Middelhoff, F. Ortanderl, A. Thrun, H. Faulstich, W. Schiebler, and H. Fasold.** 1988. Artificial dimers of native actin: preparation and properties in biological functions. *J. Protein. Chem.* **7**:571–580.
1737. **Begg, D. A., R. Rodewald, and L. I. Rebhun.** 1978. The visualization of actin filament polarity in thin sections. Evidence for the uniform polarity of membrane-associated filaments. *J. Cell Biol.* **79**:846–852.
1738. **Belagyi, J. and P. Grof.** 1983. Rotational motion of actin monomer at low and high salt concentration. *Eur. J. Biochem.* **130**:353–358.
1739. **Benyamin, Y., M. Roger, J. Gabrion, Y. Robin, and N. Van Thoai.** 1979. Immunological comparison of muscle actins from mammal, bird and fish: a quantitative approach. *FEBS Lett.* **102**:69–74.
1740. **Benyamin, Y., M. Roger, Y. Robin, and N. V. Thoai.** 1980. A competitive radioimmunoassay on a magnetic phase for actin detection. *FEBS Lett.* **110**:327–329.
1741. **Benyamin, Y., C. Roustan, and M. Boyer.** 1983. Induction by chemically modified actin derivatives of antibody specificity. A relation between modified sites and antibody interactions with monomeric and filamentous actins. *FEBS Lett.* **160**:41–45.
1742. **Benyamin, Y., C. Roustan, and M. Boyer.** 1986. Anti-actin antibodies. Chemical modification allows the selective production of antibodies to the N-terminal region. *J. Immunol. Methods* **86**:21–29.
1743. **Bochsler, P. N., N. R. Neilsen, D. F. Dean, and D. O. Slauson.** 1992. Stimulus-dependent actin polymerization in bovine neutrophils. *Inflammation.* **16**:383–392.
1744. **Bonder, E. M. and M. S. Mooseker.** 1983. Direct electron microscopic visualization of barbed end capping and filament cutting by intestinal microvillar 95-kdalton protein (villin): a new actin assembly assay using the *Limulus* acrosomal process. *J. Cell Biol.* **96**:1097–1107.
1745. **Booyse, F. M., T. P. Hoveke, and M. E. J. Rafelson.** 1973. Human platelet actin. Isolation and properties. *J. Biol. Chem.* **248**:4083–4091.
1746. **Borejdo, J. and S. Burlacu.** 1991. Distribution of actin filament lengths and their orientation measured by gel electrophoresis in capillaries. *J. Muscle Res. Cell Motil.* **12**:394–407.
1747. **Bottomley, R. C. and I. P. Trayer.** 1975. Affinity chromatography of immobilized actin and myosin. *Biochem. J.* **149**:365–379.
1748. **Boyles, J., L. Anderson, and P. Hutcherson.** 1985. A new fixative for the preservation of actin filaments: fixation of pure actin filament pellets. *J. Histochem. Cytochem.* **33**:1116–1128.
1749. **Bray, D. and C. Thomas.** 1975. The actin content of fibroblasts. *Biochem. J.* **147**:221–228.
1750. **Bray, D. and C. Thomas.** 1976. Unpolymerized actin in fibroblasts and brain. *J. Mol. Biol.* **105**:527–544.
1751. **Bulinski, J. C., S. Kumar, K. Titani, and S. D. Hauschka.** 1983. Peptide antibody specific for the amino terminus of skeletal muscle α-actin. *Proc. Natl Acad. Sci. U. S. A.* **80**:1506–1510.
1752. **Cavadore, J. C., C. Axelrud Cavadore, P. Berta, M. C. Harricane, and J. Haiech.** 1985. Preparation and characterization of bovine aortic actin. *Biochem. J.* **228**:433–441.
1753. **Cavadore, J. C., F. Martin, B. Calas, J. Mery, P. Berta, Y. Benyamin, and C. Roustan.** 1987. Actin antibodies. Preparation and characterization of antibodies specific for smooth-muscle actin isoforms. *Biochem. J.* **242**:51–54.
1754. **Chang, C. M. and R. D. Goldman.** 1973. The localization of actin-like fibers in cultured neuroblastoma cells as revealed by heavy meromyosin binding. *J. Cell Biol.* **57**:867–874.
1755. **Chantler, P. D. and W. B. Gratzer.** 1973. Interaction of myosin with monomeric matrix-bound actin. *FEBS Lett.* **34**:10–14.
1756. **Charlemagne, D., K. Schwartz, S. Bernard, and A. Paraf.** 1979. Induction of specific anti-actin antibodies and their measurement by ELISA technique. *J. Immunol. Methods* **29**:145–153.
1757. **Cheitlin, R. A. and J. Ramachandran.** 1986. Purification of rat adrenocortical actin and its use in an immunoprecipitation assay to quantitate cellular actin. *Biochim. Biophys. Acta* **883**:383–387.
1758. **Corsi, A., I. Ronchetti, and C. Cigognetti.** 1966. Observations on the actin content of the rabbit myofibril. *Biochem. J.* **100**:110–113.
1759. **Coue, M., F. Landon, and A. Olomucki.** 1982. Comparison of the properties of two kinds of preparations of human blood platelet actin with sarcomeric actin. *Biochimie.* **64**:219–226.
1760. **Cove, D. H. and N. Crawford.** 1975. Platelet contractile proteins: separation and characterization of the actin and myosin-like components. *J. Mechanochem. Cell Motil.* **3**:123–133.
1761. **Craig, S. W., S. K. Sanders, and J. V. Pardo.** 1986. Purification of isoform-selective actin antibody from polyclonal antiserum. *Methods Enzymol.* **134**:460–467.
1762. **De Couet, H. G.** 1983. Studies on the antigenic sites of actin: a comparative study of the immunogenic crossreactivity of invertebrate actins. *J. Muscle Res. Cell Motil.* **4**:405–427.

1763. **De Couet, H. G., K. D. Mazander, and U. Groschel Stewart.** 1980. A study of invertebrate actins by isoelectric focusing and immunodiffusion. *Experientia.* **36**:404–405.

1764. **Dedman, J. R. and B. G. Harris.** 1975. *Ascaris suum* actin: properties and similarities to rabbit actin. *Biochem. Biophys. Res. Commun.* **65**:170–175.

1765. **Dhermy, D., O. Bournier, and P. Boivin.** 1978. Partially polymerized erythrocyte actin obtained by affinity chromatography on DNAse sepharose. Comparison with rabbit skeletal muscle actin and role of calcium. *Biochem. Biophys. Res. Commun.* **85**:906–915.

1766. **Dosseto, M. and C. Goridis.** 1980. Differential recognition of monomeric and polymeric forms of actin by anti-actin antibodies. *Mol. Immunol.* **17**:1219–1230.

1767. **Ebashi, S.** 1985. A new simple method of preparing actin from chicken gizzard. *J. Biochem. Tokyo* **97**:693–695.

1768. **Edgar, A. J.** 1989. Gel electrophoresis of native actin and the actin-deoxyribonuclease I complex. *Electrophoresis.* **10**:722–725.

1769. **Elce, J. S., A. S. Elbrecht, M. U. Middlestadt, E. J. McIntyre, and P. J. Anderson.** 1981. Actin from pig and rat uterus. *Biochem. J.* **193**:891–898.

1770. **Fagraeus, A. and R. Norberg.** 1978. Anti-actin antibodies. *Curr. Top. Microbiol. Immunol.* **82**:1–13.

1771. **Faulstich, H., S. Zobeley, U. Bentrup, and B. M. Jockusch.** 1989. Biotinylphallotoxins: preparation and use as actin probes. *J. Histochem. Cytochem.* **37**:1035–1045.

1772. **Feuer, G., F. Molnar, E. Pettko, and F. B. Straub.** 1948.. *Hung. Acta Physiol.* **1**:150–163.

1773. **Fiss, E. and H. R. Buckley.** 1987. Purification of actin from *Candida albicans* and comparison with the *Candida* 48,000-M_r protein. *Infect. Immun.* **55**:2324–2326.

1774. **Flanagan, M. D. and S. Lin.** 1979. Comparative studies on the characteristic properties of two forms of brain actin separable by isoelectric focussing. *J. Neurochem.* **32**:1037–1046.

1775. **Forer, A.** 1982. Actin localization within cells by electron microscopy. *Methods Cell Biol.* 25 Pt B:131–142.

1776. **Fox, J. E., M. E. Dockter, and D. R. Phillips.** 1981. An improved method for determining the actin filament content of nonmuscle cells by the DNase I inhibition assay. *Anal. Biochem.* **117**:170–177.

1777. **Franklin, R. M., L. R. Emmons, R. P. Emmons, A. Oommen, J. R. Pink, A. M. Rijnbeek, M. Schnetzler, L. Tuderman, and E. Vainio.** 1983. Monoclonal antibody which recognizes a common antigenic determinant on intermediate filament proteins, actin, and myosin. *Hybridoma.* **2**:275–285.

1778. **Fujime, S., M. Takasaki Ohsita, and S. Ishiwata.** 1987. Dynamic light-scattering study of muscle F-actin. II. *Biophys. Chem.* **27**:211–224.

1779. **Fulton, C., E. Y. Lai, E. Lamoyi, and D. J. Sussman.** 1986. Naegleria actin elicits species-specific antibodies. *J. Protozool.* **33**:322–327.

1780. **Garrels, J. I. and W. Gibson.** 1976. Identification and characterization of multiple forms of actin. *Cell* **9**:793–805.

1781. **Gendry, P., J. F. Launay, and M. T. Vanier.** 1983. Rat pancreas actin: purification and characterization. *Biochem. Biophys. Res. Commun.* **113**:163–170.

1782. **Gimona, M., J. Vandekerckhove, M. Goethals, M. Herzog, Z. Lando, and J. V. Small.** 1994. β-Actin-specific monoclonal antibody. *Cell Motil. Cytoskeleton* **27**:108–116.

1783. **Gittes, F., B. Mickey, J. Nettleton, and J. Howard.** 1993. Flexural rigidity of microtubules and actin filaments measured from thermal fluctuations in shape. *J. Cell Biol.* **120**:923–934.

1784. **Goldmann, W. H. and G. Isenberg.** 1992. Temperature dependent specific heat capacity (Cp) of G-actin and talin or talin-vinculin bound to G-actin. *Biochem. Soc. Trans.* **20**:273S.

1785. **Gordon, D. J., E. Eisenberg, and E. D. Korn.** 1976. Characterization of cytoplasmic actin isolated from *Acanthamoeba castellaniii* by a new method. J. Biol. Chem. **251**:4778–4786.

1786. **Govindan, V. M. and T. Wieland.** 1975. Isolation and identification of an actin from rat liver. *FEBS Lett.* **59**:117–119.

1787. **Gown, A. M., A. M. Vogel, D. Gordon, and P. L. Lu.** 1985. A smooth muscle-specific monoclonal antibody recognizes smooth muscle actin isozymes. *J. Cell Biol.* **100**:807–813.

1788. **Grandmont Leblanc, A. and J. Gruda.** 1977. Affinity chromatography of myosin, heavy meromyosin, and heavy meromyosin subfragment one on F-actin columns stabilized by phalloidin. *Can. J. Biochem.* **55**:949–957.

1789. **Grazi, E. and E. Magri.** 1981. Studies on the polymerisation of actin: a rapid method for the separation of the monomeric from the polymeric species. *FEBS Lett.* **123**:193–194.

1790. **Grazi, E. and G. Trombetta.** 1986. [γ-^{32}P]ATP as a tracer of the fragmentation of Ca-F-actin. *Biochem. J.* **239**:801–803.

1791. **Greer, C. and R. Schekman.** 1982. Actin from *Saccharomyces cerevisiae*. *Mol. Cell Biol.* **2**:1270–1278.

1792. **Groschel Stewart, U., S. Ceurremans, I. Lehr, C. Mahlmeister, and E. Paar.** 1977. Production of specific antibodies to contractile proteins, and their use in immunofluorescence microscopy. II. Species-specific and species-non-specific antibodies to smooth and striated chicken muscle actin. *Histochemistry* **50**:271–279.

1793. **Hambly, B. D., R. L. Raison, and C. G. dos Remedios.** 1983. Monoclonal antibodies directed against skeletal muscle actin. *Biochem. Int.* **7**:739–746.

1794. **Harris, H. E., M. Y. Tso, and H. F. Epstein.** 1977. Actin and myosin-linked calcium regulation in the nematode *Caenorhabditis elegans*. Biochemical and structural properties of native filaments and purified proteins. *Biochemistry* **16**:859–865.

1795. **Hartwig, J. H. and T. P. Stossel.** 1975. Isolation and properties of actin, myosin, and a new actinbinding protein in rabbit alveolar macrophages. *J. Biol. Chem.* **250**:5696–5705.

1796. **Hatano, S., H. Kondo, and T. Miki Noumura.** 1969. Purification of sea urchin egg actin. *Exp. Cell Res.* **55**:275–277.

1797. **Hatano, S. and F. Oosawa.** 1966. Isolation and characterization of plasmodium actin. *Biochim. Biophys. Acta* **127**:488–498.

1798. **Hatano, S. and F. Oosawa.** 1966. Extraction of an actin-like protein from the plasmodium of a myxomycete and its interaction with myosin A from rabbit striated muscle. *J. Cell Physiol.* **68**:197–202.

1799. **Hatano, S. and K. Owaribe.** 1977. A simple method for the isolation of actin from myxomycete plasmodia. *J. Biochem. Tokyo* **82**:201–205.

1800. **Heggeness, M. H. and J. F. Ash.** 1977. Use of the avidin-biotin complex for the localization of actin and myosin with fluorescence microscopy. *J. Cell Biol.* **73**:783–788.

1801. **Henney, H. R. J. and C. Hubbell.** 1978. Actin from diploid and haploid cells of the myxomycete *Physarum flavicomum* – comparison with skeletal muscle actin. *Microbios.* **23**:25–34.

1802. **Henney, H. R. J. and M. Yee.** 1979. Nuclear actin and histones from the myxomycete *Physarum flavicomum*. *Cytobios.* **24**:103–116.

1803. **Herman, I. M. and T. D. Pollard.** 1979. Comparison of purified anti-actin and fluorescent-heavy meromyosin staining patterns in dividing cells. *J. Cell Biol.* **80**:509–520.

1804. **Hilmo, A. and T. H. Howard.** 1987. F-actin content of neonate and adult neutrophils. *Blood.* **69**:945–949.

1805. **Hirabayashi, T. and Y. Hayashi.** 1972. Antibody specific for actin from frog skeletal muscles. *J. Biochem. Tokyo* **71**:153–156.

1806. **Hirabayashi, T., R. Tamura, I. Mitsui, and Y. Watanabe.** 1983. Investigation of actin in *Tetrahymena* cells. A comparison with skeletal muscle actin by a devised two-dimensional gel electrophoresis method. *J. Biochem. Tokyo* **93**:461–468.

1807. **Hirono, M., Y. Kumagai, O. Numata, and Y. Watanabe.** 1989. Purification of *Tetrahymena* actin reveals some unusual properties. *Proc. Natl Acad. Sci. U. S. A.* **86**:75–79.

1808. **Hiruma, S. and S. Hashimoto.** 1988. Monoclonal antibody to rat brain actin antigenically enhanced with HVJ (Sendai virus) M protein. *Histochemistry* **90**:1–8.

1809. **Houk, T. W. J. and K. Ue.** 1974. The measurement of actin concentration in solution: a comparison of methods. *Anal. Biochem.* **62**:66–74.

1810. **Huang, Z., S. Yue, W. You, and R. P. Haugland.** 1993. A fluorometric microplate-based assay of submicrogram monomeric actin by inhibition of deoxyribonuclease I. *Anal. Biochem.* **214**:272–277.

1811. **Huang, Z. J., R. P. Haugland, and W. M. You.** 1992. Phallotoxin and actin binding assay by fluorescence enhancement. *Anal. Biochem.* **200**:199–204.

1812. **Ireland, M., N. Lieska, and H. Maisel.** 1983. Lens actin: purification and localization. *Exp. Eye Res.* **37**:393–408.

1813. **Ito, T., A. Suzuki, and T. P. Stossel.** 1992. Regulation of water flow by actin-binding protein-induced actin gelatin. *Biophys. J.* **61**:1301–1305.

1814. **Jackson, P. and N. Crawford.** 1976. The isolation and partial characterization of polymorphonuclear-leucocyte actin. *Biochem. Soc. Trans.* **4**:333–336.

1815. **Janmey, P. A., J. Peetermans, K. S. Zaner, T. P. Stossel, and T. Tanaka.** 1986. Structure and mobility of actin filaments as measured by quasi-elastic light scattering, viscometry, and electron microscopy. *J. Biol. Chem.* **261**:8357–8362.

1816. **Jockusch, B. M., M. Becker, I. Hindennach, and E. Jockusch.** 1974. Slime mould actin: homology to vertebrate actin and presence in the nucleus. *Exp. Cell Res.* **89**:241–246.

1817. **Jockusch, B. M., K. H. Kelley, R. K. Meyer, and M. M. Burger.** 1978. An efficient method to produce specific anti-actin. *Histochemistry* **55**:177–184.

1818. **Kaehn, K. and P. Bachmann.** 1986. Different immunofluorescence staining patterns of the thin filament revealed by use of three monoclonal antibodies specific to actin. *Bibl. Anat.* 202–208.

1819. **Kakar, S. K. and F. A. Bettelheim.** 1991. Birefringence of actin. *Biopolymers.* **31**:1283–1287.

1820. **Kane, R. E.** 1975. Preparation and purification of polymerized actin from sea urchin egg extracts. *J. Cell Biol.* **66**:305–315.

1821. **Kas, J., H. Strey, and E. Sackmann.** 1994. Direct imaging of reptation for semiflexible actin filaments. *Nature* **368**:226–229.

1822. **Kersey, Y. M., P. K. Hepler, B. A. Palevitz, and N. K. Wessells.** 1976. Polarity of actin filaments in Characean algae. *Proc. Natl Acad. Sci. U. S. A.* **73**:165–167.

1823. **Koffer, A. and M. J. Dickens.** 1987. Isolation and characterization of actin from cultured BHK cells. *J. Muscle Res. Cell Motil.* **8**:397–406.

1824. **Kondo, H., H. Hayashi, and K. Mihashi.** 1972. Preparation of F-actin: Sepharose 4B matrix. *J. Biochem. Tokyo* **72**:759–760.

1825. **Koteliansky, V. E., M. A. Glukhova, M. V. Bejanian, A. P. Surguchov, and V. N. Smirnov.** 1979. Isolation and characterization of actin-like protein from yeast *Saccharomyces cerevisiae*. *FEBS Lett.* **102**:55–58.

1826. **Kreis, T. E.** 1986. Preparation, assay, and microinjection of fluorescently labeled cytoskeletal proteins: actin, α-actinin, and vinculin. *Methods Enzymol.* **134**:507–519.

1827. **Krizanova, O., E. Zavodska, L. Solarikova, F. Ciampor, and D. Kociskova.** 1984. Actin organization in chick embryo fibroblasts after influenza virus infection. I. Isolation and characterization of actin from chick embryo cells. *Acta Virol. Praha.* **28**:177–184.

1828. **Kron, S. J., D. G. Drubin, D. Botstein, and J. A. Spudich.** 1992. Yeast actin filaments display ATP-dependent sliding movement over surfaces coated with rabbit muscle myosin. *Proc. Natl Acad. Sci. U. S. A.* **89**:4466–4470.

1829. **Kuczmarski, E. R. and J. L. Rosenbaum.** 1979. Chick brain actin and myosin. Isolation and characterization. *J. Cell Biol.* **80**:341–355.
1830. **Kumar, A., Raziuddin;, T. H. Finlay, J. O. Thomas, and W. Szer.** 1984. Isolation of a minor species of actin from the nuclei of *Acanthamoeba castellaniii. Biochemistry* **23**:6753–6757.
1831. **Lachapelle, M.** 1990. Phalloidin directly complexed to colloidal gold as useful markers for F-actin. *J. Histochem. Cytochem.* **38**:135.
1832. **Lachapelle, M. and H. C. Aldrich.** 1988. Phalloidin-gold complexes: a new tool for ultrastructural localization of F-actin. *J. Histochem. Cytochem.* **36**:1197–1202.
1833. **LaFountain, J. R. J., C. R. Zobel, H. R. Thomas, and C. Galbreath.** 1977. Fixation and staining of F-actin and microfilaments using tannic acid. *J. Ultrastruct. Res.* **58**:78–86.
1834. **Landon, F., C. Huc, F. Thome, C. Oriol, and A. Olomucki.** 1977. Human platelet actin. Evidence of β and γ forms and similarity of properties with sarcomeric actin. *Eur. J. Biochem.* **81**:571–577.
1835. **Lawrence, J. B. and R. H. Singer.** 1985. Quantitative analysis of in situ hybridization methods for the detection of actin gene expression. *Nucl. Acids Res.* **13**:1777–1799.
1836. **Lazarides, E.** 1982. Antibody production and immunofluorescent characterization of actin and contractile proteins. *Methods Cell Biol.* **24**:313–331.
1837. **Lazarides, E. and K. Weber.** 1974. Actin antibody: the specific visualization of actin filaments in non-muscle cells. *Proc. Natl Acad. Sci. U. S. A.* **71**:2268–2272.
1838. **Lee, R. W., W. E. Mushynski, and J. M. Trifaro.** 1979. Two forms of cytoplasmic actin in adrenal chromaffin cells. *Neuroscience.* **4**:843–852.
1839. **Lessard, J. L.** 1988. Two monoclonal antibodies to actin: one muscle selective and one generally reactive. *Cell Motil. Cytoskeleton.* **10**:349–362.
1840. **Lessard, J. L., D. Carlton, D. C. Rein, and R. Akeson.** 1979. A solid-phase assay for antiactin antibody and actin using protein-A. *Anal. Biochem.* **94**:140–149.
1841. **Levilliers, N., M. Peron Renner, G. Coffe, and J. Pudles.** 1984. Actin purification from a gel of rat brain extracts. *Biochimie* **66**:531–537.
1842. **Liebes, L. F., R. Stark, D. Nevrla, G. Grusky, D. Zucker Franklin, and R. Silber.** 1983. Purification and characterization of actin from normal and chronic lymphocytic leukemia lymphocytes. *Cancer Res.* **43**:4966–4973.
1843. **Lin, E. C. and H. F. Cantiello.** 1993. A novel method to study the electrodynamic behavior of actin filaments. Evidence for cable-like properties of actin. *Biophys. J.* **65**:1371–1378.
1844. **Lubit, B. W. and J. H. Schwartz.** 1980. An antiactin antibody that distinguishes between cytoplasmic and skeletal muscle actins. *J. Cell Biol.* **86**:891–897.
1845. **Lubit, B. W. and J. H. Schwartz.** 1983. Immunological characterization of an anti-actin antibody specific for cytoplasmic actins and its use for the immunocytological localization of actin in *Aplysia* nervous tissue. *J. Histochem. Cytochem.* **31**:728–736.
1846. **Luna, E. J., Y. L. Wang, E. W. J. Voss, D. Branton, and D. L. Taylor.** 1982. A stable, high capacity, F-actin affinity column. *J. Biol. Chem.* **257**:13095–13100.
1847. **Mabuchi, I. and J. A. Spudich.** 1980. Purification and properties of soluble actin from sea urchin eggs. *J. Biochem. Tokyo* **87**:785–802.
1848. **MacLean Fletcher, S. and T. D. Pollard.** 1980. Identification of a factor in conventional muscle actin preparations which inhibits actin filament self-association. *Biochem. Biophys. Res. Commun.* **96**:18–27.
1849. **Maruyama, K., M. Kaibara, and E. Fukada.** 1974. Rheology of F-actin. I. Network of F-actin in solution. *Biochim. Biophys. Acta* **371**:20–29.
1850. **Maupin Szamier, P. and T. D. Pollard.** 1978. Actin filament destruction by osmium tetroxide. *J. Cell Biol.* **77**:837–852.
1851. **Maupin, P. and T. D. Pollard.** 1983. Improved preservation and staining of HeLa cell actin filaments, clathrin-coated membranes, and other cytoplasmic structures by tannic acid–glutaraldehyde–saponin fixation. *J. Cell Biol.* **96**:51–62.
1852. **Mejean, C., C. Roustan, and Y. Benyamin.** 1987. Anti-actin antibodies. Detection and quantitation of total and skeletal muscle actin in human plasma using a competitive ELISA. *J. Immunol. Methods* **99**:129–135.
1853. **Meza, I., M. Sabanero, F. Cazares, and J. Bryan.** 1983. Isolation and characterization of actin from *Entamoeba histolytica. J. Biol. Chem.* **258**:3936–3941.
1854. **Mihashi, K., H. Yoshimura, T. Nishio, A. Ikegami, and K. J. Kinosita.** 1983. Internal motion of F-actin in 10^{-6}–10^{-3} s time range studied by transient absorption anisotropy: detection of torsional motion. *J. Biochem. Tokyo* **93**:1705–1707.
1855. **Miki Noumura, T.** 1969. An actin-like protein of the sea urchin eggs. II. Direct isolation procedure. *Dev. Growth. Differ. Nagoya.* **11**:219–231.
1856. **Miki Noumura, T. and F. Oosawa.** 1969. An actin-like protein of the sea urchin eggs. I. Its interaction with myosin from rabbit striated muscle. *Exp. Cell Res.* **56**:224–232.
1857. **Miller, K. G., C. M. Field, B. M. Alberts, and D. R. Kellogg.** 1991. Use of actin filament and microtubule affinity chromatography to identify proteins that bind to the cytoskeleton. *Methods Enzymol.* **196**:303–319.
1858. **Minakata, A.** 1966. Dielectric dispersion of G-actin. *Biochim. Biophys. Acta* **126**:570–577.
1859. **Minkoff, L. and R. Damadian.** 1975. Biological ion exchanger resins: VIII. A preliminary report on actin-like protein in *E. coli* and the cytotonus concept. *Physiol. Chem. Phys.* **7**:385–389.
1860. **Mommaerts, W. F. H. M.** 1951. Reversible polymerisation and ultracentrifugal purification of actin. *J. Biol. Chem.* **188**:559–565.
1861. **Moring, S., M. Ruscha, P. Cooke, and F. Samson.** 1975. Isolation and polymerization of brain actin. *J. Neurobiol.* **6**:245–255.
1862. **Mossakowska, M., J. Belagyi, and H. Strzelecka Golaszewska.** 1988. An EPR study of the rotational dynamics of actins from striated and smooth muscle and their complexes with heavy meromyosin. *Eur. J. Biochem.* **175**:557–564.
1863. **Nagy, B.** 1972. Difference spectral method for determination of native G-actin concentration. *Anal. Biochem.* **47**:371–377.
1864. **Nakamura, T., M. Yamaguchi, and T. Yanagisawa.** 1979. Comparative studies on actins from various sources. Fragments of actins from *Ascaris* muscle cleaved at cysteinyl residues in comparison with those of other actins. *J. Biochem. Tokyo* **85**:627–631.
1865. **Nakashima, K. and E. Beutler.** 1979. Comparison of structure and function of human erythrocyte and human muscle actin. *Proc. Natl Acad. Sci. U. S. A.* **76**:935–938.
1866. **Nefsky, B. and A. Bretscher.** 1992. Yeast actin is relatively well behaved. *Eur. J. Biochem.* **206**:949–955.
1867. **Neimark, H. C.** 1977. Extraction of an actin-like protein from the prokaryote *Mycoplasma pneumoniae. Proc. Natl Acad. Sci. U. S. A.* **74**:4041–4045.
1868. **Nishioka, M., T. Aibiki, M. Shirai, S. Terada, H. Kagawa, and S. Watanabe.** 1986. Rabbit autoantibodies to actin induced by immunization with modified homologous actins. *Microbiol. Immunol.* **30**:1291–1297.
1869. **Ohnishi, T.** 1977. Isolation and characterization of an actin-like protein from membranes of human red cells. *Br. J. Haematol.* **35**:453–458.
1870. **Okamoto, Y., I. Komatsu, E. Kinjo, and T. Sekine.** 1983. Actin and actin-associated proteins of rabbit liver cell. *J. Biochem. Tokyo* **94**:645–653.
1871. **Oplatka, A., A. Muhlrad, and R. Lamed.** 1976. Immobilized ATP and actin columns as a tool for the characterization and separation of different myosins and active myosin fragments. *J. Biol. Chem.* **251**:3972–3976.
1872. **Ostap, E. M., T. Yanagida, and D. D. Thomas.** 1992. Orientational distribution of spin-labeled actin oriented by flow. *Biophys. J.* **63**:966–975.
1873. **Osung, O. A., B. H. Toh, A. Gray, J. Sotelo, A. Yildiz, T. Diggle, and E. J. Holborow.** 1980. Immunoperoxidase EM localisation of cytoplasmic actin in cultured fibroblasts. *J. Immunol. Methods* **34**:303–313.
1874. **Owaribe, K. and S. Hatano.** 1975. Induction of antibody against actin from myxomycete plasmodium and its properties. *Biochemistry* **14**:3024–3029.
1875. **Palevitz, B. A., J. F. Ash, and P. K. Hepler.** 1974. Actin in the green alga, *Nitella. Proc. Natl Acad. Sci. U. S. A.* **71**:363–366.
1876. **Palmer, E. and J. L. Saborio.** 1978. *In vivo* and *in vitro* synthesis of multiple forms of rat brain actin. *J. Biol. Chem.* **253**:7482–7489.
1877. **Pardee, J. D. and J. R. Bamburg.** 1976. Quantitation of actin in developing brain. *J. Neurochem.* **26**:1093–1098.
1878. **Pardee, J. D. and J. A. Spudich.** 1982. Purification of muscle actin. *Methods Enzymol.* **85** Pt B:164–181.
1879. **Pardee, J. D. and J. A. Spudich.** 1982. Purification of muscle actin. *Methods Cell Biol.* **24**:271–289.
1880. **Peckham, M., J. E. Molloy, J. C. Sparrow, and D. C. S. White.** 1990. Physiological properties of the dorsal longitudinal flight muscle and the tergal depressor of the trochanter muscle of *Drosophila melanogaster. J. Muscle Res. Cell Motil.* **11**:203–215.
1881. **Pollard, T. D.** 1983. Measurement of rate constants for actin filament elongation in solution. *Anal. Biochem.* **134**:406–412.
1882. **Pollard, T. D. and J. A. Cooper.** 1982. Methods to characterize actin filament networks. *Methods Enzymol.* 85 Pt B:211–233.
1883. **Polzar, B., E. Nowak, R. S. Goody, and H. G. Mannherz.** 1989. The complex of actin and deoxyribonuclease I as a model system to study the interactions of nucleotides, cations and cytochalasin D with monomeric actin. *Eur. J. Biochem.* **182**:267–275.
1884. **Polzar, B., A. Rosch, and H. G. Mannherz.** 1989. A simple procedure to produce monospecific polyclonal antibodies of high affinity against actin from muscular sources. *Eur. J. Cell Biol.* **50**:220–229.
1885. **Popp, D., V. V. Lednev, and W. Jahn.** 1987. Methods of preparing well-orientated sols of f-actin containing filaments suitable for X-ray diffraction. *J. Mol. Biol.* **197**:679–684.
1886. **Preston, F. B. and G. N. Graham.** 1972. Determination of the molecular weights of actins from different species by gel chromatography. *Biochem. J.* **127**:75P–76P.
1887. **Probst, E. and F. Luscher.** 1972. Studies on thrombosthenin a, the actin-like moiety of the contractile protein from blood platelets. I. Isolation, characterization and evidence for two forms of thrombosthenin A. *Biochim. Biophys. Acta* **278**:577–584.
1888. **Puszkin, S. and S. Berl.** 1972. Actomyosin-like protein from brain. Separation and characterization of the actin-like component. *Biochim. Biophys. Acta* **256**:695–709.
1889. **Raghavan, M., C. K. Smith, and C. E. Schutt.** 1989. Analytical determination of methylated histidine in proteins: actin methylation. *Anal. Biochem.* **178**:194–197.
1890. **Rees, M. K. and M. Young.** 1967. Studies on the isolation and molecular properties of homogeneous globular actin. Evidence for a single polypeptide chain structure. *J. Biol. Chem.* **242**:4449–4458.
1891. **Rockwell, M. A., M. Fechheimer, and D. L. Taylor.** 1984. A comparison of methods used to characterize gelation of actin *in vitro. Cell Motil.* **4**:197–213.
1892. **Rohr, G. and H. G. Mannherz.** 1978. Isolation and characterization of secretory actin. DNAase I complex from rat pancreatic juice. *Eur. J. Biochem.* **89**:151–157.

1893. **Rubenstein, P. A. and J. A. Spudich.** 1977. Actin microheterogeneity in chick embryo fibroblasts. *Proc. Natl Acad. Sci. U. S. A.* **74**:120–123.

1894. **Ruscha, M. F. and R. H. Himes.** 1981. A simple procedure for the isolation of brain actin. *Prep. Biochem.* **11**:351–360.

1895. **Safer, D.** 1989. An electrophoretic procedure for detecting proteins that bind actin monomers. *Anal. Biochem.* **178**:32–37.

1896. **Sakakibara, I. and K. Yagi.** 1970. Molecular weight of G-actin obtained by the light-scattering method. *Biochim. Biophys. Acta* **207**:178–183.

1897. **Sanger, J. W.** 1975. Intracellular localization of actin with fluorescently labelled heavy meromyosin. *Cell Tissue. Res.* **161**:431–434.

1898. **Sawada, S., T. Iio, Y. Hayashi, and S. Takahashi.** 1992. Fluorescent rotors and their applications to the study of G-F transformation of actin. *Anal. Biochem.* **204**:110–117.

1899. **Schachat, F. H., H. E. Harris, and H. F. Epstein.** 1977. Actin from the nematode, *Caenorhabditis elegans*, is a single electrofocusing species. *Biochim. Biophys. Acta* **493**:304–309.

1900. **Schwartz, R. J. and K. Rothblum.** 1980. Regulation of muscle differentiation: isolation and purification of chick actin messenger ribonucleic acid and quantitation with complementary deoxyribonucleic acid probes. *Biochemistry* **19**:2506–2514.

1901. **Shapiro, H. M.** 1971. A multivariate statistical method for comparing protein amino acid compositions: studies of muscle actins and proteins derived from membranes and microtubular organelles. *Biochim. Biophys. Acta* **236**:725–738.

1902. **Sharrard, R. M. and P. Banks.** 1980. Purification and properties of actins from bovine splenic nerve and brain. *Neuroscience.* **5**:1781–1791.

1903. **Shimada, Y. and T. Obinata.** 1977. Polarity of actin filaments at the initial stage of myofibril assembly in myogenic cells *in vitro*. *J. Cell Biol.* **72**:777–785.

1904. **Sikora, L. and G. A. Marzluf.** 1982. Identification and isolation of actin from *Neurospora crassa*. *J. Gen. Microbiol.* **128**:439–445.

1905. **Simpson, P. A., J. A. Spudich, and P. Parham.** 1984. Monoclonal antibodies prepared against *Dictyostelium* actin: characterization and interactions with actin. *J. Cell Biol.* **99**:287–295.

1906. **Skalli, O., P. Ropraz, A. Trzeciak, G. Benzonana, D. Gillessen, and G. Gabbiani.** 1986. A monoclonal antibody against α-smooth muscle actin: a new probe for smooth muscle differentiation. *J. Cell Biol.* **103**:2787–2796.

1907. **Small, J. V.** 1981. Organization of actin in the leading edge of cultured cells: influence of osmium tetroxide and dehydration on the ultrastructure of actin meshworks. *J. Cell Biol.* **91**:695–705.

1908. **Small, J. V., S. Zobeley, G. Rinnerthaler, and H. Faulstich.** 1988. Coumarin-phalloidin: a new actin probe permitting triple immunofluorescence microscopy of the cytoskeleton. *J. Cell Sci.* **89**:21–24.

1909. **Smith, J. M.** 1984. A method for staining actin-containing structures in thick plastic sections for medium voltage electron microscopy. *Tissue. Cell* **16**:43–51.

1910. **Snabes, M. C., A. E. Boyd, and J. Bryan.** 1981. Detection of actin-binding proteins in human platelets by 125I-actin overlay of polyacrylamide gels. *J. Cell Biol.* **90**:809–812.

1911. **Snabes, M. C., A. E. Boyd, R. L. Pardue, and J. Bryan.** 1981. A DNase I binding/immunoprecipitation assay for actin. *J. Biol. Chem.* **256**:6291–6295.

1912. **Spudich, J. A.** 1974. Biochemical and structural studies of actomyosin-like proteins from non-muscle cells. II. Purification, properties, and membrane association of actin from amoebae of *Dictyostelium discoideum*. *J. Biol. Chem.* **249**:6013–6020.

1913. **Spudich, J. A. and S. Watt.** 1971. The regulation of rabbit skeletal muscle contraction. I. Biochemical studies of the interaction of the tropomyosin–troponin complex with actin and the proteolytic fragments of myosin. *J. Biol. Chem.* **246**:4866–4871.

1914. **Stark, R., L. F. Liebes, D. Nevrla, and R. Silber.** 1982. The quantitation of actin in human lymphocytes by isoelectric focusing. *Biochem. Med.* **27**:200–206.

1915. **Stone, D. B., P. M. Curmi, and R. A. Mendelson.** 1987. Preparation of deuterated actin from *Dictyostelium discoideum*. *Methods Cell Biol.* **28**:215–229.

1916. **Strader, C. D., E. Lazarides, and M. A. Raftery.** 1980. The characterization of actin associated with postsynaptic membranes from *Torpedo californica*. *Biochem. Biophys. Res. Commun.* **92**:365–373.

1917. **Strzelecka Golaszewska, H., E. Prochniewicz, E. Nowak, S. Zmorzynski, and W. Drabikowski.** 1980. Chicken-gizzard actin: polymerization and stability. *Eur. J. Biochem.* **104**:41–52.

1918. **Stuewer, D. and U. Groschel Stewart.** 1990. Immune response of rabbits to native G-actins. *Biochim. Biophys. Acta* **1039**:5–11.

1919. **Taneja, K. and R. H. Singer.** 1987. Use of oligodeoxynucleotide probes for quantitative in situ hybridization to actin mRNA. *Anal. Biochem.* **166**:389–398.

1920. **Tannenbaum, J. and A. Rich.** 1979. An isoelectric focusing study of plasma membrane actin. *Anal. Biochem.* **95**:236–244.

1921. **Thorstensson, R., G. Utter, R. Norberg, and A. Fagraeus.** 1981. A radioimmunoassay for determination of anti-actin antibodies. *J. Immunol. Methods* **45**:15–26.

1922. **Tiburzy, R., H. C. Hoch, and R. C. Staples.** 1990. Isolation and identification of actin from the phytopathogenic filamentous fungus uromyces appendiculatus. *Eur. J. Cell Biol.* **53**:364–372.

1923. **Tilley, L., M. Dwyer, and G. B. Ralston.** 1986. Solubilization of native actin monomers from human erythrocyte membranes. *Aust. J. Biol. Sci.* **39**:117–124.

1924. **Tilley, L. and G. Ralston.** 1984. Purification and kinetic characterisation of human erythrocyte actin. *Biochim. Biophys. Acta* **790**:46–52.

1925. **Torbet, J. and M. J. Dickens.** 1984. Orientation of skeletal muscle actin in strong magnetic fields. *FEBS Lett.* **173**:403–406.

1926. **Tsukada, T., D. Tippens, D. Gordon, R. Ross, and A. M. Gown.** 1987. HHF35, a muscle-actin-specific monoclonal antibody. I. Immunocytochemical and biochemical characterization. *Am. J. Pathol.* **126**:51–60.

1927. **Turoverov, K. K., S. Y. Haitlina, and G. P. Pinaev.** 1976. Ultra-violet fluorescence of actin. Determination of native actin content in actin preparations. *FEBS Lett.* **62**:4–6.

1928. **Uyemura, D. G., S. S. Brown, and J. A. Spudich.** 1978. Biochemical and structural characterization of actin from *Dictyostelium discoideum*. *J. Biol. Chem.* **253**:9088–9096.

1929. **Valenta, R., F. Ferreira, M. Grote, I. Swoboda, S. Vrtala, M. Duchene, P. Deviller, R. B. Meagher, E. McKinney, E. Heberle Bors,** *et al.* 1993. Identification of profilin as an actin-binding protein in higher plants. *J. Biol. Chem.* **268**:22777–22781.

1930. **Vande Berg, J. S., D. R. Disharoon, W. L. Poolman, and R. Rudolph.** 1988. Evaluation of an enzyme-linked immunosorbent assay to actin in cultured fibroblasts. *Immunol. Invest.* **17**:273–294.

1931. **Villanueva, M. A., S. C. Ho, and J. L. Wang.** 1990. Isolation and characterization of one isoform of actin from cultured soybean cells. *Arch. Biochem. Biophys.* **277**:35–41.

1932. **Wang, Y. L. and D. L. Taylor.** 1981. Probing the dynamic equilibrium of actin polymerization by fluorescence energy transfer. *Cell* **27**:429–436.

1933. **Warrick, H. M., R. M. Simmons, J. T. Finer, T. Q. Uyeda, S. Chu, and J. A. Spudich.** 1993. *In vitro* methods for measuring force and velocity of the actin–myosin interaction using purified proteins. *Methods Cell Biol.* **39**:1–21.

1934. **Water, R. D., J. R. Pringle, and L. J. Kleinsmith.** 1980. Identification of an actin-like protein and of its messenger ribonucleic acid in *Saccharomyces cerevisiae*. *J. Bacteriol.* **144**:1143–1151.

1935. **Weber, K., R. Koch, W. Herzog, and J. Vandekerckhove.** 1977. The isolation of tubulin and actin from mouse 3T3 cells transformed by simian virus 40 (SV3T3 cells), an established cell line growing in culture. *Eur. J. Biochem.* **78**:27–32.

1936. **Weihing, R. R. and E. D. Korn.** 1971. *Acanthamoeba* actin. Isolation and properties. *Biochemistry* **10**:590–600.

1937. **Weir, J. P. and D. W. Frederiksen.** 1980. The isolation and characterization of actin from porcine brain. *Arch. Biochem. Biophys.* **203**:1–10.

1938. **Weir, J. P. and D. W. Frederiksen.** 1982. Preparation of cytoplasmic actin. *Methods Enzymol.* 85 Pt B:371–373.

1939. **Wilson, F. J. and H. Finck.** 1971. Actin: immunochemical and immunofluorescence studies. *J. Biochem. Tokyo* **70**:143–148.

1940. **Wolosewick, J. J., J. De Mey, and V. Meininger.** 1983. Ultrastructural localization of tubulin and actin in polyethylene glycol-embedded rat seminiferous epithelium by immunogold staining. *Biol. Cell* **49**:219–226.

1941. **Woolley, D. E.** 1972. An actin-like protein from amoebae of *Dictyostelium discoideum*. *Arch. Biochem. Biophys.* **150**:519–530.

1942. **Work, S. S. and D. M. Warshaw.** 1992. Computer-assisted tracking of actin filament motility. *Anal. Biochem.* **202**:275–285.

1943. **Yabu, H., R. Takahashi, and E. Miyazaki.** 1969. Intestinal actin-like protein. *Jpn. J. Physiol.* **19**:722–732.

1944. **Yang, Y. Z. and J. F. Perdue.** 1972. Contractile proteins of cultured cells. I. The isolation and characterization of an actin-like protein from cultured chick embryo fibroblasts. *J. Biol. Chem.* **247**:4503–4509.

1945. **Zechel, K.** 1980. Isolation of polymerization-competent cytoplasmic actin by affinity chromatography on immobilized DNAse I using formamide as eluant. *Eur. J. Biochem.* **110**:343–348.

1946. **Zechel, K. and K. Weber.** 1978. Actins from mammals, bird, fish and slime mold characterized by isoelectric focusing in polyacrylamide gels. *Eur. J. Biochem.* **89**:105–112.

1947. **Zot, H. G. and J. D. Potter.** 1981. Purification of actin from cardiac muscle. *Prep. Biochem.* **11**:381–395.

TOXINS AND DRUGS AFFECTING ACTIN FUNCTION

1948. *Recent Advances in Cytochalasins.* London, Chapman and Hall.

1949. **Aktories, K.** 1990. Clostridial ADP-ribosyltransferases–modification of low molecular weight GTP-binding proteins and of actin by clostridial toxins. *Med. Microbiol. Immunol. Berl.* **179**:123–136.

1950. **Aktories, K.** 1990. ADP-ribosylation of actin. *J. Muscle Res. Cell Motil.* **11**:95–97.

1951. **Aktories, K., T. Ankenbauer, B. Schering, and K. H. Jakobs.** 1986. ADP-ribosylation of platelet actin by botulinum C2 toxin. *Eur. J. Biochem.* **161**:155–162.

1952. **Aktories, K., M. Barmann, I. Ohishi, S. Tsuyama, K. H. Jakobs, and E. Habermann.** 1986. Botulinum C2 toxin ADP-ribosylates actin. *Nature* **322**:390–392.

1953. **Aktories, K., K. H. Reuner, P. Presek, and M. Barmann.** 1989. Botulinum C2 toxin treatment increases the G-actin pool in intact chicken cells: a model for the cytopathic action of actin-ADP-ribosylating toxins. *Toxicon.* **27**:989–993.

1954. **Aktories, K. and A. Wegner.** 1989. ADP-ribosylation of actin by clostridial toxins. *J. Cell Biol.* **109**:1385–1387.

1955. **Aktories, K. and A. Wegner.** 1992. Mechanisms of the cytopathic action of actin-ADP-ribosylating toxins. *Mol. Microbiol.* **6**:2905–2908.

1956. **Aktories, K., M. Wille, and I. Just.** 1992. Clostridial actin-ADP-ribosylating toxins. *Curr. Top. Microbiol. Immunol.* **175**:97–113.

1957. **Barak, L. S. and R. R. Yocum.** 1981. 7-Nitrobenz-2-oxa-1,3-diazole (NBD)-phallacidin: synthesis of a fluorescent actin probe. *Anal. Biochem.* **110**:31–38.

1958. **Barak, L. S., R. R. Yocum, E. A. Nothnagel, and W. W. Webb.** 1980. Fluorescence staining of the actin cytoskeleton in living cells with 7-nitrobenz-2-oxa-1,3-diazole-phallacidin. *Proc. Natl Acad. Sci. U. S. A.* **77**:980–984.

1959. **Barak, L. S., R. R. Yocum, and W. W. Webb.** 1981. *In vivo* staining of cytoskeletal actin by autointernalization of nontoxic concentrations of nitrobenzoxadiazole-phallacidin. *J. Cell Biol.* **89**:368–372.

1960. **Blackholm, H. and H. Faulstich.** 1981. Protection of one of the two reactive thiol groups in F-actin by ATP and phalloidin. *Biochem. Biophys. Res. Commun.* **103**:125–130.

1961. **Bonder, E. M. and M. S. Mooseker.** 1986. Cytochalasin B slows but does not prevent monomer addition at the barbed end of the actin filament. *J. Cell Biol.* **102**:282–288.

1962. **Borovikov YuS,** 1984. The effect of glutaraldehyde and phalloidin on the conformation of F-actin. *Gen. Physiol. Biophys.* **3**:513–516.

1963. **Brenner, S. L. and E. D. Korn.** 1979. Substoichiometric concentrations of cytochalasin D inhibit actin polymerization. Additional evidence for an F-actin treadmill. *J. Biol. Chem.* **254**:9982–9985.

1964. **Brenner, S. L. and E. D. Korn.** 1980. Spectrin/actin complex isolated from sheep erythrocytes accelerates actin polymerization by simple nucleation. Evidence for oligomeric actin in the erythrocyte cytoskeleton. *J. Biol. Chem.* **255**:1670–1676.

1965. **Brown, S. S. and J. A. Spudich.** 1979. Cytochalasin inhibits the rate of elongation of actin filament fragments. *J. Cell Biol.* **83**:657–662.

1966. **Brown, S. S. and J. A. Spudich.** 1981. Mechanism of action of cytochalasin: evidence that it binds to actin filament ends. *J. Cell Biol.* **88**:487–491.

1967. **Cano, M. L., L. Cassimeris, M. Joyce, and S. H. Zigmond.** 1992. Characterization of tetramethylrhodaminyl-phalloidin binding to cellular F-actin. *Cell Motil. Cytoskeleton.* **21**:147–158.

1968. **Carlier, M. F., P. Criquet, D. Pantaloni, and E. D. Korn.** 1986. Interaction of cytochalasin D with actin filaments in the presence of ADP and ATP. *J. Biol. Chem.* **261**:2041–2050.

1969. **Colombo, R., I. Dalle Donne, and A. Milzani.** 1990. Metal ions modulate the effect of doxorubicin on actin assembly. *Cancer Biochem. Biophys.* **11**:217–226.

1970. **Colombo, R. and A. Milzani.** 1988. How does doxorubicin interfere with actin polymerization? *Biochim. Biophys. Acta* **968**:9–16.

1971. **Colombo, R., A. Necco, G. Vailati, and A. Milzani.** 1988. Dose-dependence of doxorubicin effect on actin assembly *in vitro*. *Exp. Mol. Pathol.* **49**:297–304.

1972. **Colombo, R., A. Necco, G. Vailati, B. Saracco, A. Milzani, and G. Scari.** 1984. Doxorubicin affects actin assembly *in vitro*. *Cell Biol. Int. Rep.* **8**:127–135.

1973. **Coluccio, L. M. and L. G. Tilney.** 1984. Phalloidin enhances actin assembly by preventing monomer dissociation. *J. Cell Biol.* **99**:529–535.

1974. **Cooper, J. A.** 1987. Effects of cytochalasin and phalloidin on actin. *J. Cell Biol.* **105**:1473–1478.

1975. **Coue, M., S. L. Brenner, I. Spector, and E. D. Korn.** 1987. Inhibition of actin polymerization by latrunculin A. *FEBS Lett.* **213**:316–318.

1976. **Cramb, G. and J. W. Dow.** 1983. Uptake of bepridil into isolated ventricular myocytes. Retention by actin. *Biochem. Pharmacol.* **32**:227–231.

1977. **Cribbs, D. H., J. R. J. Glenney, P. Kaulfus, K. Weber, and S. Lin.** 1982. Interaction of cytochalasin B with actin filaments nucleated or fragmented by villin. *J. Biol. Chem.* **257**:395–399.

1978. **Dabrowska, R., E. Prochniewicz, and W. Drabikowski.** 1983. The effect of cytochalasin and glutaraldehyde on F-actin filaments containing muscle and non-muscle tropomyosin. *J. Muscle Res. Cell Motil.* **4**:83–93.

1979. **Dancker, P., L. Hess, and K. Ritter.** 1991. Product release is not the rate-limiting step during cytochalasin B-induced ATPase activity of monomeric actin. *Z. Naturforsch. C.* **46**:139–144.

1980. **Dancker, P. and I. Low.** 1979. Complex influence of cytochalasin B on actin polymerization. *Z. Naturforsch. C.* **34**:555–557.

1981. **Dancker, P., I. Low, W. Hasselbach, and T. Wieland.** 1975. Interaction of actin with phalloidin: polymerization and stabilization of F-actin. *Biochim. Biophys. Acta* **400**:407–414.

1982. **De Vries, J. X., A. J. Schafer, H. Faulstich, and T. Wieland.** 1976. Protection of actin from heat denaturation by various phallotoxins. *Hoppe Seylers Z. Physiol. Chem.* **357**:1139–1143.

1983. **Drubin, D. G., H. D. Jones, and K. F. Wertman.** 1993. Actin structure and function – roles in mitochondrial organisation and morphogenesis in budding yeast and identification of the phalloidin-binding site. *Mol. Biol. Cell* **4**:1277–1294.

1984. **Elias, E. and J. L. Boyer.** 1979. Chlorpromazine and its metabolites alter polymerization and gelation of actin. *Science* **206**:1404–1406.

1985. **Estes, J. E., L. A. Selden, and L. C. Gershman.** 1981. Mechanism of action of phalloidin on the polymerization of muscle actin. *Biochemistry* **20**:708–712.

1986. **Faulstich, H., A. Buku, H. Bodenmuller, and T. Wieland.** 1980. Virotoxins: actin-binding cyclic peptides of *Amanita virosa* mushrooms. *Biochemistry* **19**:334–343.

1987. **Faulstich, H., A. J. Schafer, and M. Weckauf.** 1977. The dissociation of the phalloidin-actin complex. *Hoppe Seylers Z. Physiol. Chem.* **358**:181–184.

1988. **Faulstich, H., S. Zobeley, U. Bentrup, and B. M. Jockusch.** 1989. Biotinylphallotoxins: preparation and use as actin probes. *J. Histochem. Cytochem.* **37**:1035–1045.

1989. **Faulstich, H., S. Zobeley, D. Heintz, and G. Drewes.** 1993. Probing the phalloidin binding site of actin. *FEBS Lett.* **318**:218–222.

1990. **Faulstich, H., S. Zobeley, G. Rinnerthaler, and J. V. Small.** 1988. Fluorescent phallotoxins as probes for filamentous actin. *J. Muscle Res. Cell Motil.* **9**:370–383.

1991. **Fesus, L., L. Muszbek, and K. Laki.** 1981. The effect of methylglyoxal on actin. *Biochem. Biophys. Res. Commun.* **99**:617–622.

1992. **Flanagan, M. D. and S. Lin.** 1980. Cytochalasins block actin filament elongation by binding to high affinity sites associated with F-actin. *J. Biol. Chem.* **255**:835–838.

1993. **Forer, A., J. Emmersen, and O. Behnke.** 1972. Cytochalasin B: does it affect actin-like filaments? *Science* **175**:774–776.

1994. **Forscher, P. and S. J. Smith.** 1988. Actions of cytochalasins on the organization of actin filaments and microtubules in a neuronal growth cone. *J. Cell Biol.* **107**:1505–1516.

1995. **Furukawa, K., K. Sakai, S. Watanabe, K. Maruyama, M. Murakami, K. Yamaguchi, and Y. Ohizumi.** 1993. Goniodomin A induces modulation of actomyosin ATPase activity mediated through conformational change of actin. *J. Biol. Chem.* **268**:26026–26031.

1996. **Fussmann, B. and P. Dancker.** 1986. Polymerization of actin in the absence and presence of cytochalasin B: problems of determining "critical concentration". *Z. Naturforsch. C.* **41**:781–786.

1997. **Geipel, U., I. Just, and K. Aktories.** 1990. Inhibition of cytochalasin D-stimulated G-actin ATPase by ADP-ribosylation with *Clostridium perfringens* iota toxin. *Biochem. J.* **266**:335–339.

1998. **Gicquaud, C., A. Turcotte, and S. St Pierre.** 1983. *Amanita virosa* peptides: viroidin and viroisin are more effective than phalloidin for the *in vitro* protection of actin against the effects of osmic acid. *Eur. J. Cell Biol.* **32**:171–173.

1999. **Goddette, D. W. and C. Frieden.** 1985. The binding of cytochalasin D to monomeric actin. *Biochem. Biophys. Res. Commun.* **128**:1087–1092.

2000. **Goddette, D. W. and C. Frieden.** 1986. The kinetics of cytochalasin D binding to monomeric actin. *J. Biol. Chem.* **261**:15970–15973.

2001. **Goddette, D. W. and C. Frieden.** 1986. Actin polymerization. The mechanism of action of cytochalasin D. *J. Biol. Chem.* **261**:15974–15980.

2002. **Grundmann, E., U. Wissemann, and C. B. Boschek.** 1980. The effect of phalloidin on polymerisation of actin isolated from rat hepatocytes and AS-30d ascites hepatoma cells. *Hoppe Seylers Z. Physiol. Chem.* **361**:457–460.

2003. **Hartwig, J. H. and T. P. Stossel.** 1979. Cytochalasin B and the structure of actin gels. *J. Mol. Biol.* **134**:539–553.

2004. **Himes, R. H. and L. L. Houston.** 1976. The action of cytochalasin A on the *in vitro* polymerization of brain tubulin and muscle G-actin. *J. Supramol. Struct.* **5**:81–90.

2005. **Himes, R. H., R. N. Kersey, M. Ruscha, and L. L. Houston.** 1976. Cytochalasin A inhibits the *in vitro* polymerization of brain tubulin and muscle actin. *Biochem. Biophys. Res. Commun.* **68**:1362–1370.

2006. **Hirata, M., T. Inamitsu, T. Hashimoto, and T. Koga.** 1984. An inhibitor of lipoxygenase, nordihydroguaiaretic acid, shortens actin filaments. *J. Biochem. Tokyo* **95**:891–894.

2007. **Hirono, M., Y. Kumagai, O. Numata, and Y. Watanabe.** 1989. Purification of *Tetrahymena* actin reveals some unusual properties. *Proc. Natl Acad. Sci. U. S. A.* **86**:75–79.

2008. **Hirose, T., Y. Izawa, K. Koyama, S. Natori, K. Iida, I. Yahara, S. Shimaoka, and K. Maruyama.** 1990. The effects of new cytochalasins from *Phomopsis* sp. and the derivatives on cellular structure and actin polymerization. *Chem. Pharm. Bull. Tokyo* **38**:971–974.

2009. **Howard, T. H. and S. Lin.** 1979. Specific interaction of cytochalasins with muscle and platelet actin filaments *in vitro*. *J. Supramol. Struct.* **11**:283–293.

2010. **Just, I., E. S. Hennessey, D. R. Drummond, K. Aktories, and J. C. Sparrow.** 1993. ADP-ribosylation of *Drosophila* indirect-flight-muscle actin and arthrin by *Clostridium botulinum* C2 toxin and *Clostridium perfringens* iota toxin. *Biochem. J.* **291**:409–412.

2011. **Just, I., M. Wille, C. Chaponnier, and K. Aktories.** 1993. Gelsolin-actin complex is target for ADP-ribosylation by *Clostridium botulinum* C2 toxin in intact human neutrophils. *Eur. J. Pharmacol.* **246**:293–297.

2012. **Kahl, J. U., G. P. Vlasov, A. Seeliger, and T. Wieland.** 1984. Analogs of viroidin. Synthesis of four virotoxin-like F-actin binding heptapeptides with one less hydroxyl monomer dissociation. *J. Cell Biol.* **99**:529–535.

1974. **Cooper, J. A.** 1987. Effects of cytochalasin and phalloidin on actin. *J. Cell Biol.* **105**:1473–1478.

1975. **Coue, M., S. L. Brenner, I. Spector, and E. D. Korn.** 1987. Inhibition of actin polymerization by latrunculin A. *FEBS Lett.* **213**:316–318.

1976. **Cramb, G. and J. W. Dow.** 1983. Uptake of bepridil into isolated ventricular myocytes. Retention by actin. *Biochem. Pharmacol.* **32**:227–231.

1977. **Cribbs, D. H., J. R. J. in gei-sol transformation.** *FEBS Lett.* **153**:311–314.

2016. **Le Bihan, T. and C. Gicquaud.** 1991. Stabilization of actin by phalloidin: a differential scanning calorimetric study. *Biochem. Biophys. Res. Commun.* **181**:542–547.

2017. **Lengsfeld, A. M., I. Low, T. Wieland, P. Dancker, and W. Hasselbach.** 1974. Interaction of phalloidin with actin. *Proc. Natl Acad. Sci. U. S. A.* **71**:2803–2807.

2018. **Lin, D. C. and S. Lin.** 1980. A rapid assay for actin-associated high-affinity cytochalasin binding sites based on isoelectric precipitation of soluble protein. *Anal. Biochem.* **103**:316–322.

2019. **Lin, D. C., K. D. Tobin, and D. H. Cribbs.** 1984. On the mechanism for inactivation of cytochalasin binding activity associated with F-actin and spectrin-band 4.1-actin complex by sulfhydryl reagents. *Biochem. Biophys. Res. Commun.* **122**:244–251.

2020. **Lin, D. C., K. D. Tobin, M. Grumet, and S. Lin.** 1980. Cytochalasins inhibit nuclei-induced actin polymerization by blocking filament elongation. *J. Cell Biol.* **84**:455–460.

2021. **Lorenz, M., D. Popp, and K. C. Holmes.** 1993. Refinement of the F-actin model against X-ray fiber diffraction data by the use of a directed mutation algorithm. *J. Mol. Biol.* **234**:826–836.

2022. **Low, I. and P. Dancker.** 1976. Effect of cytochalasin B on formation and properties of muscle F-actin. *Biochim. Biophys. Acta* **430**:366–374.

2023. **Low, I., P. Dancker, and Wieland Th.** 1975. Stabilization of F-actin by phalloidin. Reversal of the destabilizing effect of cytochalasin B. *FEBS Lett.* **54**:263–265.

2024. **Low, I., P. Dancker, and T. Wieland.** 1976. Stabilization of actin polymer structure by phalloidin: ATPase activity of actin induced by phalloidin at low pH. *FEBS Lett.* **65**:358–360.

2025. **Low, I., W. Jahn, T. Wieland, S. Sekita, K. Yoshihira, and S. Natori.** 1979. Interaction between rabbit muscle actin and several chaetoglobosins or cytochalasins. *Anal. Biochem.* **95**:14–18.

2026. **Low, I. and T. Wieland.** 1974. The interaction of phalloidin. Some of its derivatives, and of other cyclic peptides with muscle actin as studied by viscosimetry. *FEBS Lett.* **44**:340–343.

2027. **Mabuchi, I.** 1983. Electron microscopic determination of the actin filament end at which cytochalasin B blocks monomer addition using the acrosomal actin bundle from horseshoe crab sperm. *J. Biochem. Tokyo* **94**:1349–1352.

2028. **MacLean Fletcher, S. and T. D. Pollard.** 1980. Mechanism of action of cytochalasin B on actin. *Cell* **20**:329–341.

2029. **Mariano, R., B. Gonzalez, and W. Lewis.** 1986. Cardiac actin interactions with doxorubicin *in vitro*. *Exp. Mol. Pathol.* **44**:7–13.

2030. **Maruyama, K., J. H. Hartwig, and T. P. Stossel.** 1980. Cytochalasin B and the structure of actin gels. II. Further evidence for the splitting of F-actin by cytochalasin B. *Biochim. Biophys. Acta* **626**:494–500.

2031. **Maruyama, K., M. Oosawa, A. Tashiro, T. Suzuki, M. Tanikawa, M. Kikuchi, S. Sekita, and S. Natori.** 1986. Effects of chaetoglobosin J on the G-F transformation of actin. *Biochim. Biophys. Acta* **874**:137–143.

2032. **Mauss, S., C. Chaponnier, I. Just, K. Aktories, and G. Gabbiani.** 1990. ADP-ribosylation of actin isoforms by *Clostridium botulinum* C2 toxin and *Clostridium perfringens* iota toxin. *Eur. J. Biochem.* **194**:237–241.

2033. **Miki, M.** 1987. The recovery of the polymerizability of Lys-61-labelled actin by the addition of phalloidin. Fluorescence polarization and resonance-energy-transfer measurements. *Eur. J. Biochem.* **164**:229–235.

2034. **Miki, M., J. A. Barden, C. G. dos Remedios, L. Phillips, and B. D. Hambly.** 1987. Interaction of phalloidin with chemically modified actin. *Eur. J. Biochem.* **165**:125–130.

2035. **Mitchell, M. J., B. E. Laughon, and S. Lin.** 1987. Biochemical studies on the effect of *Clostridium difficile* toxin B on actin *in vivo* and *in vitro*. *Infect. Immun.* **55**:1610–1615.

2036. **Miyamoto, Y., M. Kuroda, E. Munekata, and T. Masaki.** 1986. Stoichiometry of actin and phalloidin binding: one molecule of the toxin dominates two actin subunits. *J. Biochem. Tokyo* **100**:1677–1680.

2037. **Morris, A. and J. Tannenbaum.** 1980. Cytochalasin D does not produce net depolymerization of actin filaments in HEp-2 cells. *Nature* **287**:637–639.

2038. **Moss, M. L., R. E. Palmer, P. Kuzmic, B. E. Dunlap, W. Henzel, J. L. Kofron, W. S. Mellon, C. A. Royer, and D. H. Rich.** 1992. Identification of actin and HSP 70 as cyclosporin A binding proteins by photoaffinity labeling and fluorescence displacement assays. *J. Biol. Chem.* **267**:22054–22059.

2039. **Nakamura, S., K. Ohmi, and Y. Nonomura.** 1988. Low concentration of reserpine accelerates actin polymerization via interaction with G-actin. *Mol. Pharmacol.* **33**:604–610.

2040. **Natori, S.** 1986. Cytochalasins-actin filament modifiers as a group of mycotoxins. *Dev. Toxicol. Environ. Sci.* **12**:291–299.

2041. **Necco, A., G. Vailati, and R. Colombo.** 1983. Actin behaviour in doxorubicin treatment. *Cell Biol. Int. Rep.* **7**:129–134.

2042. **Norgauer, J., E. Kownatzki, R. Seifert, and K. Aktories.** 1988. Botulinum C2 toxin ADP-ribosylates actin and enhances O_2-production and secretion but inhibits migration of activated human neutrophils. *J. Clin. Invest.* **82**:1376–1382.

2043. **Ohishi, I., Y. Morikawa, and T. Baba.** 1990. ADP-ribosylation of nonmuscle actin by component I of botulinum C2 toxin inactivates the ability to interact with unmodified actin. *J. Biochem. Tokyo* **107**:420–425.

2044. **Ohishi, I. and S. Tsuyama.** 1986. ADP-ribosylation of nonmuscle actin with component I of C2 toxin. *Biochem. Biophys. Res. Commun.* **136**:802–806.

2045. **Ohmori, H. and S. Toyama.** 1992. Direct proof that the primary site of action of cytochalasin on cell motility processes is actin. *J. Cell Biol.* **116**:933–941.

2046. **Ottlinger, M. E. and S. Lin.** 1988. Clostridium difficile toxin B induces reorganization of actin, vinculin, and talin in cultured cells. *Exp. Cell Res.* **174**:215–229.

2047. **Patterson, G. M., C. D. Smith, L. H. Kimura, B. A. Britton, and S. Carmeli.** 1993. Action of tolytoxin on cell morphology, cytoskeletal organization, and actin polymerization. *Cell Motil. Cytoskeleton* **24**:39–48.

2048. **Pollender, J. M. and J. Gruda.** 1979. Effect of phalloidin on actin proteolysis as measured by viscometry and fluorimetry. *Can. J. Biochem.* **57**:49–55.

2049. **Popoff, M. R. and P. Boquet.** 1988. Clostridium spiroforme toxin is a binary toxin which ADP-ribosylates cellular actin. *Biochem. Biophys. Res. Commun.* **152**:1361–1368.

2050. **Popoff, M. R., E. J. Rubin, D. M. Gill, and P. Boquet.** 1988. Actin-specific ADP-ribosyltransferase produced by a *Clostridium difficile* strain. *Infect. Immun.* **56**:2299–2306.

2051. **Rampal, A. L., H. B. Pinkofsky, and C. Y. Jung.** 1980. Structure of cytochalasins and cytochalasin B binding sites in human erythrocyte membranes. *Biochemistry* **19**:679–683.

2052. **Reuner, K. H., P. Presek, C. B. Boschek, and K. Aktories.** 1987. Botulinum C2 toxin ADP-ribosylates actin and disorganizes the microfilament network in intact cells. *Eur. J. Cell Biol.* **43**:134–140.

2053. **Rosqvist, R., A. Forsberg, and H. Wolf Watz.** 1991. Intracellular targeting of the *Yersinia* YopE cytotoxin in mammalian cells induces actin microfilament disruption. *Infect. Immun.* **59**:4562–4569.

2054. **Rosqvist, R., A. Forsberg, and H. Wolf Watz.** 1991. Microinjection of the *Yersinia* YopE cytotoxin in mammalian cells induces actin microfilament disruption. *Biochem. Soc. Trans.* **19**:1131–1132.

2055. **Runnegar, M. T. and I. R. Falconer.** 1986. Effect of toxin from the cyanobacterium *Microcystis aeruginosa* on ultrastructural morphology and actin polymerization in isolated hepatocytes. *Toxicon.* **24**:109–115.

2056. **Sampath, P. and T. D. Pollard.** 1991. Effects of cytochalasin, phalloidin, and pH on the elongation of actin filaments. *Biochemistry* **30**:1973–1980.

2057. **Schafer, A., J. X. De Vries, H. Faulstich, and T. Wieland.** 1975. Phalloidin counteracts the inhibitory effect of actin on deoxyribonuclease I. *FEBS Lett.* **57**:51–54.

2058. **Schafer, A. J. and H. Faulstich.** 1977. A protein-binding assay for phallotoxins using muscle actin. *Anal. Biochem.* **83**:720–723.

2059. **Schering, B., M. Barmann, G. S. Chatwal, U. Geipel, and K. Aktories.** 1988. ADP-ribosylation of skeletal muscle and non-muscle actin by *Clostridium perfringens* iota toxin. *Eur. J. Biochem.* **171**:225–229.

2060. **Sekita, S., K. Yoshihira, S. Natori, F. Harada, K. Iida, and I. Yahara.** 1985. Structure-activity relationship of thirty-nine cytochalasans observed in the effects on cellular structures and cellular events and on actin polymerization *in vitro*. *J. Pharmacobiodyn.* **8**:906–916.

2061. **Selden, L. A., L. C. Gershman, and J. E. Estes.** 1980. A proposed mechanism of action of cytochalasin D on muscle actin. *Biochem. Biophys. Res. Commun.* **95**:1854–1860.

2062. **Selden, L. A., H. J. Kinosian, L. C. Gershman, and J. E. Estes.** 1991. Cytochalasin D-activation of Mg-ATP-actin ATPase activity. *Biophys. J.* **59**:53.

2063. **Shanahan, M. F.** 1983. Characterization of cytochalasin B photoincorporation into human erythrocyte D-glucose transporter and F-actin. *Biochemistry* **22**:2750–2756.

2064. **Spudich, J. A. and S. Lin.** 1972. Cytochalasin B, its interaction with actin and actomyosin from muscle (cell movement-microfilaments-rabbit striated muscle). *Proc. Natl Acad. Sci. U. S. A.* **69**:442–446.

2065. **Suzuki, N. and K. Mihashi.** 1991. Binding mode of cytochalasin B to F-actin is altered by lateral binding of regulatory proteins. *J. Biochem. Tokyo.* **109**:19–23.

2066. **Tannenbaum, J. and J. G. Brett.** 1985. Evidence for regulation of actin synthesis in cytochalasin D-treated HEp-2 cells. *Exp. Cell Res.* **160**:435–448.

2067. **Terashima, M., K. Mishima, K. Yamada, M. Tsuchiya, T. Wakutani, and M. Shimoyama.** 1992. ADP-ribosylation of actins by arginine-specific ADP-ribosyltransferase purified from chicken heterophils. *Eur. J. Biochem.* **204**:305–311.

2068. **Toyama, S.** 1984. A variant form of β-actin in a mutant of KB cells resistant to cytochalasin B. *Cell* **37**:609–614.

2069. **Toyama, S.** 1988. Functional alterations in β'-actin from a KB cell mutant resistant to cytochalasin B. *J. Cell Biol.* **107**:1499–1504.

2070. **Turcotte, A., C. Gicquaud, M. Gendreau, and S. St Pierre.** 1984. Separation of the virotoxins of the mushroom *Amanita virosa* and comparative study of their interaction on actin *in vitro*. *Can. J. Biochem. Cell Biol.* **62**:1327–1334.

2071. **Urbanik, E. and B. R. Ware.** 1989. Actin filament capping and cleaving activity of cytochalasins B, D, E, and H. *Arch. Biochem. Biophys.* **269**:181–187.

2072. **Vandekerckhove, J., A. Deboben, M. Nassal, and T. Wieland.** 1985. The phalloidin binding site of F-actin. *EMBO J.* **4**:2815–2818.

2073. **Vandekerckhove, J., B. Schering, M. Barmann, and K. Aktories.** 1987. *Clostridium perfringens* iota toxin ADP-ribosylates skeletal muscle actin in Arg-177. *FEBS Lett.* **225**:48–52.

2074. **Vandekerckhove, J., B. Schering, M. Barmann, and K. Aktories.** 1988. Botulinum C2 toxin ADP-ribosylates cytoplasmic β/γ-actin in arginine 177. *J. Biol. Chem.* **263**:696–700.

2075. **Walling, E. A., G. A. Krafft, and B. R. Ware.** 1988. Actin assembly activity of cytochalasins and cytochalasin analogs assayed using fluorescence photobleaching recovery. *Arch. Biochem. Biophys.* **264**:321–332.

2076. **Wendel, H. and P. Dancker.** 1987. Influence of phalloidin on both the nucleation and the elongation phase of actin polymerization. *Biochim. Biophys. Acta* **915**:199–204.

2077. **Wendel, H. and P. Dancker.** 1987. Influence of phalloidin on ATP hydrolysis during actin polymerization. *Biochim. Biophys. Acta* **915**:205–209.

2078. **Wieland, T.** 1976. Interaction of phallotoxins with actin. *Adv. Enzyme. Regul.* **15**:285–300.

2079. **Wieland, T.** 1977. Modification of actins by phallotoxins. *Naturwissenschaften.* **64**:303–309.

2080. **Wieland, T., J. X. De Vries, A. Schafer, and H. Faulstich.** 1975. Spectroscopic evidence for the interaction of phalloidin with actin. *FEBS Lett.* **54**:73–75.

2081. **Wieland, T. and V. M. Govindan.** 1974. Phallotoxins bind to actins. *FEBS Lett.* **46**:351–353.
2082. **Wieland, T., T. Miura, and A. Seeliger.** 1983. Analogs of phalloidin. D-Abu2-Lys7-phalloin, an F-actin binding analog, its rhodamine conjugate (RLP) a novel fluorescent F-actin-probe, and D-Ala2-Leu7-phalloin, an inert peptide. *Int. J. Pept. Protein. Res.* **21**:3–10.
2083. **Williamson, R. E. and U. A. Hurley.** 1986. Growth and regrowth of actin bundles in *Chara*: bundle assembly by mechanisms differing in sensitivity to cytochalasin. *J. Cell Sci.* **85**:21–32.
2084. **Wilson, P., E. Fuller, and A. Forer.** 1987. Irradiations of rabbit myofibrils with an ultraviolet microbeam. II. Phalloidin protects actin in solution but not in myofibrils from depolymerization by ultraviolet light. *Biochem. Cell Biol.* **65**:376–385.
2085. **Wulf, E., A. Deboben, F. A. Bautz, H. Faulstich, and T. Wieland.** 1979. Fluorescent phallotoxin, a tool for the visualization of cellular actin. *Proc. Natl Acad. Sci. U. S. A.* **76**:4498–4502.

NEW REFERENCES FOR 3rd EDITION

BOOKS

Estes, J.E. and P.J. Higgins (eds). 1994. Actin: biophysics, biochemistry and cell biology. *Adv. Exp. Med. Biol.* **358**.
Lloyd, C.W. (ed.). 1982. *The Cytoskeleton in Plant Growth and Development*. Cambridge University Press, Cambridge.
Lloyd, C.W. (ed.). 1991. *The Cytoskeletal Basis of Plant Growth and Form*. Academic Press, London.
Menzell, D. (ed.). 1991. *The Cytoskeleton of the Algae*. CRC Press, Boca Raton.
Perry, S.V. and A. Margreth (eds). 1976. *Contractile Systems in Non-muscle Cells*. North Holland, Amsterdam.

REVIEWS

2086. **Estes, J. E. and P. J. Higgins** (eds). 1994. Actin: biophysics, biochemistry and cell biology. *Adv. Exp. Med. Biol.* **358**.
2087. **Ampe, C. and J. Vandekerckhove.** 1994. Actin–actin binding protein interfaces. *Semin. Cell Biol.* **5**:175–182.
2088. **Bassell, G. J., K. L. Taneja, E. H. Kislauskis, C. L. Sundell, C. M. Powers, A. Ross, and R. H. Singer.** 1994. Actin filaments and the spatial positioning of mRNAS. *Adv. Exp. Med. Biol.* **358**:183–189.
2089. **Bretscher, A.** 1993. Microfilaments and membranes. *Curr. Opin. Cell Biol.* **5**:653–660.
2090. **Bretscher, A., B. Drees, E. Harsay, D. Schott, and T. Wang.** 1994. What are the basic functions of microfilaments? Insights from studies in budding yeast. *J. Cell Biol.* **126**:821–825.
2091. **Carlier, M. F. and D. Pantaloni.** 1994. Actin assembly in response to extracellular signals: role of capping proteins, thymosin beta 4 and profilin. *Semin. Cell Biol.* **5**:183–191.
2092. **Carlier, M. F., C. Valentin Ranc, C. Combeau, S. Fievez, and D. Pantoloni.** 1994. Actin polymerization: regulation by divalent metal ion and nucleotide binding, ATP hydrolysis and binding of myosin. *Adv. Exp. Med. Biol.* **358**:71–81.
2093. **Chrzanowska Wodnicka, M. and K. Burridge.** 1992. Rho, rac and the actin cytoskeleton. *Bioessays* **14**:777–778.
2094. **Cossart, P. and C. Kocks.** 1994. The actin-based motility of the facultative intracellular pathogen Listeria monocytogenes. *Mol. Microbiol.* **13**:395–402.
2095. **Cramer, L. P., T. J. Mitchison, and J. A. Theriot.** 1994. Actin-dependent motile forces and cell motility. *Curr. Opin. Cell Biol.* **6**:82–86.
2096. **Crawford, L. E., R. W. Tucker, A. W. Heldman, and P. J. Goldschmidt Clermont.** 1994. Actin regulation and surface catalysis. *Adv. Exp. Med. Biol.* **358**:105–112.
2097. **Doonan, J. H. and L. Clayton.** 1987. Immunofluorescent studies on the plant cytoskeleton. *Soc. Exp. Biol. Sem. Ser.* **29**:111–136.
2098. **Doonan, J. H. and C. W. Lloyd.** 1988. The bryophyte cytoskeleton: experimental and immunofluorescence studies. *Adv. Bryol.* **3**:1–34.
2099. **dos Remedios, C. G. and P. D. Moens.** 1995. Actin and the actomyosin interface: a review. *Biochim. Biophys. Acta* 1228:99–124.
2100. **Egelman, E. H.** 1994. Actin filament structure. The ghost of ribbons past. *Curr. Biol.* **4**:79–81.
2101. **Fang, H. and B. P. Brandhorst.** 1994. Evolution of actin gene families of sea urchins. *J. Mol. Evol.* **39**:347–356.
2102. **Fechheimer, M. and S. H. Zigmond.** 1993. Focusing on unpolymerized actin. *J. Cell Biol.* **123**:1–5.
2103. **Fyrberg, C., L. Ryan, M. Kenton, and E. Fyrberg.** 1994. Genes encoding actin-related proteins of *Drosophila melanogaster*. *J. Mol. Biol.* **241**:498–503.
2104. **Fyrberg, C., L. Ryan, L. McNally, M. Kenton, and E. Fyrberg.** 1994. The actin protein superfamily. *Soc. Gen. Physiol. Ser.* **49**:173–178.
2105. **Gershman, L. C., L. A. Selden, H. J. Kinosian, and J. E. Estes.** 1994. Actin-bound nucleotide/divalent cation interactions. *Adv. Exp. Med. Biol.* **358**:35–49.
2106. **Gips, S. J., D. E. Kandzari, and P. J. Goldschmidt Clermont.** 1994. Growth factor receptors, phospholipases, phospholipid kinases and actin reorganization. *Semin. Cell Biol.* **5**:201–208.
2107. **Hall, A.** 1994. Small GTP-binding proteins and the regulation of the actin cytoskeleton. *Annu. Rev. Cell Biol.* **10**:31–54.
2108. **Hatano, S.** 1994. Actin-binding proteins in cell motility. *Int. Rev. Cytol.* **156**:199–273.
2109. **Hitchcock DeGregori, S. E.** 1994. Structural requirements of tropomyosin for binding to filamentous actin. *Adv. Exp. Med. Biol.* **358**:85–96.
2110. **Hori, K. and F. Morita.** 1994. Molecular mechanism of actin–actin contact. *Tanpakushitsu. Kakusan. Koso.* **39**:871–876.
2111. **Janmey, P.** 1993. Cell biology. A slice of the actin. *Nature* **364**:675–676.
2112. **Janmey, P. A.** 1994. Phosphoinositides and calcium as regulators of cellular actin assembly and disassembly. *Annu. Rev. Physiol.* **56**:169–191.
2113. **Jockusch, B. M., C. Wiegand, C. J. Temm Grove, and G. Nikolai.** 1993. Dynamic aspects of microfilament-membrane attachments. *Symp. Soc. Exp. Biol.* **47**:253–266.
2114. **Kabsch, W. and K. C. Holmes.** 1995. Protein motifs **2**: the actin fold. *FASEB J.* **9**:167–174.
2115. **Kocks, C.** 1994. Intracellular motility. Profilin puts pathogens on the actin drive. *Curr. Biol.* **4**:465–468.
2116. **Lefebvre, P., J. G. White, M. D. Krumwiede, and I. Cohen.** 1993. Role of actin in platelet function. *Eur. J. Cell Biol.* **62**:194–204.
2117. **Lepault, J., J. L. Ranck, I. Erk, and M. F. Carlier.** 1994. Small angle X-ray scattering and electron cryomicroscopy study of actin filaments: role of the bound nucleotide in the structure of F-actin. *J. Struct. Biol.* **112**:79–91.
2118. **Matsudaira, P.** 1994. Actin crosslinking proteins at the leading edge. *Semin. Cell Biol.* **5**:165–174.
2119. **Mendelson, R. and E. Morris.** 1994. Combining electron microscopy and X-ray crystallography data to study the structure of F-actin and its implications for thin-filament regulation in muscle. *Adv. Exp. Med. Biol.* **358**:13–23.
2120. **Mendelson, R. A. and E. Morris.** 1994. The structure of F-actin. Results of global searches using data from electron microscopy and X-ray crystallography. *J. Mol. Biol.* **240**:138–154.
2121. **Ohm, T. and A. Wegner.** 1994. Mechanism of ATP hydrolysis by polymeric actin. *Biochim. Biophys. Acta* 1208:8–14.
2122. **Pollard, T. D., S. Almo, S. Quirk, V. Vinson, and E. E. Lattman.** 1994. Structure of actin binding proteins: insights about function at atomic resolution. *Annu. Rev. Cell Biol.* **10**:207–249.
2123. **Rayment, I., H. M. Holden, M. Whittaker, C. B. Yohn, M. Lorenz, K. C. Holmes, and R. A. Milligan.** 1993. Structure of the actin–myosin complex and its implications for muscle contraction. *Science* **261**:58–65.
2124. **Reedy, M. K.** 1993. Myosin–actin motors: the partnership goes atomic. *Structure.* **1**:1–5.
2125. **Schroer, T. A., E. Fyrberg, J. A. Cooper, R. H. Waterston, D. Helfman, T. D. Pollard, and D. I. Meyer.** 1994. Actin-related protein nomenclature and classification. *J. Cell Biol.* **127**:1777–1778.
2126. **Sellers, J. R. and H. V. Goodson.** 1995. Motor proteins **2**: myosin. *Protein Profile* **2**:960–1051.
2127. **Small, J. V.** 1994. Lamellipodia architecture: actin filament turnover and the lateral flow of actin filaments during motility. *Semin. Cell Biol.* **5**:157–163.
2128. **Sohn, R. H. and P. J. Goldschmidt Clermont.** 1994. Profilin: at the crossroads of signal transduction and the actin cytoskeleton. *Bioessays* **16**:465–472.
2129. **Southwick, F. S. and D. L. Purich.** 1994. Dynamic remodeling of the actin cytoskeleton: lessons learned from *Listeria* locomotion. *Bioessays* **16**:885–891.
2130. **Theriot, J. A.** 1994. Regulation of the actin cytoskeleton in living cells. *Semin. Cell Biol.* **5**:193–199.
2131. **Theriot, J. A.** 1994. Actin filament dynamics in cell motility. *Adv. Exp. Med. Biol.* **358**:133–145.
2132. **Theurkauf, W. E.** 1994. Actin cytoskeleton. Through the bottleneck. *Curr. Biol.* **4**:76–78.
2133. **Wegner, A., K. Aktories, A. Ditsch, I. Just, B. Schoepper, N. Selve, and M. Wille.** 1994. Actin–gelsolin interaction. *Adv. Exp. Med. Biol.* **358**:97–104.
2134. **Yamakita, Y., S. Yamashiro, and F. Matsumura.** 1993. Microfilament system in cell cycle. *Tanpakushitsu. Kakusan. Koso.* **38**:2392–2395.

ACTIN-RELATED PROTEINS

2135. **Adams, M. D., J. M. Kelley, J. D. Gocayne, M. Dubnick, M. H. Polymeropoulos, H. Xiao, C. R. Merril, A. Wu, B. Olde, R. F. Moreno, A. R. Kerlavage, W. R. McCombie, and J. C. Venter.** 1991. Complementary DNA sequencing: expressed sequence tags and human genome project. *Science* **252**:1651–1656.
2136. **Clark, S. W. and D. I. Meyer.** 1992. Centractin is an actin homologue associated with the centrosome. *Nature* **359**:246–250.
2137. **Clark, S. W. and D. I. Meyer.** 1994. *ACT3*: a putative centractin homologue in *S. cerevisiae* is required for proper orientation of the mitotic spindle. *J. Cell Biol.* **127**:129–138.
2138. **Frankel, S., M. B. Heintzelman, S. Artavanis Tsakonas, and M. S. Mooseker.** 1994. Identification of a divergent actin-related protein in *Drosophila*. *J. Mol. Biol.* **235**:1351–1356.
2139. **Fyrberg, C. and E. Fyrberg.** 1993. A *Drosphila* homologue of the *Schizosaccharomyces pombe act2* gene. *Biochem. Genet.* **31**:329–341.
2140. **Fyrberg, C., L. Ryan, M. Kenton, and E. Fyrberg.** 1994. Genes encoding actin-related proteins of *Drosophila melanogaster*. *J. Mol. Biol.* **241**:498–503.

2141. **Fyrberg, C., L. Ryan, L. McNally, M. Kenton, and E. Fyrberg.** 1994. The actin protein superfamily. *Soc. Gen. Physiol. Ser.* **49**:173–178.

2142. **Harata, M., A. Karwan, and U. Wintersberger.** 1994. An essential gene of *Saccharomyces cerevisiae* coding for an actin-related protein. *Proc. Natl. Acad. Sci. U. S. A.* **91**:8258–8262.

2143. **Lees-Miller, J. P., D. M. Helfman, and T. A. Schroer.** 1992. A vertebrate actin-related protein is a component of a multisubunit complex involved in microtubule-based motility. *Nature* **359**:244–246.

2144. **Lees-Miller, J. P., G. Henry, and D. M. Helfman.** 1992. Identification of act2, an essential gene in the fission yeast *Schizosaccharomyces pombe* that encodes a protein related to actin. *Proc. Natl. Acad. Sci. U. S. A.* **89**:80–83.

2145. **Machesky, L. M., S. J. Atkinson, C. Ampe, J. Vandekerckhove, and T. D. Pollard.** 1994. Purification of a cortical complex containing two unconventional actins from *Acanthamoeba* by affinity chromatography on profilin-agarose. *J. Cell Biol.* **127**:107–115.

2146. **McCombie, W. R., M. D. Adams, J. M. Kelley, M. G. Fitzgerald, T. R. Utterback, M. Khan, M. Dubnick, A. R. Kerlavage, J. C. Venter, and C. Fields.** 1992. *Caenorhabditis elegans* expressed sequence tags identify gene families and potential gene disease homologues. *Nature Gen.* **1**:124–131.

2147. **Melki, R., I. E. Vainberg, R. L. Chow, and N. J. Cowan.** 1993. Chaperonin-mediated folding of vertebrate actin-related protein and gamma-tubulin. *J. Cell Biol.* **122**:1301–1310.

2148. **Michaille, J. J., M. Gouy, S. Blanchet, and L. Duret.** 1995. Isolation and characterization of a cDNA encoding a chicken actin-like protein. *Gene* **154**:205–209.

2149. **Muhua, L., T. S. Karpova, and J. A. Cooper.** 1994. A yeast actin-related protein homologous to that in vertebrate dynactin complex is important for spindle orientation and nuclear migration. *Cell* **78**:669–679.

2150. **Murgia, I., S. K. Maciver, and P. Morandini.** 1995. An actin-related protein from *Dictyostelium discoideum* is developmentally regulated and associated with mitochondria. *FEBS Lett.* **360**:235–241.

2151. **Paschal, B. M., E. L. Holzbaur, K. K. Pfister, S. W. Clark, D. I. Meyer, and R. B. Vallee.** 1993. Characterization of a 50-kDa polypeptide in cytoplasmic dynein preparations reveals a complex with p150GLUED and a novel actin. *J. Biol. Chem.* **268**:15318–15323.

2152. **Plamann, M., P. F. Minke, J. H. Tinsley, and K. S. Bruno.** 1994. Cytoplasmic dynein and actin-related protein Arp1 are required for normal nuclear distribution in filamentous fungi. *J. Cell Biol.* **127**:139–149.

2153. **Schafer, D. A., S. R. Gill, J. A. Cooper, J. E. Heuser, and T. A. Schroer.** 1994. Ultrastructural analysis of the dynactin complex: an actin-related protein is a component of a filament that resembles F-actin. *J. Cell Biol.* **126**:403–412.

2153a. **Schroer, T. A., E. Fyrberg, J. A. Cooper, R. H. Waterston, D. Helfman, T. D. Pollard, and D. I. Meyer.** 1994. Actin-related protein nomenclature and classification. *J. Cell Biol.* **127**:1777–1778.

2154. **Schwob, E. and R. P. Martin.** 1992. New yeast actin-like gene required late in the cell cycle. *Nature* **355**:179–182.

2155. **Tanaka, T., F. Shibasaki, M. Ishikawa, N. Hirano, R. Sakai, J. Nishida, T. Takenawa, and H. Hirai.** 1992. Molecular cloning of bovine actin-like protein, actin2. *Biochem. Biophys. Res. Commun.* **187**:1022–1028.

2156. **Waterman Storer, C. M., S. Karki, and E. L. Holzbaur.** 1995. The p150Glued component of the dynactin complex binds to both microtubules and the actin-related protein centractin (Arp-1). *Proc. Natl. Acad. Sci. U. S. A.* **92**:1634–1638.

2157. **Waterston, R., C. Martin, M. Craxton, C. Huynh, A. Coulson, L. Hillier, R. Durbin, P. Green, R. Shownkeen, N. Halloran, M. Metzstein, T. Hawkins, R. Wilson, M. Berks, Z. Du, K. Thomas, J. Thiery-Mieg, and J. Sulston.** 1992. A survey of expressed genes in *Caenorhabditis elegans*. *Nature Gen.* **1**:114–123.

ACTIN POLYMERIZATION

2158. **Ballweber, E., E. Hannappel, B. Niggemeyer, and H. G. Mannherz.** 1994. Induction of the polymerization of actin from the actin:thymosin β4 complex by phalloidin, skeletal myosin subfragment 1, chicken intestinal myosin I and free ends of filamentous actin. *Eur. J. Biochem.* **223**:419–426.

2159. **Cao, L. G., D. J. Fishkind, and Y. L. Wang.** 1993. Localization and dynamics of nonfilamentous actin in cultured cells. *J. Cell Biol.* **123**:173–181.

2160. **Carlier, M. F. and D. Pantaloni.** 1994. Actin assembly in response to extracellular signals: role of capping proteins, thymosin beta 4 and profilin. *Semin. Cell Biol.* **5**:183–191.

2161. **Carlier, M. F., C. Valentin Ranc, C. Combeau, S. Fievez, and D. Pantoloni.** 1994. Actin polymerization: regulation by divalent metal ion and nucleotide binding, ATP hydrolysis and binding of myosin. *Adv. Exp. Med. Biol.* **358**:71–81.

2162. **Fechheimer, M. and S. H. Zigmond.** 1993. Focusing on unpolymerized actin. *J. Cell Biol.* **123**:1–5.

2163. **Finkel, T., J. A. Theriot, K. R. Dise, G. F. Tomaselli, and P. J. Goldschmidt Clermont.** 1994. Dynamic actin structures stabilized by profilin. *Proc. Natl. Acad. Sci. U. S. A.* **91**:1510–1514.

2164. **Friederich, E., T. E. Kreis, and D. Louvard.** 1993. Villin-induced growth of microvilli is reversibly inhibited by cytochalasin D. *J. Cell Sci.* **105**:765–775.

2165. **Ha, K. S., E. J. Yeo, and J. H. Exton.** 1994. Lysophosphatidic acid activation of phosphatidylcholine-hydrolysing phospholipase D and actin polymerization by a pertussis toxin-sensitive mechanism. *Biochem. J.* **303**:55–59.

2166. **Haugwitz, M., A. A. Noegel, J. Karakesisoglou, and M. Schleicher.** 1994. Dictyostelium amoebae that lack G-actin-sequestering profilins show defects in F-actin content, cytokinesis, and development. *Cell* **79**:303–314.

2167. **Hori, K., T. Itoh, K. Takahashi, and F. Morita.** 1994. Different modes of interaction of two peptide fragments from subdomain 4 of rabbit skeletal muscle actin with actin promoters. *Biochim. Biophys. Acta* 1186:35–42.

2168. **Huang, C. and N. C. Liang.** 1994. Increase in cytoskeletal actin induced by inositol 1,4-bisphosphate in saponin-permeated pig platelets. *Cell Biol. Int.* **18**:797–804.

2169. **Janmey, P. A.** 1994. Phosphoinositides and calcium as regulators of cellular actin assembly and disassembly. *Annu. Rev. Physiol.* **56**:169–191.

2170. **Janmey, P. A., S. Hvidt, J. Kas, D. Lerche, A. Maggs, E. Sackmann, M. Schliwa, and T. P. Stossel.** 1994. The mechanical properties of actin gels. Elastic modulus and filament motions. *J. Biol. Chem.* **269**:32503–32513.

2171. **Koch, G., J. Norgauer, and K. Aktories.** 1994. ADP-ribosylation of the GTP-binding protein Rho by *Clostridium limosum* exoenzyme affects basal, but not *N*-formyl-peptide-stimulated, actin polymerization in human myeloid leukaemic (HL60) cells. *Biochem. J.* **299**:775–779.

2172. **Koyama, Y. and A. Baba.** 1994. Endothelins are extracellular signals modulating cytoskeletal actin organization in rat cultured astrocytes. *Neuroscience.* **61**:1007–1016.

2173. **Kozma, R., S. Ahmed, A. Best, and L. Lim.** 1995. The Ras-related protein Cdc42Hs and bradykinin promote formation of peripheral actin microspikes and filopodia in Swiss 3T3 fibroblasts. *Mol. Cell Biol.* **15**:1942–1952.

2174. **Li, R., Y. Zheng, and D. G. Drubin.** 1995. Regulation of cortical actin cytoskeleton assembly during polarized cell growth in budding yeast. *J. Cell Biol.* **128**:599–615.

2175. **Nanavati, D., F. T. Ashton, J. M. Sanger, and J. W. Sanger.** 1994. Dynamics of actin and alpha-actinin in the tails of *Listeria monocytogenes* in infected PtK2 cells. *Cell Motil. Cytoskeleton.* **28**:346–358.

2176. **Norman, J. C., L. S. Price, A. J. Ridley, A. Hall, and A. Koffer.** 1994. Actin filament organization in activated mast cells is regulated by heterotrimeric and small GTP-binding proteins. *J. Cell Biol.* **126**:1005–1015.

2177. **Redmond, T., M. Tardif, and S. H. Zigmond.** 1994. Induction of actin polymerization in permeabilized neutrophils. Role of ATP. *J. Biol. Chem.* **269**:21657–21663.

2178. **Rijken, P. J., S. M. Post, W. J. Hage, and P. M. P. V. E. Henegouwen.** 1995. Actin polymerisation localises to the activated growth factor receptor in the plasma membrane, independent of the cytosolic free calcium transient. *Exp. Cell Res.* **218**:223–232.

2179. **Small, J. V.** 1994. Lamellipodia architecture: actin filament turnover and the lateral flow of actin filaments during motility. *Semin. Cell Biol.* **5**:157–163.

2180. **Southwick, F. S. and D. L. Purich.** 1994. Arrest of *Listeria* movement in host cells by a bacterial ActA analogue: implications for actin-based motility. *Proc. Natl. Acad. Sci. U. S. A.* **91**:5168–5172.

CATION BINDING TO ACTIN

2181. **DalleDonne, I., A. Milzani, U. Fascio, A. Ratti, and R. Colombo.** 1993. Lithium preserves F-actin from the disarrangement induced by either DNase I or cytochalasin D. *Biochem. Cell Biol.* **71**:440–446.

2182. **Gershman, L. C., L. A. Selden, H. J. Kinosian, and J. E. Estes.** 1994. Actin-bound nucleotide/divalent cation interactions. *Adv. Exp. Med. Biol.* **358**:35–49.

2183. **Perelroizen, I., M. F. Carlier, and D. Pantaloni.** 1995. Binding of divalent cation and nucleotide to G-actin in the presence of profilin. *J. Biol. Chem.* **270**:1501–1508.

2184. **Selden, L. A., H. J. Kinosian, J. E. Estes, and L. C. Gershman.** 1994. Influence of the high affinity divalent cation on actin tryptophan fluorescence. *Adv. Exp. Med. Biol.* **358**:51–57.

NUCLEOTIDE BINDING AND ACTIN ATPASE

2185. **Angelastro, J. M. and D. L. Purich.** 1994. Phosphorylation states of actin filament adenine nucleotides in detergent-extracted neuronal cytoskeletal fractions. *Biochem. Biophys. Res. Commun.* **201**:1490–1494.

2186. **Carlier, M. F., C. Jean, K. J. Rieger, M. Lenfant, and D. Pantaloni.** 1993. Modulation of the interaction between G-actin and thymosin β4 by the ATP/ADP ratio: possible implication in the regulation of actin dynamics. *Proc. Natl. Acad. Sci. U. S. A.* **90**:5034–5038.

2187. **Chen, X., J. Peng, M. Pedram, C. A. Swenson, and P. A. Rubenstein.** 1995. The effect of the S14A mutation on the conformation and thermostability of *Saccharomyces cerevisiae* G-actin and its interaction with adenine nucleotides. *J. Biol. Chem.* **270**:11415–11423.

2188. **Chen, X. and P. A. Rubenstein.** 1995. A mutation in the ATP-binding loop of *Saccharomyces cerevisiae* (S14A) causes a temperature-sensitive phenotype *in vivo* and *in vitro*. *J. Biol. Chem.* **270**:11406–11414.

2189. **Dufort, P. A. and C. J. Lumsden.** 1993. High microfilament concentration results in barbed-end ADP caps. *Biophys. J.* **65**:1757–1766.

2190. **Kasprzak, A. A.** 1994. Myosin subfragment 1 activates ATP hydrolysis on Mg^{2+}-G-actin. *Biochemistry* **33**:12456–12462.
2191. **Lepault, J., J. L. Ranck, I. Erk, and M. F. Carlier.** 1994. Small angle X-ray scattering and electron cryomicroscopy study of actin filaments: role of the bound nucleotide in the structure of F-actin. *J. Struct. Biol.* **112**:79–91.
2192. **Maciver, S. K. and A. G. Weeds.** 1994. Actophorin preferentially binds monomeric ADP-actin over ATP-bound actin: consequences for cell locomotion. *FEBS Lett.* **347**:251–256.
2193. **Nikolaeva, O. P., I. V. Dedova, I. S. Khvorova, and D. I. Levitsky.** 1994. Interaction of F-actin with phosphate analogues studied by differential scanning calorimetry. *FEBS Lett.* **351**:15–18.
2194. **Ohm, T. and A. Wegner.** 1994. Mechanism of ATP hydrolysis by polymeric actin. *Biochim. Biophys. Acta* 1208:8–14.
2195. **Oishi, N. and H. Sugi.** 1994. In vitro ATP-dependent F-actin sliding on myosin is not influenced by substitution or removal of bound nucleotide. *Biochim. Biophys. Acta* 1185:346–349.
2196. **Perelroizen, I., M. F. Carlier, and D. Pantaloni.** 1995. Binding of divalent cation and nucleotide to G-actin in the presence of profilin. *J. Biol. Chem.* **270**:1501–1508.

BINDING OF PROTEINS AND OTHER LIGANDS TO ACTIN

2197. **Ampe, C. and J. Vandekerckhove.** 1994. Actin–actin binding protein interfaces. *Semin. Cell Biol.* **5**:175–182.
2198. **Asada, H., K. Uyemura, and T. Shirao.** 1994. Actin-binding protein, drebrin, accumulates in submembranous regions in parallel with neuronal differentiation. *J. Neurosci. Res.* **38**:149–159.
2199. **Bektas, M., R. Nurten, Z. Gurel, Z. Sayers, and E. Bermek.** 1994. Interactions of eukaryotic elongation factor 2 with actin: a possible link between protein synthetic machinery and cytoskeleton. *FEBS Lett.* **356**:89–93.
2200. **Benedetti, H., S. Raths, F. Crausaz, and H. Riezman.** 1994. The END3 gene encodes a protein that is required for the internalization step of endocytosis and for actin cytoskeleton organization in yeast. *Mol. Biol. Cell* **5**:1023–1037.
2201. **Bertrand, R., J. Derancourt, and R. Kassab.** 1994. The covalent maleimidobenzoyl–actin–myosin head complex. Cross-linking of the 50 kDa heavy chain region to actin subdomain-2. *FEBS Lett.* **345**:113–119.
2202. **Bonet Kerrache, A. and D. Mornet.** 1995. Importance of the *C*-terminal part of actin in interactions with calponin. *Biochem. Biophys. Res. Commun.* **206**:127–132.
2203. **Bryan, J., R. Edwards, P. Matsudaira, J. Otto, and J. Wulfkuhle.** 1993. Fascin, an echinoid actin-bundling protein, is a homolog of the *Drosophila* singed gene product. *Proc. Natl. Acad. Sci. U. S. A.* **90**:9115–9119.
2204. **Bubb, M. R., J. R. Knutson, D. K. Porter, and E. D. Korn.** 1994. Actobindin induces the accumulation of actin dimers that neither nucleate polymerization nor self-associate. *J. Biol. Chem.* **269**:25592–25597.
2205. **Bubb, M. R., M. S. Lewis, and E. D. Korn.** 1994. Actobindin binds with high affinity to a covalently cross-linked actin dimer. *J. Biol. Chem.* **269**:25587–25591.
2206. **Carraway, C. A., M. E. Carvajal, Y. Li, and K. L. Carraway.** 1993. Association of p185neu with microfilaments via a large glycoprotein complex in mammary carcinoma microvilli. Evidence for a microfilament-associated signal transduction particle. *J. Biol. Chem.* **268**:5582–5587.
2207. **Cary, R. B., M. W. Klymkowsky, R. M. Evans, A. Domingo, J. A. Dent, and L. E. Backhus.** 1994. Vimentin's tail interacts with actin-containing structures *in vivo*. *J. Cell Sci.* **107**:1609–1622.
2208. **Chilcote, T. J., Y. L. Siow, E. Schaeffer, P. Greengard, and G. Thiel.** 1994. Synapsin IIa bundles actin filaments. *J. Neurochem.* **63**:1568–1571.
2209. **Chuang, J. Z., D. C. Lin, and S. Lin.** 1995. Molecular cloning, expression and mapping of the high-affinity actin-capping domain of chicken cardiac tensin. *J. Cell Biol.* **128**:1095–1109.
2210. **dos Remedios, C. G. and P. D. Moens.** 1995. Actin and the actomyosin interface: a review. *Biochim. Biophys. Acta* 1228:99–124.
2211. **Duong, A. M. and E. Reisler.** 1994. C-terminus on actin: spectroscopic and immunochemical examination of its role in actomyosin interactions. *Adv. Exp. Med. Biol.* **358**:59–70.
2212. **Freeman, N. L., Z. Chen, J. Horenstein, A. Weber, and J. Field.** 1995. An actin monomer binding activity localizes to the carboxyl-terminal half of the *Saccharomyces cerevisiae* cyclase-associated protein. *J. Biol. Chem.* **270**:5680–5685.
2213. **Fujimoto, T., A. Miyawaki, and K. Mikoshiba.** 1995. Inositol 1,4,5-trisphosphate receptor-like protein in plasmalemmal caveolae is linked to actin filaments. *J. Cell Sci.* **108**:7–15.
2214. **Gicquaud, C. and P. Wong.** 1994. Mechanism of interaction between actin and membrane lipids: a pressure-tuning infrared spectroscopy study. *Biochem. J.* **303**:769–774.
2215. **Giebing, T., H. Hinssen, and J. D'Haese.** 1994. The complete sequence of a 40-kDa actin-modulating protein from the earthworm *Lumbricus terrestris*. *Eur. J. Biochem.* **225**:773–779.
2216. **Giehl, K., R. Valenta, M. Rothkegel, M. Ronsiek, H. G. Mannherz, and B. M. Jockusch.** 1994. Interaction of plant profilin with mammalian actin. *Eur. J. Biochem.* **226**:681–689.
2217. **Goldmann, W. H., A. Bremer, M. Haner, U. Aebi, and G. Isenberg.** 1994. Native talin is a dumbbell-shaped homodimer when it interacts with actin. *J. Struct. Biol.* **112**:3–10.
2218. **Goldmann, W. H., J. Kas, and G. Isenberg.** 1994. Talin decreases the bending elasticity of actin filaments. *Biochem. Soc. Trans.* **22**:46S.
2219. **Goldmann, W. H., R. Senger, and G. Isenberg.** 1994. Analysis of filamin-actin binding and cross-linking/bundling by kinetic method. *Biochem. Biophys. Res. Commun.* **203**:338–343.
2220. **Greengard, P., F. Benfenati, and F. Valtorta.** 1994. Synapsin I, an actin-binding protein regulating synaptic vesicle traffic in the nerve terminal. *Adv. Second. Messenger. Phosphoprotein. Res.* **29**:31–45.
2221. **Hall, A. K.** 1994. Molecular interactions between G-actin, DNase I and the beta-thymosins in apoptosis: a hypothesis. *Med. Hypotheses.* **43**:125–131.
2222. **Hatano, S.** 1994. Actin-binding proteins in cell motility. *Int. Rev. Cytol.* **156**:199–273.
2223. **Hayashibara, T. and T. Miyanishi.** 1994. Binding of the amino-terminal region of myosin alkali 1 light chain to actin and its effect on actin–myosin interaction. *Biochemistry* **33**:12821–12827.
2224. **Hitchcock DeGregori, S. E.** 1994. Structural requirements of tropomyosin for binding to filamentous actin. *Adv. Exp. Med. Biol.* **358**:85–96.
2225. **Hitt, A. L., J. H. Hartwig, and E. J. Luna.** 1994. Ponticulin is the major high affinity link between the plasma membrane and the cortical actin network in *Dictyostelium*. *J. Cell Biol.* **126**:1433–1444.
2226. **Holthuis, J. C., V. T. Schoonderwoert, and G. J. Martens.** 1994. A vertebrate homolog of the actin-bundling protein fascin. *Biochim. Biophys. Acta* **1219**:184–188.
2227. **Hu, G. F., D. J. Strydom, J. W. Fett, J. F. Riordan, and B. L. Vallee.** 1993. Actin is a binding protein for angiogenin. *Proc. Natl. Acad. Sci. U. S. A.* **90**:1217–1221.
2228. **Ishikawa, R., K. Hayashi, T. Shirao, Y. Xue, T. Takagi, Y. Sasaki, and K. Kohama.** 1994. Drebrin, a development-associated brain protein from rat embryo, causes the dissociation of tropomyosin from actin filaments. *J. Biol. Chem.* **269**:29928–29933.
2229. **Ito, J., M. Masuda, and R. Tanaka.** 1994. Sialosylcholesterol effects on reconstitution of microfilament and glia filament. *J. Neurochem.* **62**:235–239.
2230. **Jarrett, H. W. and J. L. Foster.** 1995. Alternate binding of actin and calmodulin to multiple sites on dystrophin. *J. Biol. Chem.* **270**:5578–5586.
2231. **Jean, C., K. Rieger, L. Blanchoin, M. F. Carlier, M. Lenfant, and D. Pantaloni.** 1994. Interaction of G-actin with thymosin β4 and its variants thymosin β9 and thymosin βmet9. *J. Muscle. Res. Cell Motil.* **15**:278–286.
2232. **Jin, J. P.** 1995. Cloned rat cardiac titin class I and class II motifs. Expression, purification, characterization, and interaction with F-actin. *J. Biol. Chem.* **270**:6908–6916.
2233. **Johara, M., Y. Y. Toyoshima, A. Ishijima, H. Kojima, T. Yanagida, and K. Sutoh.** 1993. Charge-reversion mutagenesis of *Dictyostelium* actin to map the surface recognized by myosin during ATP-driven sliding motion. *Proc. Natl. Acad. Sci. U. S. A.* **90**:2127–2131.
2234. **Johnson, R. P. and S. W. Craig.** 1995. F-actin binding site masked by the intramolecular association of vinculin head and tail domains. *Nature* **373**:261–264.
2235. **Juang, S. H., J. Huang, Y. Li, P. J. Salas, N. Fregien, C. A. Carraway, and K. L. Carraway.** 1994. Molecular cloning and sequencing of a 58-kDa membrane- and microfilament-associated protein from ascites tumor cell microvilli with sequence similarities to retroviral Gag proteins. *J. Biol. Chem.* **269**:15067–15075.
2236. **Krawczenko, A., L. Ciszak, and M. Malicka Boaszkiewicz.** 1994. Does a natural DNase-actin complex exist in the carp liver cytosol? *Acta Biochim. Pol.* **41**:113–116.
2237. **Kreft, J., M. Dumbsky, and S. Theiss.** 1995. The actin-polymerisation protein from *Listeria ivanovii* is a large repeat protein which shows only limited amino acid sequence homology to ActA from *Listeria monocytogenes*. *FEMS. Microbiol. Lett.* **126**:113–121.
2238. **Kuhlman, P. A., J. Ellis, D. R. Critchley, and C. R. Bagshaw.** 1994. The kinetics of the interaction between the actin-binding domain of α-actinin and F-actin. *FEBS Lett.* **339**:297–301.
2239. **Leenders, F., B. Husen, H. H. Thole, and J. Adamski.** 1994. The sequence of porcine 80 kDa 17 β-estradiol dehydrogenase reveals similarities to the short chain alcohol dehydrogenase family, to actin binding motifs and to sterol carrier protein 2. *Mol. Cell Endocrinol.* **104**:127–131.
2240. **Lo, S. H., P. A. Janmey, J. H. Hartwig, and L. B. Chen.** 1994. Interactions of tensin with actin and identification of its three distinct actin-binding domains. *J. Cell Biol.* **125**:1067–1075.
2241. **Malicka Blaszkiewicz, M., I. Majcher, and D. Nowak.** 1994. DNase-I-like enzyme from the carp liver–inhibition by muscle and endogenous actin. *Int. J. Biochem.* **26**:1147–1155.
2242. **Marcu, M. G., A. Rodriguez Del Castillo, M. L. Vitale, and J. M. Trifaro.** 1994. Molecular cloning and functional expression of chromaffin cell scinderin indicates that it belongs to the family of Ca^{2+}-dependent F-actin severing proteins. *Mol. Cell Biochem.* **141**:153–165.
2243. **Matsudaira, P.** 1994. The fimbrin and alpha-actinin footprint on actin. *J. Cell Biol.* **126**:285–287.
2244. **McGough, A., M. Way, and D. DeRosier.** 1994. Determination of the α-actinin-binding site on actin filaments by cryoelectron microscopy and image analysis. *J. Cell Biol.* **126**:433–443.
2245. **Menkel, A. R., M. Kroemker, P. Bubeck, M. Ronsiek, G. Nikolai, and B. M. Jockusch.** 1994. Characterization of an F-actin-binding domain in the cytoskeletal protein vinculin. *J. Cell Biol.* **126**:1231–1240.

2246. **Miller, C. J. and E. Reisler.** 1995. Role of charged amino acid pairs in sub-domain-1 of actin in interactions with myosin. *Biochemistry* **34**:2694–2700.
2247. **Mishra, V. S., E. P. Henske, D. J. Kwiatkowski, and F. S. Southwick.** 1994. The human actin-regulatory protein cap G: gene structure and chromosome location. *Genomics.* **23**:560–565.
2248. **Molloy, J. E., J. E. Burns, J. C. Sparrow, R. T. Tregear, J. Kendrick-Jones, and D. C. S. White.** 1995. Single molecule mechanics of heavy meromyosin and S-1 interacting with rabbit or *Drosophila* actins using optical tweezers. *Biophys. J.* **68**:298s–305s.
2249. **Mornet, D., A. Bonet Kerrache, G. M. Strasburg, V. B. Patchell, S. V. Perry, P. A. Huber, S. B. Marston, D. A. Slatter, J. S. Evans, and B. A. Levine.** 1995. The binding of distinct segments of actin to multiple sites in the *C*-terminus of caldesmon: comparative aspects of actin interaction with troponin-I and caldesmon. *Biochemistry* **34**:1893–1901.
2250. **Mosialos, G., S. Yamashiro, R. W. Baughman, P. Matsudaira, L. Vara, F. Matsumura, E. Kieff, and M. Birkenbach.** 1994. Epstein–Barr virus infection induces expression in B lymphocytes of a novel gene encoding an evolutionarily conserved 55-kilodalton actin-bundling protein. *J. Virol.* **68**:7320–7328.
2251. **Pedrotti, B., R. Colombo, and K. Islam.** 1994. Microtubule associated protein MAP1A is an actin-binding and crosslinking protein. *Cell Motil. Cytoskeleton.* **29**:110–116.
2252. **Perelroizen, I., J. B. Marchand, L. Blanchoin, D. Didry, and M. F. Carlier.** 1994. Interaction of profilin with G-actin and poly(L-proline). *Biochemistry* **33**:8472–8478.
2253. **Pfuhl, M., S. J. Winder, and A. Pastore.** 1994. Nebulin, a helical actin binding protein. *EMBO J.* **13**:1782–1789.
2254. **Prinjha, R. K., C. E. Shapland, J. J. Hsuan, N. F. Totty, I. J. Mason, and D. Lawson.** 1994. Cloning and sequencing of cDNAs encoding the actin cross-linking protein transgelin defines a new family of actin-associated proteins. *Cell Motil. Cytoskeleton.* **28**:243–255.
2255. **Rayment, I., H. M. Holden, M. Whittaker, C. B. Yohn, M. Lorenz, K. C. Holmes, and R. A. Milligan.** 1993. Structure of the actin–myosin complex and its implications for muscle contraction. *Science* **261**:58–65.
2256. **Root, D. D. and K. Wang.** 1994. Calmodulin-sensitive interaction of human nebulin fragments with actin and myosin. *Biochemistry* **33**:12581–12591.
2257. **Safer, D. and V. T. Nachmias.** 1994. Beta thymosins as actin binding peptides. *Bioessays* **16**:473–479.
2258. **Samuelsson, S. J., P. W. Luther, D. W. Pumplin, and R. J. Bloch.** 1993. Structures linking microfilament bundles to the membrane at focal contacts. *J. Cell Biol.* **122**:485–496.
2259. **Schejter, E. D. and E. Wieschaus.** 1993. Bottleneck acts as a regulator of the microfilament network governing cellularization of the *Drosophila* embryo. *Cell* **75**:373–385.
2260. **Schenkels, L. C., J. Schaller, E. Walgreen Weterings, I. L. Schadee Eestermans, E. C. Veerman, and A. V. Nieuw Amerongen.** 1994. Identity of human extra parotid glycoprotein (EP-GP) with secretory actin binding protein (SABP) and its biological properties. *Biol. Chem. Hoppe. Seyler.* **375**:609–615.
2261. **Sellers, J. R. and H. V. Goodson.** 1995. Motor proteins **2**: myosin. *Protein Profile* **2**:960–1051.
2262. **Shirao, T., K. Hayashi, R. Ishikawa, K. Isa, H. Asada, K. Ikeda, and K. Uyemura.** 1994. Formation of thick, curving bundles of actin by drebrin A expressed in fibroblasts. *Exp. Cell Res.* **215**:145–153.
2263. **Swanljung Collins, H. and J. H. Collins.** 1994. Brush border myosin I has a calmodulin/phosphatidylserine switch and tail actin-binding. *Adv. Exp. Med. Biol.* **358**:205–213.
2264. **Szczesna, D. and P. G. Fajer.** 1995. The tropomyosin domain is flexible and disordered in reconstituted thin filaments. *Biochemistry* **34**:3614–3620.
2265. **Turunen, O., T. Wahlstrom, and A. Vaheri.** 1994. Ezrin has a COOH-terminal actin-binding site that is conserved in the ezrin protein family. *J. Cell Biol.* **126**:1445–1453.
2266. **Van Etten, R. A., P. K. Jackson, D. Baltimore, M. C. Sanders, P. T. Matsudaira, and P. A. Janmey.** 1994. The COOH terminus of the c-Abl tyrosine kinase contains distinct F- and G-actin binding domains with bundling activity. *J. Cell Biol.* **124**:325–340.
2267. **Varkey, J. P., P. J. Muhlrad, A. N. Minniti, B. Do, and S. Ward.** 1995. The *Caenorhabditis elegans spe-26* gene is necessary to form spermatids and encodes a protein similar to the actin-associated proteins kelch and scruin. *Genes. Dev.* **9**:1074–1086.
2268. **Watanabe, Y., N. Usada, H. Minami, T. Morita, S. Tsugane, R. Ishikawa, K. Kohama, Y. Tomida, and H. Hidaka.** 1993. Calvasculin, as a factor affecting the microfilament assemblies in rat fibroblasts transfected by src gene. *FEBS Lett.* **324**:51–55.
2269. **Waterman Storer, C. M., S. Karki, and E. L. Holzbaur.** 1995. The p150Glued component of the dynactin complex binds to both microtubules and the actin-related protein centractin (Arp-1). *Proc. Natl. Acad. Sci. U. S. A.* **92**:1634–1638.
2270. **Weber, A., C. R. Pennise, G. G. Babcock, and V. M. Fowler.** 1994. Tropomodulin caps the pointed ends of actin filaments. *J. Cell Biol.* **127**:1627–1635.
2271. **Wegner, A., K. Aktories, A. Ditsch, I. Just, B. Schoepper, N. Selve, and M. Wille.** 1994. Actin–gelsolin interaction. *Adv. Exp. Med. Biol.* **358**:97–104.
2272. **Xiao, M., O. A. Andreev, and J. Borejdo.** 1995. Rigor cross-bridges bind to two actin monomers in thin filaments of rabbit psoas muscle. *J. Mol. Biol.* **248**:294–307.

SEQUENCE ANALYSIS OF ACTIN

2273. **Anson, M., D. R. Drummond, M. A. Geeves, E. S. Hennessey, M. D. Ritchie, and J. C. Sparrow.** 1995. Actomyosin kinetics and *in vitro* motility of wild type *Drosophila* actin and effects of two mutations in the *Act88F* gene. *Biophys. J.* **68**:1991–2003.
2274. **Bhattacharya, D. and S. K. Stickel.** 1994. Sequence analysis of duplicated actin genes in *Lagenidium giganteum* and *Pythium irregulare* (Oomycota). *J. Mol. Evol.* **39**:56–61.
2275. **Chen, X., J. Peng, M. Pedram, C. A. Swenson, and P. A. Rubenstein.** 1995. The effect of the S14A mutation on the conformation and thermostability of *Saccharomyces cerevisiae* G-actin and its interaction with adenine nucleotides. *J. Biol. Chem.* **270**:11415–11423.
2276. **Chen, X. and P. A. Rubenstein.** 1995. A mutation in the ATP-binding loop of *Saccharomyces cerevisiae* (S14A) causes a temperature-sensitive phenotype *in vivo* and *in vitro*. *J. Biol. Chem.* **270**:11406–11414.
2277. **Clark, S. W. and D. I. Meyer.** 1992. Centractin is an actin homologue associated with the centrosome. *Nature* **359**:246–250.
2278. **Clark, S. W. and D. I. Meyer.** 1994. ACT3: a putative centractin homologue in *S. cerevisiae* is required for proper orientation of the mitotic spindle. *J. Cell Biol.* **127**:129–138.
2279. **da Silva, C. M., Henrique;, B. Ferreira, M. Picon, N. Gorfinkiel, R. Ehrlich, and A. Zaha.** 1993. Molecular cloning and characterization of actin genes from *Echinococcus granulosus*. *Mol. Biochem. Parasitol.* **60**:209–219.
2280. **Fang, H. and B. P. Brandhorst.** 1994. Evolution of actin gene families of sea urchins. *J. Mol. Evol.* **39**:347–356.
2281. **Frankel, S., M. B. Heintzelman, S. Artavanis Tsakonas, and M. S. Mooseker.** 1994. Identification of a divergent actin-related protein in *Drosophila*. *J. Mol. Biol.* **235**:1351–1356.
2282. **Fyrberg, C. and E. Fyrberg.** 1993. A *Drosophila* homologue of the *Schizosaccharomyces pombe* act2 gene. *Biochem. Genet.* **31**:329–341.
2283. **Hadden, T. J. and A. Sodja.** 1994. An oligogene family encodes actins in the housefly, *Musca domestica*. *Biochem. Biophys. Res. Commun.* **203**:523–531.
2284. **Hahn, J. H., J. C. Kissinger, and R. A. Raff.** 1995. Structure and evolution of CyI cytoplasmic actin-encoding genes in the indirect- and direct-developing sea urchins *Heliocidaris tuberculata* and *Heliocidaris erythrogramma*. *Gene* **153**:219–224.
2285. **Harata, M., A. Karwan, and U. Wintersberger.** 1994. An essential gene of *Saccharomyces cerevisiae* coding for an actin-related protein. *Proc. Natl. Acad. Sci. U. S. A.* **91**:8258–8262.
2286. **He, M. and D. S. Haymer.** 1992. A muscle-specific actin gene from the Mediterranean fruit fly, *Ceratitis capitata*. *Insect. Mol. Biol.* **1**:15–24.
2287. **He, M. and D. S. Haymer.** 1994. The actin gene family in the oriental fruit fly *Bactrocera dorsalis*. Muscle specific actins. *Insect. Biochem. Mol. Biol.* **24**:891–906.
2288. **Holtzman, D. A., K. F. Wertman, and D. G. Drubin.** 1994. Mapping actin surfaces required for functional interactions *in vivo*. *J. Cell Biol.* **126**:423–432.
2289. **Holtzman, D. A., S. Yang, and D. G. Drubin.** 1993. Synthetic-lethal interactions identify two novel genes, *SLA1* and *SLA2*, that control membrane cytoskeleton assembly in *Saccharomyces cerevisiae*. *J. Cell Biol.* **122**:635–644.
2290. **Honts, J. E., T. S. Sandrock, S. M. Brower, J. L. O'Dell, and A. E. Adams.** 1994. Actin mutations that show suppression with fimbrin mutations identify a likely fimbrin-binding site on actin. *J. Cell Biol.* **126**:413–422.
2291. **Johara, M., Y. Y. Toyoshima, A. Ishijima, H. Kojima, T. Yanagida, and K. Sutoh.** 1993. Charge-reversion mutagenesis of *Dictyostelium* actin to map the surface recognized by myosin during ATP-driven sliding motion. *Proc. Natl. Acad. Sci. U. S. A.* **90**:2127–2131.
2292. **Lees-Miller, J. P., G. Henry, and D. M. Helfman.** 1992. Identification of act2, an essential gene in the fission yeast *Schizosaccharomyces pombe* that encodes a protein related to actin. *Proc. Natl. Acad. Sci. U. S. A.* **89**:80–83.
2293. **Michaille, J. J., M. Gouy, S. Blanchet, and L. Duret.** 1995. Isolation and characterization of a cDNA encoding a chicken actin-like protein. *Gene* **154**:205–209.
2294. **Mitcham, J. L. and D. M. Prescott.** 1994. The actin II-encoding gene in the macronucleus of *Oxytricha nova*. *Gene* **144**:119–122.
2295. **Muhua, L., T. S. Karpova, and J. A. Cooper.** 1994. A yeast actin-related protein homologous to that in vertebrate dynactin complex is important for spindle orientation and nuclear migration. *Cell* **78**:669–679.
2296. **Murgia, I., S. K. Maciver, and P. Morandini.** 1995. An actin-related protein from *Dictyostelium discoideum* is developmentally regulated and associated with mitochondria. *FEBS Lett.* **360**:235–241.
2297. **Plamann, M., P. F. Minke, J. H. Tinsley, and K. S. Bruno.** 1994. Cytoplasmic dynein and actin-related protein Arp1 are required for normal nuclear distribution in filamentous fungi. *J. Cell Biol.* **127**:139–149.
2298. **Salazar, C. E., D. M. Hamm, D. M. Wesson, C. B. Beard, V. Kumar, and F. H. Collins.** 1994. A cytoskeletal actin gene in the mosquito *Anopheles gambiae*. *Insect. Mol. Biol.* **3**:1–13.
2299. **Sanchez, M., A. Valencia, M. J. Ferrandiz, C. Sander, and M. Vicente.** 1994. Correlation between the structure and biochemical activities of FtsA, an essential cell division protein of the actin family. *EMBO J.* **13**:4919–4925.

2300. **Schwartz, L. M., M. E. Jones, L. Kosz, and K. Kuah.** 1993. Selective repression of actin and myosin heavy chain expression during the programmed death of insect skeletal muscle. *Dev. Biol.* **158**:448–455.
2301. **Schwob, E. and R. P. Martin.** 1992. New yeast actin-like gene required late in the cell cycle. *Nature* **355**:179–182.
2302. **Tanaka, T., F. Shibasaki, M. Ishikawa, N. Hirano, R. Sakai, J. Nishida, T. Takenawa, and H. Hirai.** 1992. Molecular cloning of bovine actin-like protein, actin2. *Biochem. Biophys. Res. Commun.* **187**:1022–1028.
2303. **Wahlberg, M. H., K. A. Karlstedt, and G. I. Paatero.** 1994. Cloning, sequencing and characterization of an actin cDNA in *Diphyllobothrium dendriticum* (Cestoda). *Mol. Biochem. Parasitol.* **65**:357–360.
2304. **Wang, A. V., L. M. Angerer, G. J. Dolecki, R. Lum, G. V. Wang, R. Carlos, R. C. Angerer, and T. Humphreys.** 1994. Distinct pattern of embryonic expression of the sea urchin CyI actin gene in *Tripneustes gratilla*. *Dev. Biol.* **165**:117–125.
2305. **Wolf, S. J. and S. P. Timko.** 1994. Characterisation of actin gene family members and their expression during development in witchweed (*Striga asiatica* L.). *Planta* **192**:61–68.

STRUCTURAL ANALYSIS OF ACTIN

2306. **Adams, S. B. and E. Reisler.** 1994. Sequence 18–29 on actin: antibody and spectroscopic probing of conformational changes. *Biochemistry* **33**:14426–14433.
2307. **Borejdo, J. and S. Burlacu.** 1994. Orientation of actin filaments during motion in *in vitro* motility assay. *Biophys. J.* **66**:1319–1327.
2308. **Bremer, A., C. Henn, K. N. Goldie, A. Engel, P. R. Smith, and U. Aebi.** 1994. Towards atomic interpretation of F-actin filament three-dimensional reconstructions. *J. Mol. Biol.* **242**:683–700.
2309. **Censullo, R. and H. C. Cheung.** 1994. Tropomyosin length and two-stranded F-actin flexibility in the thin filament. *J. Mol. Biol.* **243**:520–529.
2310. **Chen, X., J. Peng, M. Pedram, C. A. Swenson, and P. A. Rubenstein.** 1995. The effect of the S14A mutation on the conformation and thermostability of *Saccharomyces cerevisiae* G-actin and its interaction with adenine nucleotides. *J. Biol. Chem.* **270**:11415–11423.
2311. **Crosbie, R. H., C. Miller, P. Cheung, T. Goodnight, A. Muhlrad, and E. Reisler.** 1994. Structural connectivity in actin: effect of *C*-terminal modifications on the properties of actin. *Biophys. J.* **67**:1957–1964.
2312. **Egelman, E. H.** 1994. Actin filament structure. The ghost of ribbons past. *Curr. Biol.* **4**:79–81.
2313. **Hambly, B. D., P. Kiessling, and C. G. dos Remedios.** 1994. Evidence for an F-actin like conformation in the actin:DNase I complex. *Adv. Exp. Med. Biol.* **358**:25–34.
2314. **Kabsch, W. and K. C. Holmes.** 1995. Protein motifs **2**: the actin fold. *FASEB J.* **9**:167–174.
2315. **Kojima, H., A. Ishijima, and T. Yanagida.** 1994. Direct measurement of stiffness of single actin filaments with and without tropomyosin by *in vitro* nanomanipulation. *Proc. Natl. Acad. Sci. U. S. A.* **91**:12962–12966.
2316. **Lepault, J., J. L. Ranck, I. Erk, and M. F. Carlier.** 1994. Small angle X-ray scattering and electron cryomicroscopy study of actin filaments: role of the bound nucleotide in the structure of F-actin. *J. Struct. Biol.* **112**:79–91.
2317. **Lorenz, M., K. J. Poole, D. Popp, G. Rosenbaum, and K. C. Holmes.** 1995. An atomic model of the unregulated thin filament obtained by X-ray fiber diffraction on oriented actin-tropomyosin gels. *J. Mol. Biol.* **246**:108–119.
2318. **McGough, A., M. Way, and D. DeRosier.** 1994. Determination of the alpha-actinin-binding site on actin filaments by cryoelectron microscopy and image analysis. *J. Cell Biol* **126**:433–443.
2319. **Mendelson, R. and E. Morris.** 1994. Combining electron microscopy and X-ray crystallography data to study the structure of F-actin and its implications for thin-filament regulation in muscle. *Adv. Exp. Med. Biol.* **358**:13–23.
2320. **Mendelson, R. A. and E. Morris.** 1994. The structure of F-actin. Results of global searches using data from electron microscopy and X-ray crystallography. *J. Mol. Biol.* **240**:138–154.
2321. **Miki, M. and T. Kouyama.** 1994. Domain motion in actin observed by fluorescence resonance energy transfer. *Biochemistry* **33**:10171–10177.
2322. **Moens, P. D., D. J. Yee, and C. G. dos Remedios.** 1994. Determination of the radial coordinate of Cys-374 in F-actin using fluorescence resonance energy transfer spectroscopy: effect of phalloidin on polymer assembly. *Biochemistry* **33**:13102–13108.
2323. **Ng, C. M. and R. D. Ludescher.** 1994. Microsecond rotational dynamics of F-actin in ActoS1 filaments during ATP hydrolysis. *Biochemistry* **33**:9098–9104.
2324. **Orlova, A. and E. H. Egelman.** 1995. Structural dynamics of F-actin: I. Changes in the C terminus. *J. Mol. Biol.* **245**:582–597.
2325. **Orlova, A., E. Prochniewicz, and E. H. Egelman.** 1995. Structural dynamics of F-actin II. Cooperativity in structural transitions. *J. Mol. Biol.* **245**:598–607.
2326. **Owen, C. and D. DeRosier.** 1993. A 13-A map of the actin-scruin filament from the limulus acrosomal process . *J. Cell Biol.* **123**:337–344.
2327. **Rayment, I., H. M. Holden, M. Whittaker, C. B. Yohn, M. Lorenz, K. C. Holmes, and R. A. Milligan.** 1993. Structure of the actin–myosin complex and its implications for muscle contraction. *Science* **261**:58–65.
2328. **Sasaki, Y., F. Tsunomori, T. Yamashita, K. Horie, H. Ushiki, R. Ishikawa, and K. Kohama.** 1994. Local environmental change from the G- to F-form of the actin molecule detected on anisotropy decay measurement. *J. Biochem. Tokyo.* **116**:236–238.
2329. **Tirion, M. M., D. ben Avraham, and K. C. Holmes.** 1994. Vibrational modes of G-actin. *Adv. Exp. Med. Biol.* **358**:3–12.

COVALENT DERIVATIZATION OF ACTIN

2330. **Chai, Y. C., S. S. Ashraf, K. Rokutan, R. B. J. Johnston, and J. A. Thomas.** 1994. *S*-Thiolation of individual human neutrophil proteins including actin by stimulation of the respiratory burst: evidence against a role for glutathione disulfide. *Arch. Biochem. Biophys.* **310**:273–281.
2331. **Jungbluth, A., V. von Arnim, E. Biegelmann, B. Humbel, A. Schweiger, and G. Gerisch.** 1994. Strong increase in the tyrosine phosphorylation of actin upon inhibition of oxidative phosphorylation: correlation with reversible rearrangements in the actin skeleton of *Dictyostelium* cells. *J. Cell Sci.* **107**:117–125.
2332. **Just, I., P. Sehr, M. Jung, J. van Damme, M. Puype, J. Vandekerckhove, J. Moss, and K. Aktories.** 1995. ADP-ribosyltransferase type A from turkey erythrocytes modifies actin at Arg-95 and Arg-372. *Biochemistry* **34**:326–333.
2333. **Kwon, H., P. M. Hardwicke, J. H. Collins, X. Zhao, and A. G. Szent Gyorgyi.** 1994. Myosin filament ATPase is enhanced by intramolecularly cross-linked actin. *J. Muscle. Res. Cell Motil.* **15**:555–562.
2334. **Marriott, G.** 1994. Caged protein conjugates and light-directed generation of protein activity: preparation, photoactivation, and spectroscopic characterization of caged G-actin conjugates. *Biochemistry* **33**:9092–9097.
2335. **Shartava, A., C. A. Monteiro, F. A. Bencsath, K. Schneider, B. T. Chait, R. Gussio, L. A. Casoria Scott, A. K. Shah, C. A. Heuerman, and S. R. Goodman.** 1995. A posttranslational modification of β-actin contributes to the slow dissociation of the spectrin-protein 4.1-actin complex of irreversibly sickled cells. *J. Cell Biol.* **128**:805–818.
2336. **van Delft, S., A. J. Verkleij, J. Boonstra, and P. M. van Bergen en Henegouwen.** 1995. Epidermal growth factor induces serine phosphorylation of actin. *FEBS Lett.* **357**:251–254.

PROTEOLYTIC CLEAVAGE OF ACTIN

2337. **Orlova, A. and E. H. Egelman.** 1995. Structural dynamics of F-actin: I. Changes in the C terminus. *J. Mol. Biol.* **245**:582–597.

EXPRESSION AND LOCALIZATION OF ACTIN

2338. **Amberg, D. C., E. Basart, and D. Botstein.** 1995. Defining protein interactions with yeast actin *in vivo*. *Nature Struct. Biol.* **2**:28–35.
2339. **Barja, F. and G. Turian.** 1994. Cytochalasin B-sensitive actin-mediated nuclear RNA export in germinating conidia of Neurospora crassa. *Cell Biol. Int.* **18**:903–906.
2340. **Bassell, G. J., C. M. Powers, K. L. Taneja, and R. H. Singer.** 1994. Single mRNAs visualized by ultrastructural *in situ* hybridization are principally localized at actin filament intersections in fibroblasts. *J. Cell Biol.* **126**:863–876.
2341. **Baumann, O. and B. Lautenschlager.** 1994. The role of actin filaments in the organization of the endoplasmic reticulum in honeybee photoreceptor cells. *Cell Tissue. Res.* **278**:419–432.
2342. **Burns, R. G. and C. D. Surridge.** 1994. Functional role of a consensus peptide which is common to α-, β-, and γ-tubulin, to actin and centractin, to phytochrome A, and to the TCP1 α chaperonin protein. *FEBS Lett.* **347**:105–111.
2343. **Chen, X., D. S. Sullivan, and T. C. Huffaker.** 1994. Two yeast genes with similarity to TCP-1 are required for microtubule and actin function *in vivo*. *Proc. Natl. Acad. Sci. U. S. A.* **91**:9111–9115.
2344. **Czernobilsky, B., S. Remadi, and G. Gabbiani.** 1993. α-smooth muscle actin and other stromal markers in endometrial mucosa. *Virchows Arch. A Pathol. Anat. Histopathol.* **422**:313–317.
2345. **Delbruck, S. and J. F. Ernst.** 1993. Morphogenesis-independent regulation of actin transcript levels in the pathogenic yeast *Candida albicans*. *Mol. Microbiol.* **10**:859–866.
2346. **Edbladh, M., P. A. Ekstrom, and A. Edstrom.** 1994. Retrograde axonal transport of locally synthesized proteins, e.g., actin and heat shock protein 70, in regenerating adult frog sciatic sensory axons. *J. Neurosci. Res.* **38**:424–432.
2347. **Fagotti, A., F. Panara, I. Di Rosa, F. Simoncelli, G. Gabbiani, and R. Pascolini.** 1994. Actin expression in some Platyhelminthe species. *J. Submicrosc. Cytol. Pathol.* **26**:545–551.
2348. **Harrison, P. and A. J. el Haj.** 1994. Actin mRNA levels and myofibrillar growth in leg muscles of the European lobster (*Homarus gammarus*) in response to passive stretch. *Mol. Mar. Biol. Biotechnol.* **3**:35–41.

2349. **Hikosaka, A., T. Kusakabe, and N. Satoh.** 1994. Short upstream sequences associated with the muscle-specific expression of an actin gene in ascidian embryos. *Dev. Biol.* **166**:763–769.

2350. **Hill, M. A., L. Schedlich, and P. Gunning.** 1994. Serum-induced signal transduction determines the peripheral location of β-actin mRNA within the cell. *J. Cell Biol.* **126**:1221–1229.

2351. **Holliday, L. S., M. R. Bubb, and E. D. Korn.** 1993. Rabbit skeletal muscle actin behaves differently than *Acanthamoeba* actin when added to soluble extracts of *Acanthamoeba castellanii. Biochem. Biophys. Res. Commun.* **196**:569–575.

2352. **Hori, R. and R. A. Firtel.** 1994. Identification and characterization of multiple A/T-rich cis-acting elements that control expression from *Dictyostelium* actin promoters: the *Dictyostelium* actin upstream activating sequence confers growth phase expression and has enhancer-like properties. *Nucleic Acids Res.* **22**:5099–5011.

2353. **Johannessen, A. J., I. F. Pryme, and A. Vedeler.** 1995. Changes in distribution of actin mRNA in different polysome fractions following stimulation of of MPC-11 cells. *Mol. Cell Biochem.* **142**:107–115.

2354. **Kirkeeide, E. K., I. F. Pryme, and A. Vedeler.** 1993. Microfilaments and protein synthesis; effects of insulin. *Int. J. Biochem.* **25**:853–864.

2355. **Kislauskis, E. H., Z. Li, R. H. Singer, and K. L. Taneja.** 1993. Isoform-specific 3′-untranslated sequences sort α-cardiac and β-cytoplasmic actin messenger RNAs to different cytoplasmic compartments. *J. Cell Biol.* **123**:165–172.

2356. **Kislauskis, E. H., X. Zhu, and R. H. Singer.** 1994. Sequences responsible for intracellular localization of β-actin messenger RNA also affect cell phenotype. *J. Cell Biol.* **127**:441–451.

2357. **Kusakabe, T.** 1995. Expression of larval-type muscle actin-encoding genes in the ascidian *Halocynthia roretzi. Gene* **153**:215–218.

2358. **Kuznetsov, S. A., D. T. Rivera, F. F. Severin, D. G. Weiss, and G. M. Langford.** 1994. Movement of axoplasmic organelles on actin filaments from skeletal muscle. *Cell Motil. Cytoskeleton.* **28**:231–242.

2359. **Langford, G. M., S. A. Kuznetsov, D. Johnson, D. L. Cohen, and D. G. Weiss.** 1994. Movement of axoplasmic organelles on actin filaments assembled on acrosomal processes: evidence for a barbed-end-directed organelle motor. *J. Cell Sci.* **107**:2291–2298.

2360. **Latham, V. M. J., E. H. Kislauskis, R. H. Singer, and A. F. Ross.** 1994. Beta-actin mRNA localization is regulated by signal transduction mechanisms. *J. Cell Biol.* **126**:1211–1219.

2361. **Lazzarino, D. A., I. Boldogh, M. G. Smith, J. Rosand, and L. A. Pon.** 1994. Yeast mitochondria contain ATP-sensitive, reversible actin-binding activity. *Mol. Biol. Cell* **5**:807–818.

2362. **Marco, S., J. L. Carrascosa, and J. M. Valpuesta.** 1994. Reversible interaction of β-actin along the channel of the TCP-1 cytoplasmic chaperonin. *Biophys. J.* **67**:364–368.

2363. **Moroianu, J., J. W. Fett, J. F. Riordan, and B. L. Vallee.** 1993. Actin is a surface component of calf pulmonary artery endothelial cells in culture. *Proc. Natl. Acad. Sci. U. S. A.* **90**:3815–3819.

2364. **Muto, E., M. Edamatsu, M. Hirono, and R. Kamiya.** 1994. Immunological detection of actin in the 14S ciliary dynein of *Tetrahymena. FEBS Lett.* **343**:173–177.

2365. **Sauman, I. and S. J. Berry.** 1994. An actin infrastructure is associated with eukaryotic chromosomes: structural and functional significance. *Eur. J. Cell Biol.* **64**:348–356.

2366. **Schroer, T. A.** 1994. New insights into the interaction of cytoplasmic dynein with the actin-related protein, Arp1. *J. Cell Biol.* **127**:1–4.

2367. **Sternlicht, H., G. W. Farr, M. L. Sternlicht, J. K. Driscoll, K. Willison, and M. B. Yaffe.** 1993. The t-complex polypeptide 1 complex is a chaperonin for tubulin and actin *in vivo. Proc. Natl. Acad. Sci. U. S. A.* **90**:9422–9426.

2368. **Turian, G. and F. Barja.** 1995. Nuclear actin and RNA export in conidial cells of *Neurospora crassa. Cell Biol. Int.* **19**:77–78.

2369. **Vinh, D. B., M. D. Welch, A. K. Corsi, K. F. Wertman, and D. G. Drubin.** 1993. Genetic evidence for functional interactions between actin noncomplementing (Anc) gene products and actin cytoskeletal proteins in *Saccharomyces cerevisiae. Genetics* **135**:275–286.

METHODS FOR ANALYSING ACTIN

2370. **Allen, P. G. and P. A. Janmey.** 1994. Gelsolin displaces phalloidin from actin filaments. A new fluorescence method shows that both Ca^{2+} and Mg^{2+} affect the rate at which gelsolin severs F-actin. *J. Biol. Chem.* **269**:-32768.

2371. **Andersland, J. M., D. D. Fisher, C. L. Wymer, R. J. Cyr, and M. V. Parthasarathy.** 1994. Characterization of a monoclonal antibody prepared against plant actin. *Cell Motil. Cytoskeleton.* **29**:339–344.

2372. **Casella, J. F., E. A. Barroncasella, and M. A. Torres.** 1995. Quantitation of CapZ in conventional actin preparations and methods for further purification of actin. *Cell Motil. Cytoskeleton.* **30**:164–170.

2373. **Funatsu, T., Y. Harada, M. Tokanuga, K. Saito, and T. Yanagida.** 1995. Imaging of single fluorescent molecules and individual ATP turnovers by single myosin molecules in aqueous solution. *Nature* **374**:555–559.

2374. **Marriott, G.** 1994. Caged protein conjugates and light-directed generation of protein activity: preparation, photoactivation, and spectroscopic characterization of caged G-actin conjugates. *Biochemistry* **33**:9092–9097.

2375. **Morris, E. P., E. Katayama, and J. M. Squire.** 1994. Evaluation of high-resolution shadowing applied to freeze-fractured, deep-etched particles: 3-D helical reconstruction of shadowed actin filaments. *J. Struct. Biol.* **113**:47–55.

2376. **Pinder, J. C., J. A. Sleep, P. M. Bennett, and W. B. Gratzer.** 1995. Concentrated tris solutions for the preparation, depolymerisation and assay of actin: application to erythroid actin. *Anal. Biochem.* **225**:291–295.

2377. **Tilney, L. G. and M. S. Tilney.** 1994. Methods to visualize actin polymerization associated with bacterial invasion. *Methods Enzymol.* **236**:476–481.

2378. **Ziemann, F., J. Radler, and E. Sackmann.** 1994. Local measurements of viscoelastic moduli of entangled actin networks using an oscillating magnetic bead micro-rheometer. *Biophys. J.* **66**:2210–2216.

TOXINS AND DRUGS AFFECTING ACTIN FUNCTION

2379. **Abosch, A. and C. Lagenaur.** 1993. Sensitivity of neurite outgrowth to microfilament disruption varies with adhesion molecule substrate. *J. Neurobiol.* **24**:344–355.

2380. **Bubb, M. R., A. M. Senderowicz, E. A. Sausville, K. L. Duncan, and E. D. Korn.** 1994. Jasplakinolide, a cytotoxic natural product, induces actin polymerization and competitively inhibits the binding of phalloidin to F-actin. *J. Biol. Chem.* **269**:14869–14871.

2381. **Bubb, M. R., I. Spector, A. D. Bershadsky, and E. D. Korn.** 1995. Swinholide A is a microfilament-disrupting marine toxin that stabilises actin dimers and severs actin filaments. *J. Biol. Chem.* **270**:3463–3466.

2382. **Bukatina, A. E. and F. Fuchs.** 1994. Effect of phalloidin on the ATPase activity of striated muscle myofibrils. *J. Muscle. Res. Cell Motil.* **15**:29–36.

2383. **Chen, T. S., G. A. Doss, A. Hsu, R. B. Lingham, R. F. White, and R. L. Monaghan.** 1993. Microbial transformation of L-696,474, a novel cytochalasin as an inhibitor of HIV-1 protease. *J. Nat. Prod.* **56**:755–761.

2384. **Crews, P., L. V. Manes, and M. Boehler.** 1986. Jasplakinolide, a cyclodepsipeptide from the marine sponge *Jaspis* sp. *Tetrahedron Lett.* **27**:2797–2800.

2385. **De La Cruz, E. and T. D. Pollard.** 1994. Transient kinetic analysis of rhodamine phalloidin binding to actin filaments. *Biochemistry* **33**:14387–14392.

2386. **Korohoda, W., M. Michalik, and M. Pierzchalska.** 1994. Benzamide-induced changes in the actin cytoskeleton in human skin fibroblasts. *Cell Biol. Int.* **18**:791–796.

2387. **Matthews, J. B., C. S. Awtrey, R. Thompson, T. Hung, K. J. Tally, and J. L. Madara.** 1993. Na^{+}–K^{+}–$2Cl^{-}$ cotransport and Cl^{-} secretion evoked by heat-stable enterotoxin is microfilament dependent in T84 cells. *Am. J. Physiol.* **265**:G370–G378.

2388. **Miyata, H. and K. J. Kinosita.** 1994. Transformation of actin-encapsulating liposomes induced by cytochalasin D. *Biophys. J.* **67**:922–928.

2389. **Oppenheimer, D. I. and L. E. Volkman.** 1995. Proteolysis of p6.9 induced by cytochalasin D in *Autographa californica* M nuclear polyhedrosis virus-infected cells. *Virology.* **207**:1–11.

2390. **Saito, S., S. Watabe, H. Ozaki, N. Fusetani, and H. Karaki.** 1994. Mycalolide B, a novel actin depolymerizing agent. *J. Biol. Chem.* **269**:29710–29714.

2391. **Schroeder, P., I. Just, and K. Aktories.** 1994. Adenine nucleotides regulate ADP-ribosylation of membrane-bound actin and actin-binding to membranes. *Eur. J. Cell Biol.* **63**:3–9.

2392. **Smith, C. D., S. Carmeli, R. E. Moore, and G. M. Patterson.** 1993. Scytophycins, novel microfilament-depolymerizing agents which circumvent P-glycoprotein-mediated multidrug resistance. *Cancer Res.* **53**:1343–1347.

2393. **Todd, J. S., K. A. Alvi, and P. Crews.** 1992. The isolation of a monomeric carboxylic acid of swinholide A from the Indo-Pacific sponge *Theonella swinhoei. Tetrahedron Lett.* **33**:441–442.

PLANT ACTIN

2394. **Alfano, F., J. G. Duckett, and R. Gambardella.** 1993. Circular and spiral F-actin associated with chloroplasts in liverwort thalli and gametophyte cells of *Equisetum* and *Pteridium. Gior. Bot. Ital.* **127**:292–294.

2395. **Alfano, F., A. Russell, R. Gambardella, and J. G. Duckett.** 1993. The actin cytoskeleton in the liverwort *Riccia fluitans*: effects of cytochalasin B and aluminium ions on rhizoid tip growth. *J. Plant Physiol.* **142**:569–574.

2396. **Cho, S.-O. and S. M. Wick.** 1991. Actin in the developing stomatal complex of winter **rye**: a comparison of actin antibodies and RH-phalloidin labelling of control and CB-treated tissues. *Cell Motil. Cytoskeleton.* **19**:25–36.

2397. **Clayton, L. and C. W. Lloyd.** 1985. Actin organization during the cell cycle in meristematic plant cells. Actin is present in the cytokinetic phragmoplast. *Exp. Cell Res.* **156**:231–238.

2398. **Doonan, J. H. and L. Clayton.** 1987. Immunofluorescent studies on the plant cytoskeleton. *Soc. Exp. Biol. Sem. Ser.* **29**:111–136.

2399. **Doonan, J. H. and C. W. Lloyd.** 1988. Microtubules and microfilaments in tip growth: evidence that microtubules impose polarity on protonemal growth in *Physcomitrella patens. J. Cell Sci.* **89**:533–540.

2400. **Doonan, J. H. and C. W. Lloyd.** 1988. The bryophyte cytoskeleton: experimental and immunofluorescence studies. *Adv. Bryol.* **3**:1–34.

2401. **Garbary, D. J., A. R. McDonald, and J. G. Duckett.** 1992. Visualisation of the cytoskeleton in red algae using fluorescent labelling. *New Phytol.* **120**:435–444.

2402. **Goode, J. A., F. Alfano, A. D. Stead, and J. G. Duckett.** 1993. The formation of aplastidic abscission (tmema) cells and protonemal disruption in the moss *Bryum tenuisetum* Limpr. is associated with transverse arrays of microtubules and microfilaments. *Protoplasma* **174**:158–172.

2403. **Grolig, F.** 1990. Actin-based organelle movements in interphase *Spirogyra*. *Protoplasma* **155**:29–42.

2404. **Jung, G. and W. Wernicke.** 1991. Patterns of actin filaments during cell shaping in developing mesophyll of wheat (*Triticum aestivum* L.). *Eur. J. Cell Biol.* **56**:139–146.

2405. **Kadota, A. and M. Wada.** 1992. Photo-induction of formation of circular structures by microfilaments on chloroplasts during intracellular orientation in protonemal cells of the fern *Adiantum capillus-veneris*. *Protoplasma* **167**:97–107.

2406. **Kakimoto, T. and H. Shibaoka.** 1987. Actin filaments and microtubules in the preprophase band and phragmoplast of tobacco cells. *Protoplasma* **140**:151–156.

2407. **Kohno, T. and T. Shimmen.** 1987. Ca^{++}-induced fragmentation of actin filaments in pollen tubes. *Protoplasma* **141**:177–179.

2408. **Kruse, S., H. Quader, D. Volkmann, and A. Sievers.** 1989. Visualisation of actin filament pattern in plant cells without pre-fixation. A comparison of differently modified phallotoxins. *Protoplasma* **149**:178–182.

2409. **Lloyd, C. W., K. J. Pearce, D. J. Rawlins, R. W. Ridge, and P. J. Shaw.** 1987. Endoplasmic microtubules connect the advancing nucleus to the tip of legume root hairs, but F-actin is involved in basipetal migration. *Cell Motil. Cytoskeleton.* **8**:27–36.

2410. **McDonald, A. R., D. J. Garbary, and J. G. Duckett.** 1993. Rhodamine-phalloidin staining of F-actin in rhodophyta. *Biotech. Histochem.* **68**:91–98.

2411. **Palevitz, B. A. and B. Liu.** 1992. Microfilament (F-actin) in generative cells and sperm: an evaluation. *Sex. Plant Reprod.* **5**:89–100.

2412. **Quader, H. and E. Schnepf.** 1989. Actin filament array during side branch initiation in protonema cells of the moss *Funaria hygrometrica*: an actin organising centre at the plasma membrane. *Protoplasma* **151**:167–170.

2413. **Staiger, C. J., M. Yuan, R. Valenta, P. J. Shaw, R. M. Warn, and C. W. Lloyd.** 1994. Microinjected profilin affects cytoplasmic streaming in plant cells by rapidly depolymerizing actin microfilaments. *Curr. Biol.* **4**:215–219.

2414. **Wolf, S. J. and S. P. Timko.** 1994. Characterisation of actin gene family members and their expression during development in witchweed (*Striga asiatica* L.). *Planta* **192**:61–68.

FEEDBACK TO THE EDITOR

This publication is intended as a practical and accessible data resource and as an easy and comprehensive access to the literature. There will no doubt be errors and omissions, but because this series is updated and refined annually, we will consider the incorporation of suggestions that will improve its value to you. If there are improvements that you would like to see in the next updated release of this issue please send them to the Editor.

PETER SHETERLINE
Department of Human Anatomy
and Cell Biology
University of Liverpool
Liverpool L69 3BX, UK

Tel: 0151-794-5450
Fax: 0151-794-5517
Email: shep@liverpool.ac.uk

The Editor would particularly welcome comments with regard to errors, omissions and suggested improvements.

FOLD-OUT SECTION

Multiple alignments of available actin sequences. An alignment was made of the 141 unique protein sequences currently available and taken from the OWL database version 23.2 (installed June 1994) (Bleasby, A.J. & Wootton, 1990, *Protein Engineering* 3: 153–160) and additional sequences from the TREMBL database. The alignment was produced using PILEUP (part of the sequence analysis package of Genetics Computer Group Inc., Wisconsin). Sequences are divided into broad phylogenetic groups. The first block is vertebrate actins (muscle and non-muscle), the second invertebrates, the third protozoa, the fourth plants and the fifth fungi. On the header line, asterisks (*) indicate amino acids that are invariant in all sequences: dots amino acids that are only different in five or less actins; h, mainly hydrophobic amino acids, o, hydroxyl-containing (T or S), a, acidic (D or E) and b, basic (R or K) in all but five or less sequences. Unsequenced regions are indicated by long runs of dashes and uncertain amino acids by X. Further data can be obtained from Table 3. Additional sequences: the wild-type *Drosophila* sequences (Act_DROME) contain some errors, notably Act6_DROME (Sparrow, unpublished). The Act_DROME sequences used are corrected versions (Sparrow & Fyrberg, unpublished). The Act3_ARTSX complete sequence was provided by Dr Sastre (*personal comm.*) and the Act2_CAAEL and Act4_CAAEL by editing the Act_CAAEL sequence with changes indicated in the Owl:Act_CAAEL file.

Alignment of actin-related sequences. Alignment and display of available actin-related protein sequences was performed as above. Sequences were aligned with ACTL_BOVIN. The aligned sequence of ACTS_HUMAN was moved to the top for reference. The classification of Arps was as given by Schroer *et al.* [2125].

```
    7          8          9          1          1          1               1                   1          1          1          1          1
                                     0          1          2               3                   4          5          6          7          8
4567890123 4567890123 4567890123 4567890123 4567890123 4567890123 4567 8901 2345 6789 0123456789 0123456789 0123456789 0123456789 0123456789
h. h *.... *hh  ...*. .h.  .o... ...  .aa.. ..**. ...  *.*  ...  .*.       ..   .        .. *...  .*.*. * *  . ***. o.  *....*   h.*.h bha h.*ba..
ILTLKYPIEH GIITNWDDME KIWHHTFYNE LRVAPEEHPT LLTEAPLNPK ANREKMTQIM FETF --NVPA MYVA --IQAV LSLYASGRTT GIVLDSGDGV THNVPIYEGY ALPHAIMRLD LAGRDLTDYL
.......... .......... .......... .......... .......... .......... .... .... .... .... .......... .......... .......... .......... ..........
.......... .......... .....S.... .......... .......... .......... .... .... .... .... .......... .......... .......... .......... ..........
.......... .......... .....S.... .......... .......... .......... .... .... .... .... .......... .......... .......... .......... ..........
.......... ..V....... .......... .........V .......... .......... .... T... .... .... .......... ..M....... ..T....... ......L... ..........
.......... ..V....... .......... .........V .......... .......... .... T... .... .... .......... ..M....... ..T....... ......L... ..........
.......... ..V....... .......... .......R.V .......... .......... .... T... .... .... .......... ..M....... ..T....... ......L... ..........
```

```
                           1          2          3          4          5          6
               123456 7890123456 789012345 678901234 5678901234 5678901234 56789 0123

                    a h. **.o.   * *h   ..  .. .....  h. .b   hh   .  .    h  ....   ...
  S_Human  --MCDEDETT ALVCDNGSGL VKAGFAGDD- APRAVFPSI- VGRPRHQGVM VGMGQKDSYV GDEAQ-SKRG
  C_Human  .....DE... .......... .......... .......... .......... .......... ..........
  A_Human  ....E.EDS. .......... C......... .......... .......... .......... ..........
  H_Human  ....E.E-.. .......... C......... .......... .......... .......... ..........
  B_Human  ..--MD.DIA ...V.....M C......... .......... .......... .......... ..........
  G_Human  ..--M.E.IA ...I.....M C......... .......... .......... .......... ..........
  B_Rabit  ..--MD.DIA ...V.....M C......... .......... .......... .......... ..........
  A_Bovin  ..--E.EDS. .......... C......... .......... .......... .......... ..........
  A_Chick  ....E.EDS. .......... C......... .......... .......... .......... ..........
 Chkaccy5  ..-MADE.IA ...V.....M C......... .......... .......... .......... ..........
  B_Ansan  ..--M.E.IA ...I.....M C....GR... .......... .......... .......... ..........
  1_Xenla  .....DE... .......... .......... .......... .......... .......... ..........
  2_Xenla  .....D.... .......... .......... .......... .......... .......... ..........
  3_Xenla  .....D.... .......... .......... .......... .......... .......... ..........
  B_Xenbo  ..-MAD.DIA ...I.....M C......... .......... .......... .......... ..........
  2_Xentr  .....DE... .......... .......... .......... .......... .......... ..........
  S_Plewa  ..-------- ---------- ---------. ---------. ---------- ---------- -----.----
  B_Cypca  ..--MD..IA ...V.....M C......... .......... .......... .......... ..........

  M_Strpu  ..-MCDEDVA ...V.....M C.T....... .......... ......H... .......... ..........
  C_Strpu  ..-MCD.DVA ...I.....M .......... .......... .......... .......... ..........
  3_Strpu  ..-MCD.DVA ...V.....M .......... .......... .......... .......... ..........
  1_Strfn  ..-MCD.DVA ...I.....M .......... .......... .......... .......... ..........
  2_Strfn  ..-MCD.DVA ...I.....M .......... .......... .......... .......... ..........
Lp09651_1  ..-MCD.DIA ...I.....M .......... .......... .......... .......... .GQ.......
Lp09652_1  ..-MCD.DVS ...I.....M .......... .......... .......... .......... ..........
  M_Pisoc  ..-MCDEDVA ...V.....M C......... .......... .......... .......... ..........
  C_Pisoc  ..-MCDEDVA ...V.....M C......... .......... .......... .......... ..........
   A60945  ..-------- ---------- ---------. ---------. ---------- ---------- -----.----
  M_Stycl  .MSDG.EDQ. .I........ ..S....... .......... .......... .......... ..........
   A61043  ..--MD..VA ...V.....M C......... .......... .......... .......... ..........
  M_Stypl  MEDDQDE.Q. .......... .....P..A. P......LT. .......... .......... ..........
  C_Stypl  ..--MD..VA ...V.....M C....E.... .......... .......... ...V...... ..........
  M_Halro  .MSDG.ED.. .I........ ..S....... .......... .......... .......... ..........
    Aplca  ..-MCD..VA ...V.....M C......... .......... .......... .......... ..........
Cvactin_1  ..-------- ---------- ---------. ---------. ---------- ---------- -----.----
  3_Drome  ..-MCD..VA ...V.....M C......... .......... .......... .......... ..........
  4_Drome  ..-MCDE.AS ...V.....M C......... .......... .......... .......C.. ..........
  5_Drome  ..-MCD..VA ...V.....M C......... .......... .......... .......... ..........
  6_Drome  ..-MCD.DAG ...I.....M C......... .......... .......... .......... ..........
  1_Drome  ..-MCDE.VA ...V.....M C......... .......... .......... .......... ..........
  2_Drome  ..-MCDE.VA ...V.....M C......... .......... .......... .......... ..........
   Jc1246  ..-MCD.DAG ...I.....M C......... .......... .......... .......... ..........
Bccactinb  ..-MCD..VA ...V.....M C......... .......... .......... .......... ..........
Bccactinc  ..-MCD..AS ...V.....M C......... .......... .......... .......... ..........
  1_Bommo  ..-MCD.DVR ...V.....M C......... .......... ........L. .......... ..........
  2_Bommo  ..-MCD.DAG ...V.....M C......... .......... .......... .......... ..........
  3_Bommo  ..-MCDE.VA ...V.....M C......... .......... ........L. .......... ..........
Motactinx  ..-MCD.DVA ...V.....M C......... .......... .......... .......... ..........
Agct1da_1  ..-MCDE.VA ...V.....M C......... .......... .......... .......... ..........
  1_Artsx  ..-MCD..VA ...V.....M C......... .......... .......... .......... ..........
  2_Artsx  ..-MCD.DVA ...V.....M C......... .......... .......... .......... ..........
  3_Artsx  ..-------- ---------- ---------. ---------. ---------- ----...... ..........
  4_Artsx  ..-MCD..VA ...V.....M C......... .......... .......... .......... ..........
 Praactin  ..-------- ---------- ---------. ---------. ---------- ------GTS. ..........
 Lpact3_1  ..-MCD..VA ...V.....M C......... .......... .......... .......... ..........
 Lpact5_1  ..-MCDEDVA ...V.....M C......... .......... .......... .......... ..........
Lpact11_1  ..-MC.EDVA ...V.....M C......... .......... .......... .......... ..........
    Taeso  ..-MGDE.VQ ...V.....M C......... .......... .......... .......... ..........
  1_Echgr  ..--MAE.I. P..V.....M C......... S.......L. ......Q.SI. ....N..... ..........
  2_Echgr  ..-MADED.A ...I.....M C......... .......... .......... .......... ..........
Smactin_1  ..-MA.EDVA ...I.....M C......... .......... .......... .......... ..........
    Caeel  ..-MCD..VA ...V.....M C......... .......... .......... .......... ..........
  2_Caeel  ..-MCD.DVA ..IL.....M C......... .......... .......... .......... ..........
  4_Caeel  ..-MCD..VA ...V.....M C......... .......... .......... .......... ..........
  1_Oncvo  ..-MCDE.VA ...V.....M C......... .......... .......... .......... ..........
  2_Oncvo  ..-MCDE.VA ...V.....M C......... .......... .......... .......... ..........
    Hydat  ..-MAD..VA ...V.....M C......... .......... .......... .......... ..........
   Jn0832  ..-MAD.DVA ...V.....M C......... .......... .......... .......... ..........
   Jn0833  ..-MAD.DVA ...V.....M C......... .......... .......... .......... ..........

    Oxyfa  ..---MSDQQ TC.I.....V .......E.. .......... ....KNVSAL I.VDSASE.L .....Q-...
Ofaactin  ..---MSDQQ TC.I.....V .......E.. .......... ....KNVSAL I.VDSASE.L .....Q-...
    Oxyno  ..---MADKQ TV.V.....V .....S.E.. .......... I...KNVSAL I.VDSASE.I .....Q-...
    Tetth  ..--A.S.SP .I.I.....M C...I..... ....A..... ....KMP.I. ...D..EC.. .E....A...
    Tetpy  ..-MTDSDSP .I.I.....M C...I..... ....A..... I...KMP.I. ...D..EC.. .E....A...
    Eupcr  .MSEE.NAKE .I.V.....V ........EN. Q.CS....VV AKPKTK.VIV G.A.N..CF. ...R.Q-...
  1_Plafa  ..-MG.EVVQ ...V.....N ....V..... ...S...... ....KNP.I. ...EE..AF. ......T...
  2_Plafa  ..--MSE.AV ...V.....M ..S.L..... ..KC...... ....KMPNI. I..E..EC.. ......N...
Tg10429_1  ..-MADE.VQ ...V.....N ....V..... .......... ..K.KNP.I. ...EE..C.. ..........
  1_Trybb  ..-MSDE.Q. .I.......M ..S..S.... ...H...... ....KNEQA. M.SAN.KLF. ......A...
  2_Trybb  ..-MSDE.Q. .I.......M ..S..S.... ...HL..... ..P.KNEQA. M.SAKQEMF. ......A...
Lmactin_1  ..-MADN.QS SI.......M .....S.... ...H...... ....KNMQA. M.SAN.TV.. ..........
  1_Naefo  ..--MC.DVQ ...V.....M C......... .......... I...KQKSI. ....N..A.. ...V......
  2_Naefo  ..------VQ ..EV.....M C......... ..G....... I...KQKSI. ....N..A.. ...V......
Ngractini  ..--MC.DVQ ...V.....M C......... .......... I...KQKSI. ....N..A.. ...V......
    Enthi  ..-MGDE.VQ ...V.....M C......... .......... ......VS.. A.....A... ..........
 Glacti_1  ..--MT.DNP .I.I.....M C......... .......TV. ....KRET.L ..STH.EE.I ....-LA...
    Phypo  ..--EGEDVQ ...I.....M C......... .......... .......T... .......... ..........
  D_Phypo  ..-----.GE .V.I.....M C.......N. T...M..... ..H...TEA. MEI.H.H... ..........
  1_Acaca  ..--MG..VQ ...I.....M C......... .......... ......T... .......... ..........
  1_Dicdi  ..---.GEDVQ ...I.....M C......... .......... ......T... .......... ..........
  2_Dicdi  ..---.GEDVQ ...I.....M C......... .....S.... ......T... .......... .E........
  3_Dicdi  ..--ESEDVQ ...I.....M C......... .......... .....YT... .........I ........RK.
  4_Dicdi  ..--ERE.VQ .I.I....DM C......... .......Q... .....YT... DV.......I V...H..RK.
  8_Dicdi  ..---.GEDVQ ...I.....M C......... .......... ......T... .......... ..........

    Pinco  ..-------- ---------- ---------. ---------. ---------- ---------- -----.----
  1_Soybn  ...A.AEDIQ P......T.M ....L....T P......N.. ......TX.V .......A.. ......VQ..
  3_Soybn  ...A.AEDIE P......T.M .......... .......... ......T... .......X.. ..........
    1_Pea  ...A.AEDIQ P......T.M .......... .-........ ......T... .......A.. ..........
    2_Pea  ...A.AEDIQ P......T.M .......... .-........ ......T... .......A.. ..........
  1_Dauca  ...A.GEDIQ P......T.M .......... .......... ......T... .......A.. ..........
  2_Dauca  .MADGGEDIQ P......T.M .......... .........V ......T... .......A.. ..........
  1_Maize  ..-MADEDIQ PI.....T.M .......... .......... ......T... .......A.. ......A...
  1_Orysa  ...A.AEDIQ P......T.M .......... .......... ......T... .......A.. ..........
  Osrac1  ...A.AEDIQ P......T.M .......... .......... ......T... .......A.. ..........
  2_Orysa  ...A.AEDIQ P......T.M ....L..... .......... ......T... .......A.. ..........
  3_Orysa  ...A.GEMIQ P......T.M .......... .......... ..Q...T... .......A.. ..........
  7_Orysa  ..-MA.EDIQ PN.....TEM .........D .......... ......S... .......A.. ....AVE..D
  1_Arath  ...A.GEDIQ P......T.M .......... .......... ......T... .......A.. ..........
  1_Soltu  ...AEGEDIQ P......T.M .......... ......R... ......T... .......A.. ..........
  2_Soltu  ...A.VEDIQ P......T.M .......... .......... ......A... .......A.. V.........
  3_Soltu  ...A.V.DIQ P......T.M .......... S......... ......T... .......A.. ..........
  4_Soltu  ..-------- ---------- --........ .......... ......T... .......A.. ..........
  5_Soltu  ...A.GEDIQ P......T.M .......... .......... ......S... .......A.. ..........
  6_Soltu  ..-------- ---------- --........ .......... ......T... .......A.. ..........
  7_Soltu  ...A.AEDIE P......T.M .......... .......... ......T... .......A.. ..........
    Tobac  ...A.GEDIQ P......T.M .......... .......... ......T... .......A.. ..........
Cc03676_1  ..----MSKP .V.I.....R C.T.I..E.. C.K....AV. I.K.KQK.I. ..T....E.. ..A.-MAR..
    Coscs  ..-------- ---------- ---------. ---------. ---------- ...D...A.. ..........
Fd11697_1  ..-MADEDVQ ...V.....M C....-.... .......... ....K.P.I. ...D...A.. ..........
    Volca  ...AE.G.VS .........M .......... .......... ......T... .......... ..........
Acactix_1  ..-------- ---------- --........ .......... ..K.K..... .......T.. .E....R...
   A37431  ..-------- ----.....M C......... .......... I....QP... .......... ..........
   S68003  ..-------- ---------- ---------. ---------. ---------- ---------- -----.----
Svsoac1_1  ...A.AEDIQ P......T.M .......... .......... ......T... .......A.. ..........

Fnctin_1  ..--M.E.VA ...I.....M C......... .......... .......... .......... ..........
   Hcct_1  ..--M.E.VA ...I.....M C......... ...S...... ......H.I. I......... ..........
gtritact_1  ..--M.E.VA ...L.....M C......... .......... C.G....... .......... ..........
    Achbi  ..-MAD.DVQ ...V.....M C......... .......... ....K.P.I. ...D..A... ..........
    Phyme  ..--M..DIQ .V.I.....M C......... .......... ..M.K.L.I. ...N..A.I. .....A....
  1_Phyin  ..-MAD.DVQ ...V.....M C......... .......... ....K.L.I. ...D..A... ..........
  2_Phyin  ..--MD.DIQ .V.I.....M C......... .......... ..M.K.L.I. ...N..A.I. .....A....
  1_Absgl  ...SM.E.IA ...I.....M C......... .......... ........I. .......... ..........
  2_Absgl  ...SM.EDIA ...I..A..M C......... .......... ........I. .......... ..........
  G_Emeni  ..--M.E.VA ...I.....M C......... .......... ......H.I. I......... ..........
    Thela  ..--M.E.VA ...I.....M C......... .......... ......H.I. I......... ..........
    Canal  ..-M.GE.VA ..II.....M C......... ........L. ........I. .......... ..........
    Crypv  ..-MS.E..Q ...V.....M ....V..... ...C...... ....KMP... ...D..C... ..........
    Yeast  ..--MDS.VA ...I.....M C......... .......... ........I. .......... ..........
    Schpo  ..--M.E.IA ...I.....M C......... .......... ......H.I. .......... ..........
    Klula  ..--MDS.VA ...I.....M C......... .......... ........I. .......... ..........
mcactini_1  ..--M...IA ...I.....M C......... .......... ........I. .......... ..........
```

```
    3          3          3          3          3          3     3
    2          3          4          5          6          7     7
6789012345 6789012345 678 901234 5678901234 5678901234 5678901234 5
a    .* h  . *hh....*  b . h* *..   . . oh     .. . ah a  .. ... . .h
EITALAPSTM KIKIIAPPER KYS-VWIGGS ILASLSTFQQ MWITKQEYDE AGPSIVHRKC F----   S_Human
.......... .......... .......... .......... ...S...... .......... .....   C_Human
.......... .......... .......... .......... ...S...... .......... .....   A_Human
.......... .......... .......... .......... ...S.P.... .......... .....   H_Human
.......... .......... .......... .......... ...S...... S......... .....   B_Human
.......... .......... .......... .......... ...S...... S......... .....   G_Human
.......... .......... .......... .......... ...S...... S......... .....   B_Rabit
.......... .......... .......... .......... ...S...... .......... .....   A_Bovin
.......... .......... .......... .......... ...S...... .......... .....   A_Chick
.......... .......... .......... .......... ...S...... S......... .....   Chkaccy5
.......... .......... .......... .......... ...S...... S......... .....   B_Ansan
.......... .......... .......... .......... ...S...... .......... .....   1_Xenla
.......... .......... .......... .......... .......... .......... .....   2_Xenla
.......... .......... .......... .......... .......... .......... .....   3_Xenla
.......... .......... .......... ...L...... ...S...... S......... .....   B_Xenbo
.......... .......... .......... .......... .......... .......... .....   2_Xentr
.......... .......... .......... .......... .......... .......... .....   S_Plewa
...S...... .......... .......... .......... ...S...... S......... .....   B_Cypca

........SV .......... .......... .......... ...S....-- S.A....... .....   M_Strpu
.......P.. .......... .......... .......... ...S...... S......... .....   C_Strpu
..V....P.. .......... .......... .......... ...S...... S......... .....   3_Strpu
.......P.. .......... .......... .......... ...S...... S......... .....   1_Strfn
.......P.. .......... .......... .......... ...S...... S......... .....   2_Strfn
.......P.. .......... .......... .......... ...S...... S......... .....   Lp09651_1
..S....P.. .......... .......... .......... ...S...... S......... .....   Lp09652_1
..Q....P.. .......... .......... .......... ...S...... S......... .....   M_Pisoc
.V.....P.. .......... .......... .......... ...S...... S......... .....   C_Pisoc
D..G...... .......... .......... ........-- ---------- ---------- -....   A60945
.......... .......... .......... .......... ...S...... .......... .....   M_Stycl
..ST...P.. .......... .......... .......... ...S...... S......... .....   A61043
.......... .S........ ........A. .......... .......... S......... .....   M_Stypl
..S....P.. .......... .......... .......... ...S...... S......... .....   C_Stypl
.......... .......... .......... .......... ...S...... .......... .....   M_Halro
...S...... .......... .......... .......... ...S...... S......... .....   Aplca
.V.T...P.. ...V...... .......... .....----- ---------- ---------- -....   Cvactin_1
...S.....I .......... .......... .......... ...S.E.... S..G...... .....   3_Drome
.......... .......... .......... .......... ...S...... S..G...... .....   4_Drome
.........I .......... .......... .......... ...S...... S..G...... .....   5_Drome
.........I .......... .......... .......... ...S...... S..G...... .....   6_Drome
.......... .......... .......... ....S..... ..TS...... S......... .....   1_Drome
.......... ....V..... .......... .......... ...S...... S......... .....   2_Drome
.........I .......... .......... .......... ...S...... S..G...... .....   Jc1246
.........I .......... .......... .......... ...S...... S..G...... .....   Bccactinb
.........I .......... .......... .......... ...S...... S..G...... .....   Bccactinc
.........I .......... .......... .......... ...S.E.... S..G...... .....   1_Bommo
.........I .......... .......... .......... ...S...... S..G...... .....   2_Bommo
...R...... .......... .......... .......... ...S...... S......... .....   3_Bommo
.........I .......... .......... .......... ...S.E.... S..G...... .....   Motactinx
.......... .......... .......... .......... ...S...... S......... .....   Agct1da_1
.........I .......... .......... .......... ...S...... S..G...... .....   1_Artsx
.........I .......... .......... .......... ...S...... S..G...... .....   2_Artsx
...M....S. .......... .......... .......... ...S...... S......... .....   3_Artsx
.......... .......... .......... .......... ...S...... S......... .....   4_Artsx
.......A.I .......... .......... .........T .....E.... S..G...... .....   Praactin
.......... .......... .......... .......... ...S...... S......... .....   Lpact3_1
..G....... .......... .......... .......... ...S...... S......... .....   Lpact5_1
..C....... .......... .......... .......... ...S...... S......... .....   Lpact11_1
...S...... ....V..... .......... .......... ...S...... S..G...... .....   Taeso
DL.....T.. .......... .......... .......... .......... S......... .....   1_Echgr
..SS...T.V ..R.A..... .......... .......... ...S.V..E. I......... .....   2_Echgr
..S....... ....V..... .......... .......... ...S...... S..G...... .....   Smactin_1
.......... .......... .......... .......... ...S...... S......... .....   Caeel
.......... .......... .......... .......... ...S...... S......... .....   2_Caeel
.......... .......... .......... .......... ...S...... S......... .....   4_Caeel
.V........ .......... .......... .......... V..S...... S......... .....   1_Oncvo
.......... .......... .......... .......... ...S...... S......... .....   2_Oncvo
..S....P.. .......... .......... .......... ...S...... S......... .....   Hydat
..SS...P.. .......... .......... .......... ...S...... S......... .....   Jn0832
..SS...P.. .......... .......... .......... ...S...... S......... .....   Jn0833

..ENR..KSI NV.V..S.D. RFA...R... T.T.....AS .....ED... N.A....... I....   Oxyfa
..ENR..KSI NV.V..S.D. RFA...R... T.T.....AS .....ED... N.A....... I....   Ofaactin
..E.R..KSI NV.V..S.D. RFA...R... T.T.....AS .....ED... N.A....... L....   Oxyno
.VS.....S. ...VV..... R......... ..S......T .....A.... S......... .....   Tetth
.VS.....S. ...VV..... R......... ..S......T .....A.... S......... .....   Tetpy
.MKD...Q.. .V.V..S.D. ..A...R... T..K....AG ..V..ED.A. F.E....... I....   Eupcr
D..T...... ...VV..... .......... ..S....... .....E.... S......... .....   1_Plafa
.M.N....S. ...V...... .......... ..S....... .....E..ED S......... .....   2_Plafa
.L.S...... ...VV..... .......... ..S....... .....E.... S......... .....   Tg10429_1
..SN.P..SI .P.VV..... .......... ..S..T...S .....S.... S......S.. .....   1_Trybb
..SN....SI .P.VV..... .......... ..S..T...S .....S.... S......S.. .....   2_Trybb
..SN....SI .P.VV..... .......... ..S..T...T ..VK.S.... S......N.. .....   Lmactin_1
.L.NM..AS. ...VV..... .......... .......... .....E..ED ...G....ES .....   1_Naefo
.L.NM..AS. ...VV..... .......... .......... .....E..ED ...G....ES .....   2_Naefo
.L.NM..AS. ...VV..... .......... .......... .....E..ED ...G.....S .....   Ngractini
.MIQ...P.. ...V...... .......... .........N .....E.... S..A...... .....   Enthi
.LLD.I.AGK RVR.SS.ED. ....A.V... V.G..A..ES ..VSS...Q. N.A..AN... M....   Glacti_1
.L........ .......... .......... .......... ...S.E.... S......... .....   Phypo
.L..FE.-.. ........G. Q...A..... ........E. .C.S.K..N. C......... .....   D_Phypo
.L........ .......... .......... .......... ...S.E.... S......... .....   1_Acaca
.L........ .......... .......... .......... ...S.E.... S......... .....   1_Dicdi
.L........ .......... .......... .......... ...S.E.... S......... .....   2_Dicdi
.L........ .......... .......... .......... ...S.E.... S......... .....   3_Dicdi
.L........ .......... ......V... ......S..E ...S.E..N. S......... SNYLK   4_Dicdi
.L........ .......... .......... .......... ...S.E.... S......... .....   8_Dicdi

........S. ...VV..... .......... .......... ...A.T..E. S......... .....   Pinco
..S.....S. ...VV..S.. .FG....... .......... ...X.A.... S......K.. .....   1_Soybn
..S.....S. ...VV..... .......... .......... ...S.G.... S......... .....   3_Soybn
........S. ...VV..... .......... .......... ...S.A.... R......... .....   1_Pea
........S. ...VV..... R......... .......... ...S.A.... R......... .....   2_Pea
........S. ...VV..... ...DL..... .......... ...S.G.... S......... LLAG.   1_Dauca
........S. ...VV..... .......... .......... ...S.G.... S......... .....   2_Dauca
...S.V..S. .V.VV..R.. .......... .......... ...S.G.... T..G...M.. .....   1_Maize
.DHC....S. ...VV..... .......... .......... ...A.A.... S......... .....   1_Orysa
.DHC....S. ...VV..... .......... .......... ...A.A.... SD........ .....   Osrac1
.DHRA...S. ...VV..... .......... .......... V..SRA..E. S..A...... .....   2_Orysa
.DY.....S. ...VV..... .......... .......... ...S.G.... S......... .....   3_Orysa
.....C.GS. .--VV..... .........F .......... V..S.A.... S..G...M.. .....   7_Orysa
........S. ...VV..... .......... .......L.. ...A.A.... S......... .....   1_Arath
.L......S. ...VV..... .......... .......... ...A.A.... S......... .....   1_Soltu
..Q.....S. ...VV..... .......... .......... ...A.A.... S......... .....   2_Soltu
........S. ...VV..... .......... .......... ...A.A.F.. S......... .....   3_Soltu
---------- ---------- ---------- ---------- ---------- ---------- -....   4_Soltu
........S. ...VV..... .......... .......... ...A.A.... S......... .....   5_Soltu
........S. ...VV..... .......... .......... .....G.... S......... .....   6_Soltu
........S. ...VV..... .......... .......... .....G.... S......... .....   7_Soltu
........S. ...VV..... .......... .......... ...A.A.... S......... .....   Tobac
..ET..PS.. ...V...... Q......... ..S....... ..VS.T.... M...V..... .....   Cc03676_1
.L........ ...VV..... .......... .......... ...S.A.... S......... .....   Coscs
.L........ ...VV..... .......... .......... ...S.A.... S......... .....   Fd11697_1
........A. ...VV..... .......... .......... ...A.S.... S......... .....   Volca
...T....S. ...VV.S... .F........ ..S....... ...S.E.... ...A...... .....   Acactix_1
.......AS. .V........ .......... ..S....... ...S...... S......... .....   A37431
........S. ...VV..... .......... .......... ...A.A.... S......... .....   S68003
GNHC....S. ...VV..... .......... .......... ...A.A.... S......... .....   Svsoac1_1

........S. .V..VS.... .......... .......... ...A.S.... S......... .....   Fnctin_1
...R....S. .V........ .......... .......... ...S...... S......... .....   Hcct_1
........S. .V..V..... .......... .......... ...S...... S......... .....   gtritact_1
.L........ ...VV..... .......... ..S....... ...S.A.... S......... .....   Achbi
.M.K...TA. .V...T.... .......... .....A...H ...S.TD... S......... .....   Phyme
.L........ ...VV..... .......... .QS....... ...S.A.... S......... .....   1_Phyin
.VIK...TA. .V...T.... .......... .....A...H ...S.T.... S......... .....   2_Phyin
---------- ---------- ---------- ---------- ---------- ---------- -....   1_Absgl
........S. ...V...... .......... .......... ...S...... S......... .....   2_Absgl
........S. .V........ .......... .......... ...S...... S......... .....   G_Emeni
.L......S. .......... .......... .......... ..VS...... S......... .....   Thela
........S. .V........ .......... .......... ...S...... S......H.. .....   Canal
.L.S...... ...VV..... .......... ..S....... .....E.... S......... .....   Crypv
........S. .V........ .......... .....T.... ...S...... S......H.. .....   Yeast
..Q.....S. .V..V..... .......... .......... ...S...... S..G..Y... .....   Schpo
........S. .V........ .......... .....T.... ...S...... S......L.. .....   Klula
........S. ...V...... .......... .......... ...S...... N.....Y... .....   mcactini_1
```

```
          2          2          2          2          2          2          2          2          2          2          3          3
          0          1          2          3          4          5          6          7          8          9          0          1
123456789 0123456789 0123456789 0123456789 012 345678 9012345678 90123456 7 8901234567 89 0123456 7890123456 7890123456 7890123456 7890 12345
          .  o*. .   .. *..h h  .                    b  h.h..*    h  ..      .  h  h    a  h        o.  ..  h.h*ba.h .  hh**. o*     .  .h b
KILTERGYS FVTTAEREIV RDIKEKLCYV ALDFENEMAT AAS-SSSLEK SYELPDGQVI TIGNERFR-C PETLFQPSFI GM-ESAGIHE TTYNSIMKCD IDIRKDLYAN NVMSGGTTMY PGIA-DRMQK
......... .......... .......... .......... .......... .......... .......... .......... .......... .......... .......... ..L....... ..........
......... .......... .......... .......... .......... .......... .......... .......... .......... .......... .......... ..L....... ..........
......... .......... .......... .......... .......... .......... .......... .......... .......... .......... .......... ..L....... ..........
......... .T........ .......... ....Q..... .......... .......... .......... ..A......L ....C..... ..F....... V......... T.L....... ..........
......... .T........ .......... ....Q..... .......... .......... .......... ..A......L ....C..... ..F....... V......... T.L....... ..........
......... .T........ .......... ....Q..... .......... .......... .......... ..A......L ....C..... ..F....... V......... T.L....... ..........
......... .......... .......... .......... .......... .......... .......... .......... .......... .......... .......... ..L....... ..........
...S..... .......... .......... .......... .......... .......... .......... .......... .......... .......... .......... ..L....... ..........
......... .T........ .......... ....Q..... .......... .......... .......... ..AI.....L ....C..... ..F....... V......... T.L....... ..........
......... .T........ .......... ....Q..... .......... .......... .......... ..A......L ....C..... ..F....... V......... T.L....... ..........
......... .......... .......... .......... .......... .......... .......... .......... .......... .......... .......... ..L....... ..........
......... .......... .......... .......... .......... .......... .......... .......... .......... .A........ .......... ..L....... ..........
......... .......... ........A. .......... .......... .......... .......... .......... .......... .......... .......... ..L....... ..........
......... .T........ .......... ....Q..... .......... .......... .......... ..A......L ....C..... ..F....... V......... T.L....... ..........
......... .......... .......... .......... .......... .......... .......... .......... .......... .......... .......... ..L....... ..........
--------- ---------- ---------- ---------- ---.------ ---------- --........ .......... .......... .......... .......... .......... ..........
......... .T........ .......... ....Q..G.. .......... .......... .......... ..A......L ....C..... ..F....... V......... T.L....... ..........

......... .T........ .........T ..N..R.... ..A....... .......... .......... ....E.A... .......... .......... .......... T.L...S... ..........
......... .T........ .......... ....Q..Q.. .......... .......... .......... ..A....A.L .......... .C........ V......... T.L...S..F ..........
......... .T........ .......... ....E..Q.. .......... .......... .......... S...L..... .......... .C........ V......... T.L..AS..F ..........
......... .T........ .......... ....Q..Q.. .......... .......... .........A ..A....P.L .......... .C........ V......... C.L...S..F ..........
......... .T........ .......... ....Q..Q.. .......... .......... .........A ..A....A.L .......... .C........ V......... T.L...S..F ..........
........T .T........ .......... ....Q..Q.. .......... .......... .......... ..A......L .......... .C........ V......... T.L...S..F ..........
........T .T........ ..V....... ....Q..L.. .......... .......... .......... ...L...... .......... .C........ V......... T.L...S..F ..........
......... .T........ .........T ....Q..Q.. .S........ .......... .......... .......A.. .......... .......... .......... T.L...S... .........R
......... .T........ .......... ....Q..Q.. .......... .......... .......... ..A......L .......... .......... V......... T.L...S..F ..........
--------- ---------- ---------- ....Q..... ---..T---- ---------- --------.- ---------- --.----L.. .C........ V......... V.L...S... ..........
--------- ---------- ---------- ---------- ---------- ---------- --------.- ---------- --.------- .......... V...P..... T.L....... ..........
......... .......... .......... ....Q..... .......... .......... .......... .......... ....S.V... .......... .......... ..L....... ..........
......... .T........ .......... G...S..G.. .S....V... .......... .......... ..AM.....L .......... .......... V......... T.L....... ..........
......... .......... .......... ....Q..... .....T.... .......... .......... .......... .......... .......... .......... ..L....... ..........
......... .T........ .......... ....Q..... .......... .......... .......... ..S....ILL .......... .......... V......... T.L......F ..........
......... .T........ .......... ....Q..... .......... .......... .......... ..AM.....L .......... .S.Q...... V......... I.L......F ..........
......... .T........ .......... ....Q..... ..A..T.... .......... .......... ..S......L ....C..... .V........ V......... I......... ..........
......... .T........ .......... ....Q..... ..A..T.... .......... .........T ..A......L ....C..... .V.Q...... V......... ..L....... ..........
......... .T........ .......... ....Q..... ..A..T.... .......... .......... ..S......L ....C..... .V........ V......... I......... ..........
......... .T........ .......... ....Q..... ..A..T.... .......... .......... ..A......L ....C..... .V........ V......... S.L....... ..........
......... .T........ .......... ....Q..... .......... .......... .......... ..A..H...L ....C..... .......... V......... T.L....... ..........
......... .T........ .......... ....Q..... .......... .......... .......... ..S......L ...AC.L... .......... V......... T.L....... ..........
......... .T........ .......... ....Q..... ..A..T.... .......... .......... ..A......L ....C..... .V........ V......... S.L....... ..........
......... .T........ .......... ....Q..... ..A..T.... .......... .......... ..S......L ....S..... .V........ V......... I......... ..........
......... .T........ .......... ....Q..... ..A..T.... .......... .........T ..A......L ....C..... .V.Q...... V......... ..L....... ..........
......... .T........ .......... ....Q..... ..A..T.... .......... .......... ..A......L ....C..... .V........ V......... T......... ..........
......... .T........ .......... ....Q..Q.. ..A..T.... .......... .......... ..A......L ....C..... .V........ V......... T.L....... ..........
......... .T........ .......... ....Q..... .......... .......... .......... ..A......L ...AN..... .......... V......... T.L....... ..........
......... .T........ .......... ....Q..... ..A..T.... .......... .......... ..A......L ....C..... .V........ V......... T......... ..........
......... .T........ .......... ....Q..... .......... .......... .......... ..A......L ...AC..... .......... V......... T.L....... ..........
......... .T........ .......... ....Q..... ..A..T.... .........L .......... ..A......L ....C..... .V..T..... V......... T.L....... ..........
......... .T........ .......... ....Q..... ..A..T.... .......... .......... ..A......L ....C..... .V........ V......... T.L....... ..........
........T .T........ .......... ....Q..... .....T.... ........I. .......... ..A......L ...TC..... .A........ V......... T.L......F ..........
......... .T........ .......... ....Q..... .....T.... .......... .......... ..A......L ....C..... .......... V......... T.L....... ..........
..M...... .T........ .........I .......NV. S.A....ID. .......... .......... ..A......L ....V.V... .VHS...R.. ........F. I.V....... ........L.
......... .T........ .......... ...DQ..... .......... .......... .......... ..A......L ....S..... .......... .......... T.L....... ..........
......... .T........ .......... ....H..T.. .......... .......... .......... ..AM.....L ...AC..... ..F....... V......... T.L...S..F ..........
.V....... .T........ .......... .......T.. .......... .......... .......... ..AM.....L ...AC..Q.. ..F....... V......... T.L...S..F ..........
......... .T........ .......... ....Q..... .......... .......... .......... ..S......L .......... S.F.A..... V......... T.L....... ..........
.L......A .T...L.... .......... .......... ..T....... T......... .V........ ..A.Y....L .L..CN.... .......... L........S V.L...S... ........N.
..M.....N .T.......A .......... ....Q..I.. ST...AI..R ........L. .......... ..A....G.L .L..AT.... .......... V........S T.L...S..F Q.....E..R
......... .T........ .......... ....Q..G.. ......A... .......... .......... ..A......L ...AS..... .......... V........S T.L...S... ..........
......... .T........ .......... ....Q..... .......... .......... .V........ ..AM.....L .......... .S........ .......... T.L....... ..........
......... .T........ .......... ....Q..... .......... .......... .V........ ..AI.....L .......... .S........ .......... T.L....... ..........
......... .T........ .......... ....Q..... .......... .......... .V........ ..AI.....L .......... .S........ .......... T.L....... ..........
......... .T........ .......... ....Q..... .......... .......... .V........ ..A......L .......... S......... .......... I.L....... ..........
......... .T........ .......... ....Q..... .......... .......... .V........ ..A......L .......... .......... .......... ..L...S... ..........
......... .T........ ........S. ....Q..Q.. .......... .......... .......... .......... .......... .......... V......... T.L......F ..........
......... .T........ ........A. ....Q..Q.. .VT....... .......... .......... .......A.. .......... .......... V......... T.L......F ..........
......... .T........ ........A. ....Q..Q.. .......... .......... .........D ..A....A.L .......... .......... V......... T.L......F ..........

.L......N .TSS..L... ........F. ..NY.SALKQ SHD...QF.. N......K.. ...S...... ..Y..K.LEM NGR.LDS.QD L..K..QE.. V.V.R...Q. IIL....... E..G.E.LL.
.LF.....N .TSS..L... ........FL ..NY.SALKQ SHD...QF.. N....H.K.. ...S...... ..Y..K.LEM NGR.LDS.QD L..K..QE.. V.V.R...Q. IIL....... E..G.E.LL.
.L...K..V .TSS..M... ........F. ...Y.AA.KQ SYE..TTF.. N.......R. ......A... ..Y..K.LEM NGK.LDS.QS L.....QE.. V.V.R...Q. ITL....... E..G.E.LL.
.L.Y.I.LN .SS......I .......... .I.Y.A.LK- .YKE..QND. .......NT. .VQDQ..... ..L..K.A.. .K..FP.... L.F....... V.V...P.N. I.L......F ......E.LS.
.L.Y.I.LN .SS......I .......... ...Y.S.LK- .YKE..TND. .......NT. .VQDQ..... ..L..K.A.. .K..FP.... L.F....... V.V......N. I.L......F ......E.LS.
R..KDDD.H .E.....K.T. .......... .D.Y.A.LKK .GE.GGE..E ..A....RPL K.STQ..Q.. ..F....DLG .R..CKSV.Q L..D...T.. L.V....... IIL......F ..LG.E.LY.
...H....G .S.S..K... .........I ..N.DE..K. SEQ...DI.. .......NI. .V........ ..A......L .K..A....T ..F...K... V........G. I.L....... E.TG.E.LTR
.WF....HT .T........ .........I .M.YDE.LKR SEEH.DEI.E I......NL. .V.S...... ..A..N.TL. .R..CP.L.I .A.Q...... ......E..N. I.L....... NN.G.E.LT.
...H....G .T.S..K... .........I ....DE..KA .ED...DI.. .......NI. .V........ ..A......L .K..A..V.R ..FD...... V........G. V.L....... E..G.E.LT.
...MHT.MT .T.S..K... .N...Q.... ....DE..TN S.K.-.VS.E PF.....N.M QV...Q.... ..A...K.AL. .LD.AP.F.. M.FQ...N.. ...V.R...G. I.L......F KNLP.E.LA.
...M.T.MT .T.S.QK... .N...Q.... ....DE..TN S.K.-.VS.E PF.....N.M QV...Q.... .QA..K.AL. .LD.AP.F.. M.FQ...N.. ...V.R...G. I.L......F KNLP.E.LG.
..MM.T.TT .T.....K... .NV..Q.... ....E...TN S.K.-.AN.E AF.....N.M MV...Q.... ..V...K..L. .LD.AP.FP. MV.Q...N.. ...V.RE..G. I.L...S..F LNLP.E.LA.
EDSHGT... .N.......CQ ....G.ALLY CF....Q..KI ..E.....V.. L........N. .V........ ..V....N.. ....A..V.. ...F...G.. .H.......G. V.L......F E.....E...T.
EDSHGTC.. .N.......CQ ....G.ALLY CF....Q..KI ..E.....V.. L........N. .V........ ..V....N.. ....A..V.. ...F...G.. .H.......G. V.L......F E.....E...T.
...M..... .N........ .........I ......Q..KI ..E.....V.. .........N. .V........ ..V....N.. ....A..V.. ...F...G.. .........G. V.L......F E.....E...T.
........A .T........ .......... .E..NE..QK ......E... .......... .......... ..V....... ....CN.... .......... V........G. I.L......F E.....E...T.
N.Q.N.A.. .T.S...F... .........F. ...Y.SVL.- .SME.ANYT. T......V.. .VNQA..K.T ..L...R.EL- NNSDMD...Q LC.KT.Q... .....SE...S. V.L....SS.F A.LP.E.LE.
......... .T........ ........A.. ....Q..Q.. ......A... .......... .......... ..A......L .......... .......... V........G. V.L......F ..........
M.A....A. .N.....L... ........A.. ....Q..Q.. .S.L------ N......R.V .......... S.V...SC.L ........P. .......... V.......FG. L.L......V .......V..
......... .T........ .......... ....Q..H.. ......A... .......... .........A ..A......L .......... .......... V........G. V.L......F ........N.
......... .T........ ........A.. ....Q..... ......A... .......... .......... ..A......L .......... .......... V........G. V.L......F ........N.
......... .T........ ........A.. ....A.LQ.. ......A... .......... .......... ..A......L .......... .......... V........G. V.L......F ........N.
....K.... .T.....K...I K.......S.. ....DA...L. .....TT... ........K.. .......... ..A.....LL A......... ..I....... V....RY.FG. VIL...S..F ........N.
......... .T........ ........A.. ....A..Q.. ......A... .......... .......... ..A......L .......... .......... V........G. V.L......F ........N.

--------- ---------- -----...A.. ...Y.Q.LE. SK.....Q.. N......... .V.A...... A.V......L. ....A..... .......... V........G. I.L...S..F .........S.
...Q..... .T.S.Q.G.. ..M.....A.I ...C.Q.LE. SET.....V.. .......... ....A...G.. ..GV.Y...MV ....AS.... .......... V........G. I.L...S..F .S.......S.
........M .T.S...... ..M.....A.. ...Y.Q.LE. .K.....V.. ...H...... ....A..... ..KI.....M. E...A..... .......... V........G. I.L...S..F L........S.
........M .T.S...... ..M.....A.. ...Y.Q.LE. .K......I. N......... ....A..... ..V......M. ....A..... .......... V........G. I.L......F .........S.
........M .T.S...... ..M.....A.. ...Y.Q.LE. .K......I. N......... ....A..... ..V......M. ....A..... .......... V........G. I.L......F .........S.
....KSIC. HY........ ..M.....A.S ...Y.Q.LE. .KR.R.AV.. N......... ....A..... .QV......M. ....A..... .......... V........G. I.L......F .........S.
........M .T.....--TGM SYM.....A.. ..VMSKSWRL PRA.RLLV.. N......... ....AV.GSG. ..V......M. .......... .......... V........G. I.L...S..F ..SCYAS.S.
......... LT.S...... ..M.....A.. ...Y.Q.LE. .KT.....V.. ....M..... ....S..... ..V.....LV .....PSV.. A......... V........G. V.L...F..F .........S.
......... .T........ ..M.....S.I ...YDQ...E. .KT.....V.. .......... ....A..... ..V....... ....A..... .......... V........G. I.L......F .........S.
......... .T........ ..M.....S.I ...YDQ...E. .KT.....V.. .......... ....A..... ..V....... ....A..... .......... V........G. I.L......F .........S.
......... VT.P.....A A......... R..Y.Q.LE. .K......V.. .......... ....A..... ..VM.....L. ....AP.... .......... V........G. I.L...S..F ...R.RP.S.
......... .T.....S. .......A.I ...Y.Q.LRV .K......V.. ...VA..... ...AREVQ.. ..-......L. ....AP.... .......... V........G. I.L......F .........S.
......... .T........ ........A.. ...Y.Q.LD. .R.......I.. ......L... ....A..... ..V......L. ....AP.... A.......V...E V........G. V.L...S..F ...G.....S.
......... .T........ .........I ...Y.Q.LE. .KT.....V.. N......... ....S..... ..V.Y....M. ....N..... .......... V........G. I.L......F .........S.
......... .T........ ..V.....S.I SKT.....V. .......... .......... ....A..... ..V......L. ....A..... .......... V........G. I.L......F .........S.
......... .T.....K.. .V.....A.L ....Q.LE. TKT.G.AV.. N......... ....A..... ..V.Y....L. ....X..... .......... V........G. H.L...S..F .........S.
......... .T.S...K.. ..M.....A.I ....Q.LDM VK.....V.. NF........ ....A..... ..V......T. ....A..... .......... V........G. I.L...S..F .........S.
......... .T........ ..V.------ ---------- ---.------ ---------- --------.- ---------- --.------- ---------- ---------- ---------- ----.-----
......... .T.S...... ..V.....A.I ...Y.Q.LE. SKT.....V. .......... ....A..... ..V......M. ....A..... .......... V........G. I.L......F .........S.
......... .T.S...... ..M.....A.. ...Y.Q.LE. .K.....AV. .......... ....A..... ..V.....LV ....A..... .......... V........G. I.L......F .........S.
......... .T.S...... ..M.....A.. ...Y.Q.LE. .K.....AV. .......... ....A..... ..V.....LV ....A..... .......... V........G. I.L......F .........S.
......... .T........ ..V.....S.I ....Q...E. SKT.....V. .......... ....A..... ..V......M. ....A..... .......... V.......GK I.L...S..F .........S.
L.MQ..... .T.S...L... ..V.QQ..F. .E.Y.K.L.N .S.N...-... E......... .VEA...Q.. ..A...K.ELL .I...MD.M.L .AF.....T. .........S. I.........F A.....S.V..
V........ .T........ ........T.. ....DQ...K. ..E.....Q.. ........N.. V......... ..VR...... ......S...D C.FKT..... V........S. I.L......F ...G.E..T.
V........ .T..R..... ...R....T.. ....DQ...K. .GE......Q.. ........N.. V......... ..V....... ......S...D C.FKT..... V........S. I.L......F ...G.E..T.
...M..... .T........ .......... .........Q. ......A... T........P. .......... ..V.YN...L. ...AV....D. .F........ V........N. I.L......F ...G.E...T.
L.M...... .T........ ....K..... ...I.Q..LTL LPR.---... .......... .V.S...... ..A...N.GLL ...G.-...AR YLF....... V........S. T.L.....M.F ...G.E....
L........ .T........ ........A.. ...DQ..Q.. .......... .......... .......... ..A.......L .........V.. .......... V........G. V.L......F A........S.
--------- ---------- ---------- ---------- ---------- ---------- ---------- .......M.. ....A..... .......... V........G. I.L...S..F .........S.
......... .T........ ..M.....A.I ...YDQ...E. .KT.....V.. .......... ..AAD..... ..V....... ...A...... .......... V........G. I.L......F .........RQ

..M.....L .T.S...... .......... ...C.Q.LQ. ..Q.....Q.. .......... .........A ..A.....LL .L...A.... .......... L........G. I......... N.........
..A...... .S........ .......... ....QQ.IQ. .SQ....... .........A ..A.....VL .L....G...A .......A.. L.V......G. I......... ..S.......
..M...... .T.........R G........Q. ....Q.LQ. .SQ.....A.. .........A ..A.....LL .L....AS... .......... L........G. V......... S..S.......
......... .T........ ........T.I ..MN.DE..EK ..R.....T.D. ........N.L V.........T ..V.....AL. .K...AS....D C.FQT..... V........C. I.L....... ..S.E...T.
......... .T........ ........T.I .........DQ..K. ..E.....G.. ........N. V.........T ..V...K...M. .R...CT.V.. CAFQT..... V.....R...N. V.L...S..F ...G.E...T.
......... .T........ ........T.I ....DQ..K. ..E.....G. ........N. V.........T ..V.......L. .K...AS....D C.FQT..... V........C. I.L....... ...G.E...T.
......... .T........ ........T.I ..M..DE..EK STR.....A.D. T.......N. V.........T ..V...N...M. .R...CS.V.. CAFQT..... V.....R...S. V.L...S..F ...G.E...T.
--------- ---------- ---------- ---------- ---------- ---------- ---------- ---------- ---------- ---------- ---------- ---------- ----------
R..A...H. .T........ .........T G.....Q..Q. ..Q.....A.. .......... .V.......A ..A.......L .L........ .......... V........S. I......... ..........
..A.....T .S........ .......... ....Q.IQ. ..Q.....Q.. .........A .KA......VL .H....G....V ..F....I.. V.V......G. I......... ....S.....
..A...... .S........ .......... .F....Q.LQ. ..Q.....T.. .........A ..A....N.AVL .H....G.... ..F....I.. V.V......G. I......... ..........
..S...... .T.S...... ........R.. .........Q..Q. SSQ....AI... .........A ..A.....R.ADL .L...A....DQ ..F....... M.V.....E..G. I......... ....E.....
..HD..... .T........ .........I ....Y.E...KK SQE......E.. T........H.. .V.S...... ..A......G.L .K...AV..G. ..FQ...... L......... I.L....... ...G.E...T.
..S...... .S........ .......... ....Q..Q. ..Q.....I.. .........A ..A...H...VL .L....G.... .......... V.V.....E..G. I.........F ....E.....
..M....T. .S........ .......... ....Q.LQ. ..Q.....Q.. .........A ..A.......AL .L...N.....DQ A......... V........G. V......... ..........
..A...... .S........ .......... ....Q..Q. .SQ....AV... .........A ..A...H...VL SL...A....DQ .......... V.V.....E..G. I.........F ....E.....
.......N. YT........ ........R.. ....Q.IH. .S........ .......... .........A ..A.......IV ...TC..... ..F....... V........S. I......... ..........
```

```
                                                                                                                                        1
         1          2          3             4                                5            6          7          8          9           0
    678901 2345678901 2345678901 2345678  9 012345                      6789 01234567 8 9 01234567 8901234567 8901234567 8901 23456 7890123456
----T.L.C. N.S.LV.A.F ..DDA.RAVF ..IVGRP--R HQ.VMV---- .......... ....--.MGQ K.SYV..... Q.SKRGILTL .Y..E...IT N..D..KIWH HTFY-NE..V A..E.PT...
       . * .*o.h b *  ..   *   . .  h.                                     .  a             b  h h   *  . hh   *  *.  .    h               h**
AGRLPACVVD CGTGYTKLGY AGNTEPQFII PSCIAIKESA KVGDQAQRRV ---------- ----MKGVDD LDFFIGDE-A I-E-KPTYAT KWPIRHGIVE DWDLMERFME QVIFK-YLRA EPEDHYFLLT

........I. V....S...F ...K...... ..A....... R...TNT..I .......... ....T..IE. .......... F.D.STG.SI .Y.V...L.. ........L. .CV....... ..........
.SG...V.I. N......M.. ...ND.S... .TT..TQS.K --.K.TAASQ .......... ....K...E. .......... .AN.SK..DM TN.VK..QI. N.TH..QYW. HCV......C ..........
.SFNVPIIM. N....S.... ...DA.SYVF .TV..TRSAG ASSGP.VSSK PSYMASKGSG HLSSKRATE. ......ND.. LKKASAG.SL DY.....QI. N..H....WQ .SL......C ..........

MDPHNPI.L. Q...FV.I.R ..ENF.DYTF ..IVGRPILR AEERAS---- .......... ....--VATP .KDIMIGD.E AS.VRSYLQI SY.MEN..IK N.TD..LLWD YAF.EQMKLP STSNGKI...
SKGRNVI.C. N...FV.C.. ..SNF.TH.F ..MVGRPMIR A.NKIG---- .......... ....---DIE VKDLMVGD.E ASQLRSLLEV SY.MEN.V.R N..D.CHVW. YTFGPKKMDI D.TNTKI...

NSHHA.V.I. N.S.VC.A.F SPEDT.RAVF ..IVGRP--R HLNVLL---- .......... ....--DSVI G.SV..ER.Q P..KRGILTL .Y..E..M.K N..E..MVWQ HT-Y-EL... D.M.LPA...

VIANQPV.I. N.S.VI.A.F ..DQI.KYCF .NYVGRP--K H.RVM.---- .......... ....--.ALE G.I...PK.. E..HRGLLSI RY.ME....K ..ND...IWQ Y.YS.DQ.QT FS.E.PV...
VIANQPV.I. N.S.VI.A.F ..DQI.KYCF .NYVGRP--K H.RVM.---- .......... ....--.ALE G.I...PK.. E..HRGLLSI RY.ME....K ..ND...IWQ Y.YS.DQ.QT FS.E.PV...
IIANQPV.I. N.S.VI.A.F ..DQI.KYCF .NYVGRP--K HMRVM.---- .......... ....--.ALE G.L...PK.. E..HRGLLTI RY.ME..V.R ..ND...IWQ Y.YS.DQ.QT FS.E.PV...
VVVNQPV.I. N.S.VI.A.F ..EHI.KCRF .NY.GRP--K H.RVM.---- .......... ....--.ALE G.I.V.PK.. E..HRGLLSI RY.ME....T ..ND...IWS YIYS.EQ.AT FT...PV...
SLHNAPI.L. N.S.TIRA.F ..DDV.KCHF ..FVGRP--K HLRVL.---- .......... ....--.ALE GEV...QK.. AS.LRGLLKI RY.LE....T ...D..KIWA Y.Y-DEG.KT LS.E.PV...
SLHNAPI.L. N.S.TIRA.F ..DDV.KCHF ..FVGRP--K HLRVL.---- .......... ....--.ALE GEV...QK.. AS.LRGLLKI RY.LE....T ...D..KIWA Y.Y-DEG.KT LS.E.PV...
VLTNQPICI. N.S.VI.A.F ..EDQ.KSFF ..YVGRP--K HLKIM.---- .......... ....--.AIE G.I...NK.. Q..LRGLLKI .Y..E....V ...D...IWQ FIY-TEE.KT VS.E.PV...
.LYNQPV.I. N.S.II.A.F S.EER.KALE YCLVGNT--K YDKVML---- .......... ....--EGLQ G.T...NN.. Q.KLRGLLKL RY..K..V.. ...S..LIWS Y.LNEVLQLQ NIGE.PL.I.

-MANAVV.L. N.AHTA.V.L .NQD..H-VV .N..MKAK.E RRRA------ .......... ....------ ---.V.NQID ECRDTSALYY ILAFQR.YLL N.HTQKTVWD YIFS.DGIGC SL.NRNIVI.
GDEVS.V.I. P.SYT.NI.. S.SDF..S.L ..VYGK---- ---------- .......... ....-YTA.E GNKK.FS.QS .GIPRKD.EL .PI.EN.L.I ...TAQEQWQ WALQNELYLN SNSGIPA...
```

```
        2                  2                           2                                                2                                 2
        2                  3                           4                                                5                                 6
2345678901 23456789        01234 56                789 012345678                                       901                       23 4567890123
I..KLC..AL .FEN.MAT-- -----AASSS SL........ .......EKS YELPDGQV.. .......... .......... .......... ....TI.... .......... ........N. ..RC..TL.Q
h**   .h       a                                       h  .a   h                                         .                          bh  .ah .
VKERYSYVCP DLVKEFNKYD TDGSKWIKQY TG-------- -------INA IS-KKEFSI- ---------- ---------- ---------- ----DVG--- ---------- --------YE RFLGPEIFFH

I..KHC.I.. .IA...A... .EPG...RNF S......... .......V.T VT..AP.NV. .......... .......... .......... .......... .......... .......... ..........
...Q...... .I....G... SEPD....TI NA........ .......QDS VT...P..Y. .......... .......... .......... .......... .......... .......... ......L..N
I..ECC.... .I....SRF. REPDRYL.-- YA........ .......SES .T.GHSTT.. .......... .......... .......... .......... .......... ........F. ...A.....N

I..KLC..SY ..DLDTKL-- -----ARETT AL........ .......VES YELPDGRT.. .......... .......... .......... ....K..... .......... ........Q. ..EA..CL.Q
M..KLC.IGY .IEM.QRL-- -----ALETT VL........ .......VES YTLPDGRV.. .......... .......... .......... ....K..... .......... ........G. ..EA..AL.Q

I..KLC..SM NYA..MDL-- -----HG.V- --........ .......-ET YELPDGQK.. .......... .......... .......... ....VL.... .......... ........C. ..RC..AL.Q

I...AC.LSI NPQ.DET--- --------LE .E........ .......KAQ YYLPDGST.. .......... .......... .......... ....EI.... .......... ........PS ..RA..LL.R
I...AC.LSI NPQ.DET--- --------LE .E........ .......KAQ YYLPDGST.. .......... .......... .......... ....EI.... .......... ........PS ..RA..LL.R
I...AC.LSI NPQ.DEA--- --------LE .E........ .......KVQ YTLPDGSTL. .......... .......... .......... .......... .......... ........PA ..RA..LL.Q
I..KVC.LAT NPQ..ET--- --------VE .E........ .......KF. YKLPDGKIF. .......... .......... .......... ....EI.... .......... ........PA ..RA..AV.R
I..SVT..AH .PR..EKEWA AAKMDPA--- -K........ ........AE YVLPDGNKL. .......... .......... .......... ....KI.... .......... ........A. ..RA...L.D
I..SVT..AH .PR..EKEWA AAKMDPA--- -K........ ........AE YVLPDGNKL. .......... .......... .......... ....KI.... .......... ........A. ..RA...L.D
I..KC...TL .PR..EKEWI NASISGG.D. .K........ .......EEE FKLPDGNVL. .......... .......... .......... ....RL.... .......... ........A. ..RA...L.D
M..KVC.LAK NIK..EE..L QGT------Q DL........ ........ST FKLPDGRC.. .......... .......... .......... ....E..... .......... ........ND .YRA...L.S

I..DVCF.AE .FKQAMQVHY SEEKRR---- EV........ .......TVD YVLPDFTTV. .......... .......... .......... ....KR.YVR VPGKPREDEE QQQMVSLCN. ..TV..LL.N
C..TLCHI.. TKTL.ET.TE LSSTAKRSIE SPWNEEIVFD NETRYGFAEE LFLP..DD.P ANWPRSNSGV VKTWRNDYVP LKRTKPSGVN KSDKK.TPTE EKEQEAVSKS TSPAANSADT PNETGKRPLE
```

		12345
	Acts_Human	...MCDEDET
	Actl_Bovin	---------M
HUMAN SKELETAL ACTIN	**Actl_Drome**	....
REFERENCE SEQUENCE	**Ddactlpr_1**	.......MNP
	Actl_Schpo	....
Arp 3		
	Actl_Yeast	.........-
	Dmarp14d_1	.......MDR
Arp2		
	Dmarp53d_1	...MSSEVDS
UNCLASSIFIED		
	Actl_Rabit	.....MESYD
Arp1a	**Hsacent_1**	.....MESYD
	Hsbcent_1	.....MESYD
	Dmarp87c_1	.....MEPYD
	Actl_Neucr	.......MTD
Arp1b	**Nc14008_1**	.......MTD
	Pmcactin	.....MEFND
	Sch9315_8	..MDQLSDSY
	Dmarp_1	.........-
UNCLASSIFIED	**Actr_Yeast**	MSNAALQVYG

		2 1 345678901
	Acts_Human	.TAER.IVRD . hb
	Actl_Bovin	-EQSLETAKA
HUMAN SKELETAL ACTIN	**Actl_Drome**	....
REFERENCE SEQUENCE	**Ddactlpr_1**	.AE...I..R
	Actl_Schpo	.DS..K..ER
Arp 3		
	Actl_Yeast	.TADF..VRQ
	Dmarp14d_1	.SADF..VRI
Arp2		
	Dmarp53d_1	.SAER.IVRE
UNCLASSIFIED		
	Actl_Rabit	.SSEF.IV..
Arp1a	**Hsacent_1**	.SSEF.IV..
	Hsbcent_1	.SAEF.VVRT
	Dmarp87c_1	.TAEF.IVRS
	Actl_Neucr	.SAEK.VVRL
Arp1b	**Nc14008_1**	.SAEK.VVRL
	Pmcactin	.SAEK.IVRI
	Sch9315_8	.SSER.IVRT
	Dmarp_1	.MDESHVVNQ
UNCLASSIFIED	**Actr_Yeast**	YANNRGFFQE

```
.           1          1          1                      1          1          1          1          1          2
.           2          3          4                      5          6          7          8          9          0
)123456 7890123456 7890123456 789012   34 5        6 7890123456 7890123456 7890123456 7890123456 7890123456 789012

.PKA..  .KMTQ....T ....AM.V.I ....S...Y. .--------G RT..I.L... .....NV.IY ...ALPHA.M RLDL....L. DYLMKI.TE. .GYSFVT... .......... ..........
h*.  ** a    hh**  h    hh    .*..  .  h.                o*h.h*.* .*ho  h*hh  *h h       h h. baho  hh  hh      h h

LNTPENR EYTAEIMFES FNVPGLYIAV QAVLAL--AA SWTSRQVGER TLTGTVIDSG DGVTHVIPVA EGYVIGSCIK HIPIAGRDIT YFIQQLLRDR -EVGIPP--- ---------- ----------

......  .........T .......... .......... ..A..SAE.. ....I.V... ....D..... .......... .......N.. S...S...E. .......... .......... ..........
.A....  .F.......T .......... .......... ....KNA-.K .......... ........IS .......S.. .........S SYV..IM.E. ..PN...... .......... ..........
.P....  .N........ ..CA...... .......... ....SK.TD. S.....V... .....I.... .......S.. TM.L....V. ..V.S..... .N---E.... .......... ..........

M.PLK.. .KMC.V...K YDFG.V.V.I ........Y. Q--------G LSS.V.V... .....IV..Y .SV.LSHLTR RLDV....V. RHLID..SR. .GYAFNR... .......... ..........
M.PTK.. .KMI.V...K YGFDSA...I ..AWT...Y. Q--------G LIS.V..... .....IC..Y .EFALPHLTR RLD....... RYLIK..LL. .GYAFNH... .......... ..........

.PKK..  .KMT.....H .Q..AF.V.. ....S...CT T--------G RTV.I.V... .....TV.IY ..FALPHACV RVDL....L. DYLCK..LE. .G.TMGT... .......... ..........

.PRK..  .RA..VF..T ....A.F.SM ....S...Y. T--------G RT..V.L... .....AV.IY ..FAMPHS.M R.D.....VS R.LRLY..KE .GYDFHS... .......... ..........
.PRK..  .RA..VF..T ....A.F.SM ....S...Y. T--------G RT..V.L... .....AV.IY ..FAMPHS.M R.D.....VS R.LRLY..KE .GYDFHS... .......... ..........
.PSK..  .KA..VF..T ....A.F.SM ....S...Y. T--------G RT..V.L... .....AV.IY ..FAMPHS.M RVD.....VS RYLRL...KE .G.DFHT... .......... ..........
.PRR..  .KA..FF..G I.A.A.FVSM ....S...Y. T--------G RV..V.L... .....AV.IY ..FAMPHS.M RVD.....V. RYLKT.I.RE .GFNFRS... .......... ..........
.PRA..  DTA.Q.L..T ....A..TSI ....S...Y. .--------G RT..V.L... ...S.AV..Y Q.FTVPNS.R R.DV....V. EYL.T...KS .GYVFHT... .......... ..........
.PRA..  DTA.Q.L..T ....A..TSI ....S...Y. .--------G RT..V.L... ...S.AV..Y Q.FTVPNS.R R.DV....V. EYL.T...KS .GYVFHT... .......... ..........
.PRT..  DQA.QVF..T ....A.FTSI ....S...Y. .--------G RT..V.L... .....AV.IY ..FAMP.A.R R.D.....V. EYL.L...KS .GTIFHT... .......... ..........
I.PLK.. .QM.QVL..T .D.SA..VSN P...S...Y. .--------G RT..C.V.C. E.YCSTV.IY D.FALPASMM RMD.G.A... EQL.FQ..KS AG.SLFS... .......... ..........

I.FQSIQ .A.L..L..E YK.D.V.KTT A.D..AFNYV ADSEERTTME S.NCII..V. YSF...V.FV L.RRVLQG.R R.DMG.KAL. NQLKE.ISY. .--HLNV... .......... ..........
/.ST... KKSL.VLL.G MQFEAC.L.- ------..-P TS.CVSFAAG RPNCL.V.I. HDTCS.S.IV D.MTLSKSTR RNF...KF.N HL.KKA.EPK EIIPLFAIKQ RKPEFIKKTF DYEVDKSLYD
```

```
 2          2          2          3          3                     3                   3          3          3          3          3
 7          8          9          0          1                     2                   3          4          5          6          7
8901 2   34567890 1234567890 1234567890 1234567890 12345678         90           1234 567890123    45678901 2345678901 234 567890 1234567890

GMES-A ..G.H.TTYN S.MK.D..I. KD..A.N.M. ..T..YPGIA D.M.KEIT-- ---A.----- ------A.ST MKIKI.APP. ..ERK.S..I ...I...LST .Q.-MWI..Q E.D.A....V
         h  hh    h     h.h.   .h ..... *  o     h  *h   h                       h h h                 h ..  .                  ah *

NPDFTQ --PISEVVDE VIQNCPIDVR RPLYKNIVLS GGSTMFRDFG RRLQRDLKRT VDARLKLSEE LSGGRLKPKP IDVQVITHH- --MQRYAVWF GGSMLASTPE FYQVCH-TKK DYEEIGPSIC

.....I ...L..I..N .......... ....N..... ......K... ......I..S ..T..RI..N ..E..I.... .......... .......... .......... .........A A...Y.....
SS.YLT ...LPK...D T..S....C. .G........ ......K... K.....V..S ..Y.I.R... ....KI.AV. LA.N..S.N. .......... .......... ..N......A Q.D.......
SS..LT ...LP.L..N .V.SS..... KG........ ....L.KN.. N........I ..E.IHR..M ...A--.SGG V..N..S.K. ..R..N.... ...L..Q... .GSY.....A ....Y.A..A

DVEQ-P ..GVG.LLFN TV.SADV.I. SS...A.... ...S.YPGLP S..EKE..-- ---Q.WF.RV .HNDPSRLDK FK.RIEDPP. ..RRKHM.FI ..AV...IMA DKDHMWLS.Q EWQ.S...AM
.VEG-P ..G.A.LAFN T..AAD..I. PE...H.... .....YPGLP S..E.EI.-- ---Q.Y.ERV .KNDTE.LAK FKIRIEDPP. ..RRKDM.FI ..AV..EVTK DRDGFWMS.Q E.Q.Q.LKVL

GQEV-M ..G.H.ATHH S.T..DM.L. KDM.A..... ..T....NIE H.FLQ..T-- ---EM----- ------A.PS .RIK.NASP. ..DR.FS..T ...V...LTS .QN-MWIDSL E...V.SA.V

GEES-E ..G.H..LVF A..KSDM.L. .T.FS..... ....L.KG.. D..LSEV.-- ---K.----- ------A..D VKIRISAPQ. ..ERL.ST.I ...I...LDT .KK-MWVS.. E...D.ARSI
GEES-E ..G.H..LVF A..KSDM.L. .T.FS..... ....L.KG.. D..LSEV.-- ---K.----- ------A..D VKIRISAPQ. ..ERL.ST.I ...I...LDT .KK-MWVS.. E...D.ARSI
GDES-E ..GLH...AF A.HKSDM.L. .T.FA..... ....L.KG.. D..LSEV.-- ---K.----- ------A..D .KIKISAPQ. ..ERL.ST.I ...I...LDT .KK-MWVS.. E...D.SRAI
GEEC-E ..G.HD.LMY S.EKSDM.L. KM..Q..... ....L.KG.. D..LSE..-- ---KH----- ------SA.D LKIRIAAPQ. ..ERL.ST.M ...I...LDT .KK-MWIS.R E...E.QKAV
GLEY-P ..GVHQI.VD S.NRTDL.L. .D..S..... ....LTKG.. D..LTEVQ-- ---K.----- ------AV.D MRIKIFAPP. ..ERK.ST.I ...I..GLST .RK-MWVSID .WH.NPDI.H
GLEY-P ..GVHQI.VD S.NRTDL.L. .D..S..... .--------- ---------- ---------- ------AARS QR.SA.ACS. ..RKCRSSRL RTCG------ ------.--- ----------
GSEY-S ..G.HQ..VD A.SRVDL.L. KS.FG..... ....LT.G.. D..LSEIR-- ---.------ ------AV.D VKIKIFAPP. ..ERK.ST.I ...I...LST .RK-MWVSAE E.Q.DPDI.H
GLGY-D ..GL.DMCMQ S.WKVDL.L. K..LSS.I.. ..T.TLKG.. D.MLW..E-- ---A.----- ------TKGT SKIKI.APS. ..ERK.TT.I ...I.TGLST .QR-LWTK.S .WL.DSTRVY

GVQ-QV ..G.P.A.AD CLKA..WEAH .E.LL..LIV ...AQ.PG.L P..K...RAL .--------- -------.DD LE.SL.CPE. ..DPVRYA.Y ..KEV.TS.N .EEFVY..QD ....Y.FQGI
PKENNE LIGLADL.YS S.MSSDV.L. AT.AH.V..T ..TSSIPGLS D..MTE.NKI L--------- --------PS LKFRIL.TGH TIERQ.QS.L ...I.T.LGT .H.-LWVG.. E...V.VERL
```

Acts_Human	
Actl_Bovin	
Actl_Drome	HUMAN SKELETAL ACTIN
Ddactlpr_1	REFERENCE SEQUENCE
Actl_Schpo	
	Arp 3
Actl_Yeast	
Dmarp14d_1	
	Arp2
Dmarp53d_1	
	UNCLASSIFIED
Actl_Rabit	
Hsacent_1	Arp1a
Hsbcent_1	
Dmarp87c_1	
Actl_Neucr	
Nc14008_1	Arp1b
Pmcactin	
Sch9315_8	
Dmarp_1	
Actr_Yeast	UNCLASSIFIED

3 7 12345		
HRKCF-----	Acts_Human	
RHNPVFGVMS	Actl_Bovin	
.......T.T	Actl_Drome	HUMAN SKELETAL ACTIN
.F.T.I.GIN	Ddactlpr_1	REFERENCE SEQUENCE
.RYQI..NSL	Actl_Schpo	
		Arp 3
TKFGPR----	Actl_Yeast	
QKLQKISH--	Dmarp14d_1	
		Arp2
HRKCF-----	Dmarp53d_1	
		UNCLASSIFIED
HRKTF-----	Actl_Rabit	
HRKTF-----	Hsacent_1	Arp1a
HRKTF-----	Hsbcent_1	
HRKTF-----	Dmarp87c_1	
TKFT------	Actl_Neucr	
----------	Nc14008_1	Arp1b
.KSI------	Pmcactin	
Sch9315_8	SNLM------	
Dmarp_1	NQR-------	
Actr_Yeast	LNDRFR----	UNCLASSIFIED